A.S. SPIRIN (Ed.)

Cell-Free Translation Systems

Springer

Berlin
Heidelberg
New York
Barcelona
Hong Kong
London
Milan
Paris
Tokyo

A.S. Spirin (Ed.)

Cell-Free
Translation Systems

With 91 Figures and 17 Tables

 Springer

Prof. ALEXANDER S. SPIRIN
Institute for Protein Research
Russian Academy of Sciences
142292 Pushchino, Moscow Region, Russia

ISBN-13: 978-3-642-63956-2   e-ISBN-13: 978-3-642-59379-6
DOI: 10.1007/978-3-642-59379-6

Library of Congress applied for
Die Deutsche Bibliothek – CIP-Einheitsaufnahme
Cell free translation systems / Alexander S. Spirin (ed.) – Berlin ; Heidelberg ; New York ;
Barcelona ; Budapest ; Hong Kong ; London ; Mailand ; Paris ; Santa Clara ; Singapore ;
Tokyo ; Springer, 2002

Springer-Verlag Berlin Heidelberg New York
a member of BertelsmannSpringer Science+Business Media GmbH

http://www.springer.de/medizin

Cover Design: design & production, 69121 Heidelberg, Germany
Production: ProEdit GmbH, 69126 Heidelberg, Germany
Typesetting: TBS, 69207 Sandhausen, Germany
Printed on acid free paper   SPIN 10833845   18/3130 Re – 5 4 3 2 1 0

Contents

Introduction

CHAPTER 1

Cell-Free Protein Synthesis 1

ALEXANDER S. SPIRIN*

Abbreviations

NTP	nucleoside triphosphates
ARSes	aminoacyl-tRNA synthetases
PEP	phosphoenol pyruvate
CP	creatine phosphate
AcP	acetyl phosphate
PK	pyruvate kinase
CK	creatine kinase
AcK	acetyl kinase
ME	mercaptoethanol
DTT	dithiotreitol
DHFR	dihydrofolate reductase
CAT	chloramphenicol acetyltransferase
GFP	green fluorescent protein
IL-2 and IL-6	interleukin-2 and interleukin-6
TMV	tobacco mosaic virus

Introduction: Prehistory of Cell-Free Translation Systems

As early as the beginning of the 1950s, several groups independently demonstrated that protein synthesis does not require the integrity of the cell and can continue after cell disruption. Thus, disrupted cells or their isolated fractions were reported to be capable of synthesizing proteins (Borsook 1950; Winnick 1950a, 1950b; Siekevitz and Zamecnik 1951; Siekevitz 1952; Peterson and Greenberg 1952; Khesin 1953; Gale and Folkes 1954). In the meantime, ribonucleoprotein particles were observed and identified in cells (Palade 1955) and then isolated from cells and studied with respect to their physicochemical properties (Chao and Schachman 1956; Ts'o et al. 1956; Peterman and Hamilton 1957; Tissiéres and Watson 1958; Tissiéres et al. 1959; see also papers in Roberts 1958). The protein-

* Institute of Protein Research, Russian Academy of Sciences, 142290 Pushchino,
  Moscow Region, Russia
  Tel./Fax: 007(095)924-0493, e-mail: spirin@vega.protres.ru

synthesizing ability of these particles was experimentally proved (Littlefield et al. 1955; Littlefield and Keller 1957; McQuillen et al. 1959). The word "ribosome" was proposed to designate the protein-synthesizing ribonucleoprotein particles.

In the second half of that decade, Zamecnik and his colleagues made a real ribosomal system of protein synthesis (translation) based on mitochondria-free cytoplasmic extracts of animal cells (Littlefield et al. 1955; Keller and Zamecnik 1956; Littlefield and Keller 1957). The dependence of the system on energy supply in the form of ATP and GTP was shown. Zillig's group was the first to succeed in producing a bacterial cell-free translation system (Schachtschabel and Zillig 1959). Independently, bacterial cell-free translation systems were made by American groups (Lamborg and Zamecnik 1960; Tissiéres et al. 1960). Ribosomes in all of those systems were programmed with endogenous mRNA; they were simply reading the messages to which they had already been attached at the time of cell disruption. Nevertheless, the significance of these systems was great, since they opened the door for studies of molecular mechanisms of protein biosynthesis, including activation of amino acids, involvement of tRNA, GTP requirement, ribosome functions, and participation of soluble translation factors (Zamecnik 1969).

Programming Ribosomes
of Cell Extracts with Exogenous Messages

A revolutionary step in the development of cell-free translation systems was the introduction of exogenous messages. This was first done by Nirenberg and Matthaei in 1961 with a bacterial system (Nirenberg and Matthaei 1961). Preincubation of the cell extract at physiological temperature was sufficient to remove the endogenous mRNA from ribosomes. The vacant ribosomes in the extract were found to accept either exogenous natural mRNAs or synthetic polyribonucleotides as templates for polypeptide synthesis. The Nirenberg system became a classic procedure (Table 1).

In more recent versions of the system, the ribosome-free fraction of the bacterial extract freed from any RNA, in combination with either the runoff ribosomes, or the fraction of active vacant couples, or the salt-washed ribosomes was used (Table 2). The treatment with DEAE cellulose was found to be very effective in removing nucleic acids. Certainly, in this case total tRNA should be added.

The preincubation procedure similar to that of Nirenberg-Matthaei could be applied also to some eukaryotic systems. Partially fractionated animal-cell extracts in combination with purified ribosomes or ribosomal subunits were also shown to be quite adequate for expression of an exogenous message in a cell-free system (Schreier and Staehelin 1973). At the same time, it was found that the wheat germ extract could be used directly for effective expression of exogenous messages because of the intrinsically low levels of endogenous mRNA (Roberts and Paterson 1973; Marcus et al. 1974; Anderson et al. 1983). Micrococcal Ca^{2+}-dependent RNAase treatment was demonstrated to be useful for the removal of endogenous mRNA from reticulocyte lysates (Pelham and Jackson 1976; Jackson and Hunt 1983; Merrick 1983) as well as from other animal-cell extracts (Henshaw and Panniers 1983). The avoidance of endogenous messenger activity and the expression of exogenous templates became the main ways in the use of cell-free translation systems. Principal components of a typical eukaryotic system based on rabbit reticulocyte lysate are listed in Table 3. The eukaryotic cell-free systems based on yeast and *Drosophila* extracts are described by Altmann and Trachsel (p. 66) and Bergamini and Gebauer (p. 79) in the present volume.

Table 1. Crude bacterial cell-free translation system with an exogenous message (after Nirenberg and Matthaei 1961; Nirenberg 1963)

Principal components:

Supernatant fraction S-30 of *E. coli* extract treated with DNAase, preincubated with amino acids, ATP, PEP and PK at 37°C for 80 min and then dialyzed: contains endogenous ribosomes, tRNAs, ARSes, and translation factors

Natural mRNA or synthetic polyribonucleotide

Amino acids

Energy compounds: ATP and GTP

NTP-regenerating system: PEP + PK (or CP + CK, or just AcP with AcK being endogenously present in the extract)

SH-compound: ME or DTT

Mg^{2+}, and K^+ or NH_4^2, in optimal concentrations

Table 2. Fractionated bacterial cell-free translation system

Ribosomes

Natural mRNA or synthetic polyribonucleotide

Total tRNA

Supernatant fraction S-100 deprived of all nucleic acids (e.g., by DEAE cellulose treatment): contains ARSes and translation factors

Amino acids

Formyltetrahydrofolate or its congener (e.g., methenyltetrahydrofolate or folinic acid)

Energy compounds: ATP and GTP

NTP-regenerating system: PEP + PK, or CP + CK, or AcP

SH-compound: ME or DTT

Mg^{2+}, and K^+ and/or NH_4^2, in optimal concentrations

Table 3. Crude eukaryotic cell-free translation system with an exogenous mRNA (Pelham and Jackson 1976; Jackson and Hunt 1983)

Reticulocyte lysate treated with micrococcal Ca^{2+}-dependent RNAase: contains ribosomes, tRNAs, ARSes, and translation factors

mRNA

Amino acids

Energy compounds: ATP and GTP

NTP-regenerating system: CP + CK

SH-compound: DTT

Additional components that may amend the system: hemin, glutathione, cAMP

Mg^{2+}, K^+ and polyamines (spermidine) in optimal concentrations

Transcription-Translation Cell-Free Systems Using Exogenous DNA

Coupled Bacterial Transcription-Translation Systems

Again, Nirenberg and Matthaei seem to be the first to report dependence of the bacterial cell-free system on the presence of DNA (Matthaei and Nirenberg 1961). This dependence was seen especially at the later stages of incubation of the system, when pre-existing mRNA had been presumably read out. Some time later, coupled transcription-translation systems were developed by using exogenous bacteriophage DNAs (Wood and Berg 1962; Byrne et al. 1964). These systems, however, were poorly active with cellular exogenous DNAs as well as with a number of viral DNAs; in addition, the background, due to endogenous polypeptide synthesis, was significant. Coupled transcription-translation systems came into wide use after some major improvements were made by two groups (Lederman and Zubay 1967; DeVries and Zubay 1967; Zubay 1973; Chen and Zubay 1983; Gold and Schweiger 1969, 1971; Schweiger and Gold 1969). In the Zubay system (Table 4), the bacterial crude extract was used after prolonged incubation to degrade endogenous RNA and DNA by cellular nucleases; concentrations of the components of the system were optimized. The system became very popular because of the simplicity of its preparation, the stability of the extracts during storage, and its high activity. The Gold-Schweiger system (Table 5) consists of the isolated ribosomes and the supernatant fraction specially purified of endogenous amino acids and nucleic acids by ion-exchange chromatography. This provides a very low background due to endogenous synthesis and better-controlled conditions, but it is more complicated to prepare.

In the case of prokaryotic systems, just a proper DNA species, such as a plasmid, an isolated gene, or a synthetic DNA fragment, could be added to the DNA-free extract instead of mRNA, and the corresponding mRNA was synthesized in situ by the endogenous RNA polymerase present in the extract or its supernatant

Table 4. Crude bacterial coupled transcription-translation system with an exogenous DNA (Lederman and Zubay 1967; DeVries and Zubay 1967; Zubay 1973)

Supernatant fraction S-30 of *E. coli* extract preincubated (e.g., as in Table 1) to deprive of endogenous DNA and mRNA: contains RNA polymerase, ribosomes, tRNAs, ARSes, and translation factors
DNA
Amino acids
ATP, GTP, UTP and CTP
NTP-regenerating system: PEP + PK, or CP + CK, or AcP
Formyltetrahydrofolate or its congener (e.g., methenyltetrahydrofolate or folinic acid)
cAMP
SH-compound: ME or DTT
Mg^{2+}, Ca^{2+}, K^+ and NH_4^2 in optimal concentrations

Table 5. Fractionated bacterial coupled transcription-translation cell-free system (Gold and Schweiger 1969, 1971)

Ribosomes
DNA
Total tRNA
Supernatant protein fraction S-100 deprived of all nucleic acids (e.g., by DEAE cellulose treatment): contains RNA polymerase, ARS and translation factors
Amino acids
ATP, GTP, UTP, CTP
NTP-regenerating system: PEP + PK, or CP + CK, or AcP
Formyltetrahydrofolate or its congener (e.g., methenyltetrahydrofolate or folinic acid)
SH-compound: DTT
Mg^{2+}, Ca^{2+}, K^+ and NH_4^2 in optimal concentrations

fraction. In this case, ribosomes start to translate the nascent chains of mRNA, even prior to the completion of the RNA synthesis. Thus, translation is going on while mRNA is still elongating, and the rates of transcription and translation are coordinated. That is why such systems are called *coupled transcription-translation systems*.

Combined Eukaryotic Transcription-Translation Systems

The eukaryotic extracts, however, are prepared from the cytoplasmic fraction, so they lack endogenous RNA polymerase activity. This limitation can be overcome by addition of an exogenous RNA polymerase. Purified *E. coli* RNA polymerase was used in the first attempts to link DNA and translation in eukaryotic cell-free systems (Roberts et al. 1975; Lewis et al. 1975; Coen et al. 1977; Stueber et al. 1984; Bujard et al. 1987). Virion-associated RNA polymerases of some animal viruses were also combined with eukaryotic extracts to provide transcription directly in cell-free translation systems (Ball and White 1976; Content et al. 1977; Pelham et al. 1978). Later, isolated bacteriophage T7 and SP6 RNA polymerases were found to be the most convenient and efficient for use in cell-free systems, and thus they were introduced into eukaryotic extracts to produce mRNA for translation in situ, using DNA constructs with cognate promoters (Spirin 1991; Baranov and Spirin 1993; Craig et al. 1992) (Table 6). In this case, however, no real coupling between transcription and translation takes place, since the bacteriophage RNA polymerases work much faster than the translation system. Thus, mRNA is synthesized mainly in advance. The term *combined transcription-translation* can be proposed for this situation.

Table 6. Crude eukaryotic combined transcription-translation system (Spirin 1991; Baranov and Spirin 1993; Craig et al. 1992)

Reticulocyte lysate (treated with microccocal Ca^{2+}-dependent RNAase), or wheat germ extract: contains ribosomes, tRNAs, ARSes, and translation factors

DNA (plasmid with SP6 or T7 promoter)

Bacteriophage SP6 or T7 RNA polymerase

(Total tRNA addition may be required for optimal synthesis)

Amino acids

ATP, GTP, UTP, CTP

NTP-regenerating system: CP + CK

SH-compound: DTT

(Hemin, glutathione, cAMP may be useful in the case of reticulocyte system)

Mg^{2+}, K^+ and polyamines (spermidine or spermine) in concentrations optimal for translation

(Ca^{2+} may be also recommended in the case of wheat germ system)

Purified, Modified, and Simplified Peptide-Synthesizing Systems Composed of Pure Individual Components

In all previous cases, the cell-free systems were based either on the crude cell extract, including ribosomes and all soluble enzymes, factors, and tRNAs (S-30 fraction), or on the ribosome-free extract (S-100 fraction) combined with isolated ribosomes. In the cases when the ribosome-free extract had been freed from polynucleotides, total tRNA was added to the mixture. Since major progress was achieved in the isolation and purification of different translation factors as well as of individual aminoacyl-tRNA synthetases and tRNAs, the cell-free systems could be composed from a set of pure components. Today, the cell-free protein-synthesizing systems may be reconstituted from well-characterized, highly purified components, including ribosomes, mRNA or DNA + RNA polymerase, the full set of tRNAs and aminoacyl-tRNA synthetases (ARSes), and a set of special proteins called translation factors. The mixture must be supplemented with amino acids, two (ATP and GTP) or four (ATP, GTP, UTP and CTP) ribonucleoside triphosphates (NTPs), and an NTP-regenerating system (phosphoenol pyruvate + pyruvate kinase, or creatine phosphate + creatine kinase, or acetyl phosphate + acetyl kinase).

Pure Translation and Transcription-Translation Cell-Free Systems

The poly(U)-directed cell-free translation system (Table 7) is the easiest to assemble in this way. This is a very good model system for studies of elongation (initiation and termination steps of translation are absent from the system) and for testing ribosome activity. It was exploited intensively by many workers, and much useful information concerning ribosomes, elongation factors, energetics, and other features was obtained.

Table 7. Pure poly(U)-directed cell-free translation system

Ribosomes
Poly(U)
Phe-tRNA
Elongation factors:
EF-Tu, EF-Ts and EF-G in the case of bacterial ribosomes, or eEF1 A, eEF1B and eEF2 in the case of eukaryotic ribosomes
GTP
GTP-regenerating system: PEP + PK
Mg^{2+}, and K^+ or NH_4^+, in optimal concentrations

Table 8. Pure mRNA-directed cell-free translation system with a set of pre-aminoacylated tRNAs (Ganoza et al. 1985; Green et al. 1985)

Ribosomes (bacterial)
Individual mRNA (e.g., phage RNA)
Full set of aminoacyl-tRNAs, including F-Met-tRNA
Initiation factors: IF1, IF2, IF3
Elongation factors: EF-Tu, EF-Ts, EF-G
Additional protein factors: rescue factor, EF-P, W
Termination factors: RF1 or RF2, RF3, RRF
GTP
DTT
Mg^{2+} and NH_4^+ in optimal concentrations

Using natural mRNA as a template, the purified system must be supplemented by initiation factors (three proteins in prokaryotic systems or a dozen proteins in eukaryotic systems), termination factors (two or three proteins in prokaryotic and one or two in eukaryotic systems), and a full set of aminoacyl-tRNAs. Such systems were used for studies of initiation and termination phases of translation and the role of individual factors in these processes. A nice example is the bacterial *(E. coli)* system for translation of phage RNAs (Table 8) (Ganoza et al. 1985; Green et al. 1985); several additional protein factors, such as rescue factor, elongation factor EF-P, and W were also added and shown to be required for effective translation.

A more sophisticated pure system that continuously recycles tRNA during translation can be formed. Instead of including the set of aminoacylated tRNAs in the system, the full set of tRNAs, all individual ARSes (20 proteins), and formyltetrahydrofolate-Met-tRNA$_f$ transformylase should be introduced, along with amino acids, GTP, ATP, an NTP-regenerating system, and a formyl group donor (Table 9) (Kung et al. 1978). Thirty-three individual proteins were used to construct a pure coupled transcription-translation system for gene-dependent synthesis of β-galactosidase (Kung et al. 1977) that includes RNA polymerase, formyltetrahydrofolate-Met-tRNA$_f$ transformylase, 20 ARSes, three initiation factors, three elongation factors, termination factors, and several additional factors

Table 9. Pure mRNA-directed cell-free translation system with aminoacylation of tRNAs (Kung et al. 1978)

Ribosomes (bacterial)
mRNA
Total tRNA
20 individual ARSes
Formyltetrahydrofolate Met-tRNA$_f$ transformylase
Initiation factors: IF1, IF2, IF3
Elongation factors: EF-Tu, EF-Ts, EF-G
Termination factors: RF1, RF2, (RF3), RRF
Amino acids
ATP, GTP (UTP, CTP)
PEP + PK
Methenyltetrahydrofolate
DTT
Mg^{2+}, K$^+$ and NH$_4^+$ in optimal concentrations

Table 10. Pure DNA-directed cell-free transcription-translation system (Kung et al. 1977, 1979; Zarucki-Schulz et al. 1979)

Ribosomes (bacterial)
DNA (individual gene)
Total tRNA
RNA polymerase
20 individual ARSes
Formyltetrahydrofolate Met-tRNA$_f$ transformylase
Initiation factors: IF1, IF2, IF3
Elongation factors: EF-Tu, EF-Ts, EF-G
Termination factors: RF1, RF2, (RF3), RRF
Amino acids
ATP, GTP, UTP, CTP
PEP + PK
Methenyltetrahydrofolate
DTT
Mg^{2+}, K$^+$, NH$_4^+$, and spermidine in optimal concentrations

(Table 10). This system was later improved (Kung et al. 1979) and also used for synthesis of the proteins of the transcriptional and translational machinery, such as ribosomal proteins L10 and L12, EF-Tu and EF-G, and RNA polymerase subunits (Zarucki-Schulz et al. 1979). These systems can be used for studies of the regulation mechanisms of protein synthesis and the role of individual protein factors, as well as in the search for new factors required for transcription and translation. Recently, an efficient protein synthesis system composed of pure components has been proposed by Ueda et al. (p. 53 this volume).

Factor-Free Translation Systems

With poly(U) as a message in a cell-free translation system, all protein translation factors can be omitted. Thus, the system can consist of just carefully washed bacterial ribosomes, poly(U), and phenylalanyl-tRNA. It was found that in such a factor-free system, the steps of aminoacyl-tRNA binding and translocation proceed spontaneously ("non-enzymatically"), whereas the transpeptidation step is normally catalyzed by the ribosome; the result is the slow translation of the poly(U) message into polyphenylalanine (Pestka 1969, 1974; Gavrilova and Smolyaninov 1971; Gavrilova and Spirin 1974; Gavrilova et al. 1976). The factor-free translation could be significantly enhanced by some covalent modifications (SH group blocking) or mutational alterations of ribosomes, or by removal of ribosomal protein S12 from ribosomes (Gavrilova and Spirin 1972, 1974; Gavrilova et al. 1974; Asatryan and Spirin 1975). Factor-free translation of poly(A) into oligolysines (Koteliansky and Spirin 1975) and poly(U,C) into copolymers of phenylalanine, leucine, serine, and proline (Rutkevitch and Gavrilova 1982) was also demonstrated.

This translation system is the simplest of all. It has been used for studies of energetics and accuracy of elongation and, after addition of just one of the elongation factors (either EF-Tu or EF-G), for studies of the contribution of the elongation factors to the rate and accuracy of translation (e.g., Gavrilova et al. 1981).

Template-Free Elongation System

Another simplified system of peptide synthesis deserves attention: this is the template-free system for ribosomal synthesis of some polypeptides from aminoacyl-tRNA. It was observed that in a system consisting of ribosomes, elongation factors, lysyl-tRNA, and GTP, without any message, oligolysines up to six to seven residues long were synthesized; the synthesis depended entirely on ribosome functions and strictly required EF-Tu, EF-G, and GTP (Belitsina et al. 1981). Phenylalanyl-tRNA, however, was found to be incapable of serving as a substrate in the template-free, ribosome-catalyzed elongation. At the same time, when tRNA[Lys] was misacylated with phenylalanine, the system produced polyphenylalanine from Phe-tRNA[Lys] (Yusupova et al. 1986). Hence, it is a property of aminoacylated tRNA[Lys] that allows its participation in the ribosomal elongation cycle without a message. Some other aminoacyl-tRNAs were also found to be capable of serving as substrates for the template-free elongation on ribosomes. Among 16 aminoacyl-tRNAs tested, lysyl-, seryl-, threonyl-, and aspartyl-tRNAs proved to be the best, whereas phenylalanyl-, asparaginyl-, methionyl-, isoleucyl-, and some other tRNAs could not serve as substrates in the absence of a message (Yusupova et al. 1986). It is interesting that not only was the substrate-binding reaction factor-dependent, but the translocation of peptidyl-tRNAs during synthesis was normally catalyzed by EF-G and GTP.

Immobilized Translation Systems

The idea of immobilization of the message or the ribosome on a solid matrix for the use in cell-free polypeptide synthesis was always very attractive. Such a "solid-phase translation system" was first realized by covalent coupling of the 3'-terminus of a message polynucleotide to a resin of cellulose or dextran type, with subsequent use of the resin-bound message in cell-free translation (Belitsina et al. 1973; Belitsina and Spirin 1979). Free ribosomes initiated translation from the 5'-ends of the immobilized messages and moved toward the resin surface during elongation, thus also becoming immobilized on the resin via the message polynucleotide. The suspension of the resin-bound translating ribosomes could be packed into a column, washed off from non-translating ribosomes and other components of the original translation system and reused in continuing elongation with pure components. These systems were successfully applied for studies of individual steps of the elongation cycle (Belitsina et al. 1975a, 1976, 1979). The preparation of the fraction of 100% active translating ribosomes and one-functional-state translating ribosomes for physical studies was achieved by using the modified system with a message covalently linked to the resin via a cleavable bridge (Belitsina et al. 1975b; Baranov et al. 1979; Spirin et al. 1987).

Continuously Fed Cell-Free Systems

A principal shortcoming of all test-tube cell-free translation and transcription-translation systems should be mentioned: in contrast to in vivo protein synthesis, they have short lifetimes and, as a consequence, give low yields of protein products. This makes them useful mainly for analytical purposes and inappropriate for preparative syntheses of polypeptides and proteins. Indeed, the bacterial (*E. coli*) cell-free systems are active, as a rule, only for 10–30 min at 37°C. The systems based on rabbit reticulocyte lysate or wheat germ extract are typically capable of working for 1 h, although in some cases the lifetime may be prolonged up to 3 or 4 h. Recent innovations, however, allow further prolongation of the active period of test-tube cell-free systems and raising the yield (see Kim and Swartz 1999, 2000; also Kim and Swartz, p. 41 this volume; Lamla and Erdmann (p. 23 this volume).

A decade ago, a novel principle was introduced into the methodology of cell-free protein-synthesizing systems (Spirin et al. 1988; Baranov et al. 1989; Spirin 1991). Instead of incubating the reaction mixture in a fixed volume of a test-tube, the incubation was done under conditions of *continuous removal of the reaction products* (inorganic phosphates, nucleoside monophosphates, and polypeptides) and *continuous supply with the consumable substrates* (amino acids, NTPs and energy-regenerating compounds). This can be achieved with the use of a porous barrier that retains the high-molecular-weight components of the protein-synthesizing machinery within a defined reaction compartment and, at the same time, provides the continuous feeding with the substrates and the removal of the products.

Continuous-Exchange Cell-Free Systems

„The simplest configuration is a membrane bag containing the reaction region, while retaining a solution outside the membrane, which provides for the desired level of the lower molecular weight components in the reaction region. Thus, by exchange across the membrane, the lower molecular weight products produced by the reaction will be continuously dialyzed into the external solution, while the reaction components will be continuously replenished in the reaction region".

„The simplest configuration is a membrane bag containing the reaction region, while retaining a solution outside the membrane, which provides for the desired level of the lower molecular weight components in the reaction region. Thus, by exchange across the membrane, the lower molecular weight products produced by the reaction will be continuously dialyzed into the external solution, while the reaction components will be continuously replenished in the reaction region".

In practice, both dialysis bags and flat membranes could be taken for performing the CECF system run. Using this approach, the Promega Corporation group carried out the synthesis of firefly luciferase in the bacterial (*E. coli*) extract for 20 h, yielding 120–240 μg of the protein per milliliter (Davis et al. 1996), and Kim and Choi (1996) reported on the synthesis of 1.2 mg of CAT during 14 h in wheat germ extract. More recently, Kigawa et al. (1999) succeeded in synthesizing CAT and Ras proteins in amounts up to 6 mg/ml for 18 h in their version of the bacterial CECF system, and Madin et al. (2000) reached the yields of 1–4 mg/ml for several functionally active proteins (DHFR, GFP, luciferase, and RNA replicase of TMV) in the wheat embryo CECF system incubated for 60 h. Translation of pre-synthesized mRNA (Madin et al. 2000), coupled transcription-translation with bacterial RNA polymerase (Davis et al. 1996), and combined transcription-translation with bacteriophage T7 or SP6 RNA polymerases (Kim and Choi 1996; Kigawa et al. 1999) were exploited in prokaryotic or eukaryotic versions of the CECF system (see also Madin et al. p. 109, this volume).

As the product synthesized in the CECF system accumulates in the reaction compartment (a dialysis bag or a membrane-limited chamber), the syntheses of proteins and polypeptides fused with GFP provide a direct and demonstrative way to visualize the product accumulation by GFP moiety fluorescence. The examples are the syntheses of an HIV protein, the so-called Nef antigen, fused with GFP (Chekulayeva et al. 2001) and an antibacterial polypeptide Cecropin P1 (31 amino acid long) fused with GFP (Martemyanov et al. 2001), both in the bacterial CECF T7-transcription-translation system.

Continuous-Flow Cell-Free Systems

The primary configuration of the systems under consideration employs a feeding solution containing the consumable substrates continuously introduced into the reaction region *by flow*, while the same volume of liquid containing the products is continuously withdrawn from the reaction region (Spirin et al. 1988; Baranov and Spirin 1993; Alakhov et al. 1995). This format is designated as a *continuous-flow cell-free (CFCF) system*. The principal scheme of the flow reactor is shown by Shirokov et al. (p. 91) this volume.

Typically, the porous barrier for retaining the high-molecular-weight components of the protein-synthesizing machinery in the reaction compartment is an ultrafiltration membrane (e.g., Amicon PM-30, or XM-100, or YM-300). Most often the capacity of the reaction compartment in the published experiments was 1 ml, though principally the CFCF reactors can be easily scaled up (see Spirin 1991). The rate of the flow through the reaction compartment may vary, commonly being from 1/2 to 3 volumes over that of the reaction compartment capacity per hour (e.g., from 1/2 to 3 ml of feeding solution per hour when the reaction chamber is of 1 ml capacity). A more or less constant rate of the protein synthesis is usually observed during 20–100 h. Both bacterial (*E. coli*) and eukaryotic (rabbit reticulocyte or wheat germ) extracts were successfully used for the experiments with the CFCF systems (reviewed by Spirin 1991, Baranov and Spirin 1993). The CFCF systems of this type were tested in translation, coupled transcription-translation, and combined transcription-translation formats for the syntheses of functionally active DHFR, CAT, GFP, IL-2 and IL-6, as well as virus coat proteins, globin and some other polypeptides and enzymes (Spirin et al. 1988; Baranov et al. 1989; Ryabova et al. 1989, 1994, 1998; Spirin 1991; Kigawa and Yokoyama 1991; Endo et al. 1992, 1993; Kolosov et al. 1992; Kudlicki et al. 1992; Baranov and Spirin 1993; Volyanik et al. 1993; Uzawa et al. 1993; Nishimura et al. 1993, 1995; Alexandrov et al. (1996); see also Shirokov et al., p.91, this volume).

One of the most remarkable observations made with the CFCF systems was the absence of a significant leakage of the components of the protein-synthesizing machinery, such as tRNAs, translation factors, ARSes, through 50–300 kDa cut-off pores of ultrafiltration membranes, while the synthesized proteins of comparable molecular masses were removed from the reactor by flow. The retention of the macromolecules participating in protein synthesis requires them to be in a functionally active state. The explanation of this fact seems to be obvious: large dynamic complexes are formed in the process of translation or transcription-translation, so that no free tRNA, aminoacyl-tRNA, translation factors or ARSes exist in significant amounts at any given moment of the process. For instance, tRNA is bound with ARSes (or even a big complex of several ARSes in eukaryotic extracts) and quickly aminoacylated, and the aminoacyl-tRNA is immediately picked up by elongation factor 1. A large proportion of these macromolecules are present within translating polyribosomes.

Thus, one of the principal differences between the CECF and CFCF systems is the following: whereas the synthesized protein or polypeptide accumulates in the complex reaction mixture during CECF run, in the case of the CFCF system the protein product is continuously removed from the reaction compartment (if membrane pore size permits) and therefore separated from the bulk of proteins of the protein-synthesizing machinery. Moreover, numerous odd proteins present in a crude cell extract are usually leaking out through the ultrafiltration membrane of the reactor during the first hours of the run. As a result, if the first fractions of the outflow are discarded, the protein synthesized during the CFCF run can be collected from the reactor in a relatively pure state. For example, interleukin-4 of 85% purity was collected from a scaled-up CFCF reactor by Alakhov et al. (unpublished data; cited in Spirin 1991).

Conclusion

The use of cell-free translation systems in the laboratory practice of molecular biologists was a key experimental approach to molecular mechanisms of protein biosynthesis. The major part of the knowledge about the genetic code, mRNA, ribosome functions, protein factors involved in translation, translation stages, translational control, co-translational protein folding, etc. – was obtained due to the use of cell-free systems. Now, in addition to the principal contributions of the cell-free methodology to basic science, the possibilities of technological applications of cell-free translation systems have arisen. The long lifetimes and high productivity of the continuous (CFCF and CECF) systems make them promising for practical use.

Among the most obvious applications of the cell-free gene expression technology are the following:
1. Synthesis of cytotoxic proteins and polypeptides
2. Expression of unidentified open reading frames, functionally unstable or poorly expressible genes, and genes encoding for unstable products
3. Functional mapping of genomes through direct expression of genomic libraries
4. Synthesis of proteins with unnatural, chemically modified or isotope-labeled amino acid residues, including those for NMR spectroscopy
5. Synthesis of polypeptides and proteins, including direct expression of genomic libraries, for structural analyses (e.g., by NMR spectroscopy and X-ray crystallography)
6. In vitro protein engineering
7. Screening of engineered and theoretically designed proteins.

As an example of practical applications, the CECF transcription-translation system was successfully used for stable-isotope labeling of a protein product with $^{13}C/^{15}N$-amino acids for NMR spectroscopy studies (Kigawa et al. 1999).

The cell-free technology for protein synthesis has several advantages as compared with the biotechnologies based on living organisms. These are the speed and directness of all procedures, the absence of constraints from a living cell, the ease of operator control, the purity of a product, and the wide possibilities of product modifications. These are all grounds to believe that in the near future the cell-free protein-synthesizing systems will strongly contribute to biotechnology development.

References

Alakhov YB, Baranov VI, Ovodov SJ, Ryabova LA, Spirin AS, Morozov IJ (1995) Method of preparing polypeptides in cell-free translation system. United States Patent no. 5,478,730

Alexandrov A, Kolosova I, Kolosov M (1996) mRNA stabilization in continuous flow translation system. Biochem Mol Biol Intern 38:1111–1116

Anderson CW, Straus W, Dudock BS (1983) Preparation of cell-free protein-synthesizing system from wheat germ. Methods Enzymol 101:635–644

Asatryan LS, Spirin AS (1975) Non-enzymatic translocation in ribosomes from streptomycin-resistant mutants of *Escherichia coli*. Mol Gen Genet 138:315–321

Ball LA, White CN (1976) Order of transcription of genes of vesicular stomatitis virus. Proc Natl Acad Sci USA 73:442–446

Baranov VI, Morozov IY, Ortlepp SA, Spirin AS (1989) Gene expression in a cell-free system on the preparative scale. Gene 84:463–466

Baranov VI, Spirin AS (1993) Gene expression in cell-free systems on preparative scale. Methods Enzymol 217:123–142

Baranov VI, Belitsina NV, Spirin AS (1979) The use of columns with matrix-bound polyuridylic acid for isolation of translating ribosomes. Methods Enzymol 59:382–397

Belitsina NV, Spirin AS (1979) Translation of matrix-bound polyuridylic acid by *Escherichia coli* ribosomes (solid-phase translation system). Methods Enzymol 60:745–760

Belitsina NV, Girshovich AS, Spirin AS (1973) Translation of resin-bound polynucleotide. Doklady Akad Nauk SSSR 210:214–227

Belitsina NV, Glukhova MA, Spirin AS (1975a) Translocation in ribosomes by attachment-detachment of elongation factor G without GTP cleavage Evidence from a column-bound ribosome system. FEBS Lett 54:35–38

Belitsina NV, Elizarov SM, Glukhova MA, Spirin AS, Butorin AS, Vasilenko SK (1975b) Isolation of translating ribosomes with a resin-bound polyU-column. FEBS Lett 57:262–266

Belitsina NV, Glukhova MA, Spirin AS (1976) Stepwise elongation factor G-promoted elongation of polypeptides on the ribosome without GTP cleavage. J Mol Biol 108:609–613

Belitsina NV, Glukhova MA, Spirin AS (1979) Elongation factor G-promoted translocation and polypeptide elongation in ribosomes without GTP cleavage. Use of columns with matrix-bound polyuridylic acid. Methods Enzymol 60:761–779

Belitsina NV, Tnalina GZ, Spirin AS (1981) Template-free ribosomal synthesis of polylysine from lysyl-tRNA. FEBS Lett 131:289–292

Belitsina NV, Tnalina GZ., Spirin AS (1982) Template-free ribosomal synthesis of polypeptides from aminoacyl-tRNAs. BioSystems 15:233–241

Borsook H (1950) Protein turnover and incorporation of labeled amino acids into tissue proteins in vivo and in vitro. Physiol Rev 30:206–219

Bujard H, Gentz R, Lanzer M, Stueber D, Mueller M, Ibrahimi I, Haeuptle MT, and Dobberstein B (1987) A T5 promoter-based transcription-translation system for the analysis of proteins in vitro and in vivo. Methods Enzymol 155:416–433

Byrne R, Levin JG, Bladen HA, Nirenberg MW (1964) The in vitro formation of a DNA-ribosome complex. Proc Natl Acad Sci USA 52:140–148

Chao FC, Schachman HK (1956) The isolation and characterization of a macromolecular ribonucleoprotein from yeast. Arch Biochem Biophys 61:220–230

Chekulayeva MN, Kurnasov OV, Shirokov VA, Spirin AS (2001) Continuous-exchange cell-free protein-synthesizing system: Synthesis of HIV-1 antigen Nef. Biochem Biophys Res Commun 280:914–917

Chen HZ, Zubay G (1983) Prokaryotic coupled transcription-translation. Methods Enzymol 101:674–690

Coen DM, Bedbrook JR., Bogorad L, Rich A (1977) Maize chloroplast DNA fragment encoding the large subunit of ribulosebiphosphate carboxylase. Proc Natl Acad Sci USA 74: 5487–5491

Content J, de Witt L, Horisberger M (1977) Cell-free coupling of influenza virus RNA transcription and translation. J Virol 22:247–255

Craig D, Howell MT, Gibbs CL, Hunt T, Jackson RJ (1992) Plasmid cDNA-directed synthesis in a coupled eukaryotic in vitro transcription-translation system. Nucleic Acids Res 20:4987–4995

Davis J, Thompson D, Beckler GS (1996) Large scale dialysis cell-free system. Promega Notes Magazine No 56:14–21

DeVries JK, Zubay G (1967) DNA-directed peptide synthesis. II. The synthesis of the α-fragment of the enzyme β-galactosidase. Proc Natl Acad Sci USA 57:1010–1012

Endo Y, Otsuzuki S, Ito K, Miura K (1992) Production of an enzymatic active protein using a continuous flow cell-free system. J Biotech 25:221–230

Endo Y, Oka T, Ogata K, Natori Y (1993) Production of dihydrofolate reductase by an improved continuous flow cell-free translation system using wheat germ extract. Tokishima J Exp Med 40:13–17

Gale EF, Folkes JP (1954) Effect of nucleic acids on protein synthesis and amino-acid incorporation in disrupted staphylococcal cells. Nature 173:1223–1227

Ganoza MC, Cuningham C, Green RM (1985) Isolation and point of action of a factor from Escherichia coli required to reconstruct translation. Proc Natl Acad Sci USA 82:648–1652

Gavrilova LP, Smolyaninov VV (1971) Studies on mechanism of translocation in ribosomes. I. Synthesis of polyphenylalanine in Escherichia coli ribosomes in the absence of GTP and transfer protein factors. Mol Biol (Mosk) 8:883–891

Gavrilova LP, Spirin AS (1972) A modification of the 30 S ribosomal subparticle is responsible for stimulation of "non-enzymatic" translocation by p-chloromercuribenzoate. FEBS Letters 22:91–92

Gavrilova LP, Spirin AS (1974) "Nonenzymatic" translation. Methods Enzymol 30:452–462

Gavrilova LP, Koteliansky VE, Spirin AS (1974) Ribosomal protein S12 and "non-enzymatic" translocation. FEBS Lett 45:324–328

Gavrilova LP, Kostiashkina OE, Koteliansky VE, Rutkevitch NM, Spirin, AS (1976). Factor-free ("non-enzymic") and factor-dependent systems of translation of polyuridylic acid by Escherichia coli ribosomes. J Mol Biol 101:537–552

Gavrilova LP, Perminova IN, Spirin AS (1981) Elongation factor Tu can reduce translation errors in poly(U)-directed cell-free system. J Mol Biol 16:67–84

Gold LM, Schweiger M (1969) Synthesis of phage-specific α- and β-glucosyl transferases directed by T-even DNA in vitro. Proc Natl Acad Sci USA 62:892–898

Gold LM, Schweiger M (1971) Synthesis of bacteriophage-specific enzymes directed by DNA in vitro. Methods Enzymol 20:537–542

Green RH, Glick BR, Ganoza MC (1985) Requirements for in vitro reconstruction of protein synthesis. Biochem. Biophys Res Commun 126:792–798

Henshaw EC, Panniers R (1983) Translation systems prepared from the Ehrlich ascites tumor cell. Methods Enzymol 101:616–629

Jackson RJ, Hunt T (1983) Preparation and use of nuclease-treated rabbit reticulocyte lysates for the translation of eukaryotic messenger RNA. Methods Enzymol 96:50–74

Keller EB, Zamecnik PC (1956) The effect of guanosine diphosphate and triphosphate on the incorporation of labeled amino acids in protein. J Biol Chem 221:45–59

Khesin RB (1953) Formation of amylase by cytoplasmic granules isolated from pancreas cells. Biokhimiya (USSR) 18:462–474

Kigawa T, Yokoyama S (1991) A continuous cell-free protein synthesis system for coupled transcription-translation. J Biochem (Tokyo) 110:166–168

Kigawa T, Yabuki T, Yoshida Y, Tsutsui M, Ito Y, Shibata T, Yokoyama S (1999) Cell-free production and stable-isotope labelling of milligram quantities of proteins. FEBS Lett 442:15–19

Kim DM, Choi CY (1996) A semi-continuous prokaryotic coupled transcription/translation system using a dialysis membrane. Biotechnol Prog 12:645–649

Kim DM, Swartz JR (1999) Prolonging cell-free protein synthesis with a novel ATP regeneration system. Biotech Bioengineering 66:180–188

Kim DM, Swartz JR (2000) Prolonging cell-free protein synthesis by selective reagent additions. Biotech Prog 16:385–390

Kolosov MI, Kolosova IM, Alakhov VY, Ovodov SY, Alakhov YB (1992) Preparative in vitro synthesis of bioactive human interleukin-2 in a continuous flow translation system. Biotech Appl Biochem 16:125–133

Koteliansky VE, Spirin AS (1975) "Non-enzymatic" translocation in ribosomes using polyadenylic acid as a template. Doklady Akad Nauk SSSR 221:477–480

Kudlicki W, Kramer G, Hardesty B (1992) High-efficiency cell-free synthesis of proteins: refinement of the coupled transcription/translation system. Anal Biochem 206:389–393

Kung HF, Chu F, Caldwell P, Spears C, Treadwell BV, Eskin B, Brot N, Weissbach H (1978) The mRNA-directed synthesis of the α-peptide of β-galactosidase, ribosomal protein L12 and L10, and elongation factor Tu, using purified translational factors. Arch Biochem Biophys 187:457–463

Kung HF, Redfield B, Treadwell BV, Eskin B, Spears C, Weissbach H (1977) DNA-directed in vitro synthesis of β-galactosidase. J Biol Chem 252:6889–6894

Kung HF, Redfield B, Weissbach H (1979) DNA-directed in vitro synthesis of β-galactosidase. J Biol Chem 254:8404–8408

Lamborg H, Zamecnik PC (1960) Amino acid incorporation into protein by extracts of E. coli. Biochim Biophys Acta 42:206–211

Lederman M, Zubay G (1967) DNA-directed peptide synthesis. I. A comparison of T_2 and Escherichia coli DNA-directed peptide synthesis in two cell-free systems. Biochim Biophys Acta 149:253–258

Lewis JB, Anderson CW, Atkins JF, Gesteland RF (1975) The origin and destiny of adenovirus proteins. Cold Spring Harbor Symp Quant Biol 39:581–590

Littlefield JW, Keller EB (1957) Incorporation of C^{14} amino acids into ribonucleoprotein particles from the Ehrlich mouse ascites tumor. J Biol Chem 224:13–30

Littlefield, JW, Keller EB, Gross J Zamecnik PC (1955) Studies on cytoplasmic ribonucleoprotein particles from the liver of the rat. J Biol Chem 217:111–123

Madin K, Sawasaki T, Ogasawara T, Endo Y (2000) A highly efficient and robust cell-free protein synthesis system prepared from wheat embryos: Plants apparently contain a suicide system directed at ribosomes. Proc Natl Acad Sci USA 97:559–564

Marcus A, Efron D, Weeks DP (1974) The wheat embryo cell-free system. Methods Enzymol 30:749–754

Martemyanov KA, Shirokov VA, Kurnasov OV, Gudkov AT, Spirin AS (2001) Cell-free production of biologically active polypeptides: Application to the synthesis of antibacterial polypeptide cecropin. Protein Expression and Purification 21:456–461

Matthaei JH, Nirenberg MW (1961) Characteristics and stabilization of DNAase-sensitive protein synthesis in E. coli extracts. Proc Natl Acad Sci USA 47:1580–1588

McQuillen K, Roberts RB, Britten RJ (1959) Synthesis of nascent protein by ribosomes in E. coli. Proc Natl Acad Sci USA 45:1437–1447

Merrick WC (1983) Translation of exogenous mRNAs in reticulocyte lysates. Methods Enzymol 101:606–615

Nirenberg MW (1963) Cell-free protein synthesis directed by messenger RNA. Methods Enzymol 6:17–23

Nirenberg MW, Matthaei JH (1961) The dependence of cell-free protein synthesis in E. coli upon naturally occurring or synthetic polynucleotides. Proc Natl Acad Sci USA 47:1588–1602

Nishimura N, Kitaoka Y, Mimura A, Takahara Y (1993) Continuous protein synthesis system with Escherichia coli S30 extract containing endogenous T7 RNA polymerase. Biotech Lett 15:785–790

Nishimura N, Kitaoka Y, Niwano M (1995) Cell-free system derived from heat-shocked Escherichia coli: Synthesis of enzyme protein possessing higher specific activity. J Ferment Bioeng 79:131–135

Palade GE (1955) A small particulate component of the cytoplasm. Biophys Biochem Cytol 1:59–68

Pelham HRB, Jackson RJ (1976) An efficient mRNA-dependent translation system from reticulocyte lysates. Eur J Biochem 67:47–256

Pelham HRB, Sykes JMM, Hunt T (1978) Characteristics of a coupled cell-free transcription and translation system directed by vaccinia cores. Eur J Biochem 82:199–209

Pestka S (1969) Studies on the formation of transfer ribonucleic acid-ribosome complex. VI. Oligopeptide synthesis and translocation on ribosomes in the presence and absence of soluble transfer factors. J Biol Chem 244:1533–1539

Pestka S (1974) Assays for nonenzymatic and enzymatic translocation with *Escherichia coli* ribosomes. Methods Enzymol 30:462–470

Peterman ML, Hamilton MG (1957) The purification and properties of cytoplasmic ribonucleoprotein from rat liver. J Biol Chem 224:725–736

Peterson EA, Greenberg DM (1952) Characteristics of the amino acid-incorporating system of liver homogenates. J Biol Chem 194:359–375

Roberts BE, Paterson BM (1973) Efficient translation of tobacco mosaic virus RNA and rabbit globin 9 S RNA in a cell-free system from commercial wheat germ. Proc Natl Acad Sci USA 70:2330–2334

Roberts BE, Corecki M, Mulligan RC, Danna KJ, Rozenblatt S, Rich A (1975) Simian virus 40 DNA directs synthesis of authentic viral polypeptides in a linked transcription-translation cell-free system. Proc Natl Acad Sci USA 72:1922–1926

Roberts RB (ed) (1958) Microsomal Particles and Protein Synthesis. Pergamon Press New York

Rutkevitch NM, Gavrilova LP (1982) Factor-free and one-factor-promoted poly (U, C)-dependent synthesis of polypeptides in cell-free systems from *Escherichia coli*. FEBS Lett 143:115–118

Ryabova LA, Ortlepp SA, Baranov VI (1989) Preparative synthesis of globin in a continuous cell-free translation system from rabbit reticulocytes. Nucleic Acids Res 17:4412

Ryabova L, Volianik E, Kurnasov O, Spirin AS, Wu Y, Kramer FR (1994) Coupled replication-translation of amplifiable messenger RNA: A cell-free protein synthesis system that mimics viral infection. J Biol Chem 269:1501–1505

Ryabova LA, Desplancq D, Spirin AS, Plückthun A (1997) Functional antibody production using cell-free translation: Effects of protein disulfide isomerase and chaperones. Nature Biotechnology 15:79–84

Ryabova LA, Morozov IY, Spirin AS (1998) Continuous-flow cell-free translation, transcription-translation, and replication-translation systems. In: Martin R (ed) Methods in Molecular Biology, vol 77: Protein Synthesis: Methods and Protocols. Humana Press, Totowa, NJ, pp 179–193

Schachtschabel D, Zillig W (1959) Untersuchungen zur Biosynthese der Proteine. I. Uber den Einbau C14-markierter Aminosauren ins Protein zellfreier Nucleoproteid-Enzyme-Systeme aus *E. coli* B. Hoppe-Seyler's Z Physiol Chem 314:262–275

Schreier MH, Staehelin T (1973) Initiation of mammalian protein synthesis: The importance of ribosome and initiation factor quality for the efficiency of in vitro synthesis. J Mol Biol 73:329–349

Schweiger M, Gold LM (1969) Bacteriophage T4 DNA-dependent in vitro synthesis of lysozyme. Proc Natl Acad Sci USA 63:1351–1358

Siekevitz P (1952) Uptake of radioactive alanine in vitro into proteins of rat liver fractions. J Biol Chem 195:549–565

Siekevitz P, Zamecnik PC (1951) In vitro incorporation of l-C14-DL-alanine into proteins of rat-liver granular fractions. Fed Proc 10:246–247

Spirin AS (1991) Cell-free protein synthesis bioreactor. In: Todd P, Sikdar SK, Beer M (eds) Frontiers in Bioprocessing II. American Chemical Society, Washington, DC, pp 31–43

Spirin AS (2000) Ribosomes. Kluwer Academic/Plenum Publishers, New York

Spirin AS, Baranov VI, Polubesov GS, Serdyuk IN, May RL (1987) Translocation makes the ribosome less compact. J Mol Biol 194:119–126

Spirin AS, Baranov VI, Ryabova LA, Ovodov SY, Alakhov YB (1988) A continuous cell-free translation system capable of producing polypeptides in high yield. Science 242:1162–1164

Stueber D, Ibrahimi I, Cutler D, Dobberstein B, Bujard H (1984) A novel in vitro transcription-translation system: accurate and efficient synthesis of single proteins from cloned DNA sequences. EMBO J 3:3143–3148

Tissiéres A, Watson JD (1958) Ribonucleoprotein particles from *E. coli*. Nature 182:778–780

Tissiéres A, Watson JD, Schlessinger D, Hollingworth BR (1959) Ribonucleoprotein particles from *Escherichia coli*. J Mol Biol 1:221–233

Tissiéres A, Schlessinger D, Gros F (1960) Amino acid incorporation into proteins by *E. coli* ribosomes. Proc Natl Acad Sci USA 46:1450–1463

Ts'o POP, Bonner J, Vinograd J (1956) Microsomal nucleoprotein particles from pea seedlings. J Biophys Biochem Cytol 2:451–465

Uzawa T, Yamagishi A, Ueda T, Chikazumi N, Watanabe K, Oshima T (1993) Effects of polyamines on a continuous cell-free protein synthesis system of an extreme thermophile, *Thermus thermophilus*. J Biochem (Tokyo) 114:732–734

Volyanik EV, Dalley A, McKay IA, Keigh I, Williams NS, Bustin SA (1993) Synthesis of preparative amounts of biologically active interleukin-6 using a continuous-flow cell-free translation system. Anal Biochem. 214:289–294

Winnick T (1950a) Incorporation of labeled amino acids into protein of embryonic and tumor tissue homogenates. Fed Proc 9:247

Winnick T (1950b) Studies on the mechanism of protein synthesis in embryonic and tumor tissues. II. Inactivation of fetal rat liver homogenates by dialyses and reactivation by the adenylic acids system. Arch Biochem 28:338–347

Wood WB, Berg P (1962) The effect of enzymatically synthesized ribonucleic acid on amino acid incorporation by a soluble protein-ribosome system from *Escherichia coli*. Proc Natl Acad Sci USA 48:94–104

Yamamoto YI, Nagahori H, Yao S, Zhang ST, Suzuki E (1996) Hollow fiber reactor for continuous flow cell-free protein production. J Chem Eng Japan 6:1047–1050

Yusupova GZ, Belitsina NV, Spirin AS (1986) Template-free ribosomal synthesis of polypeptides from aminoacyl-tRNA. Polyphenylalanine synthesis from phenylalanyl-tRNA^{Lys}. FEBS Lett 206:142–146

Zamecnik PC (1969) An historical account of protein synthesis, with current overtones – a personalized view. Cold Spring Harbor Symp Quant Biol 34:1–16

Zarucki-Schulz T, Jeres C, Goldberg G, Kung HF, Huan KH, Brot N, Weissbach H (1979) DNA-directed synthesis of proteins involved in bacterial transcription and translation. Proc Natl Acad Sci USA 76:6115–6119

Zubay G (1973) In vitro synthesis of protein in microbial systems. Annu Rev Genet 7:267–287

PART II

Improved Classical (Batch) Cell-Free Translation Systems

II

Improved Batch Translation System Based on *E. coli* Extract

Thorsten Lamla, Volker A. Erdmann*

Keywords: *in vitro* protein synthesis; coupled transcription/translation; Strep-tag affinity peptide; affinity chromatography; protein bioreactor

Abstract

Product removal from the continuous flow cell-free (CFCF) reactor still only partially solves some of the problems of this system, since the use of ultrafiltration membranes has some remaining limitations. These can be overcome by introducing an affinity system. A cell-free protein synthesis system has therefore been employed to produce bovine heart fatty acid binding protein (FABP) and bacterial chloramphenicol acetyltransferase (CAT) with and without fusion of the Strep-tag affinity peptide. These two fusion proteins were purified via a streptavidin and StrepTactin sepharose matrix respectively. No significant influence of the Strep-tag and the conditions during the affinity chromatography on maturation or activity of the proteins were observed. In addition, quantitative removal of the fusion proteins during cell-free synthesis from a batch reaction and a semicontinuous flow cell-free (SFCF) reactor was achieved. The results document that it is possible to avoid the limitations of the ultrafiltration membranes during product removal from a CFCF reactor. The data presented show that the affinity system is also well suited for the development of a novel protein bioreactor.

Introduction

The *in vitro* protein biosynthesis system has the potential not only to produce cellular proteins, but also to synthesize cytotoxic, regulatory or unstable proteins that cannot be expressed in living cells [39]. Another advantage is the use of incorporation of radioactively labeled amino acids, which facilitates their detec-

* Correspondence to Volker A. Erdmann, Institut für Chemie-Biochemie,
  Freie Universität Berlin, Thielallee 63, D-14195 Berlin
  Tel.: 0049-30-838-56002, Fax: 0049-30-838-56403, e-mail: erdmann@chemie.fu-berlin.de

tion. Site-directed isotope labeling allows study of their structure and dynamics by NMR [5, 9, 28, 43] and FTIR [37, 17] spectroscopy. Another way of investigating the structure and function of proteins is the site-specific incorporation of unnatural amino acids [reviewed in 4, 36] and with these it is even possible to create proteins with improved biological activities [26]. Expression PCR [reviewed in 15, 40, 20] can be used for revealing the products of the fast growing number of newly discovered genes, or for producing and analyzing proteins carrying engineered or random mutations [3]. Screening and evolution of peptides and proteins has recently been demonstrated by the use of *in vitro* translation [21, 6, 29].

Different versions of cell-free protein synthesis systems derived from rabbit reticulocyte lysates, as well as wheat germ or *Escherichia coli* are currently used. The batch configuration is sufficient to achieve analytical amounts of the desired protein [7, 11, 27, 13]. For production on a preparative scale, continuous flow cell-free (CFCF) reactors [39, 38, 2, 8, 14] and dialysis cells [24, 12, 18] have been established. Continuous product removal from the CFCF-reactor still only partially solves some of the well-known problems. The employment of ultrafiltration membranes is limited by the fact that the proteins may not pass through or might interact with the membrane. In addition, it may well be that some translation components get lost during the reaction. However, the use of a synthesized protein for further applications, such as crystallization or NMR studies, depends to a great extent upon its purity. For this purpose, the recombinant production and purification of proteins with short affinity tails have gained widespread application in biotechnology [31, 25]. In most of the examples investigated so far it has been found that these short peptide extensions, between three and twelve amino acids in length, do not interfere with the biological function of the protein and therefore need not be removed via proteolysis. One of these short affinity tags, termed Strep-tag (in this work Strep-tag I), is a nine amino acid peptide (AWRHPQFGG) with intrinsic streptavidin binding activity [K_d ca. 10^{-5} M] [32]. It has been shown that the Strep-tag I allows single-step protein purification from bacterial expression systems [33] and from *in vitro* synthesized proteins [1], but its fusion to recombinant proteins is restricted to the C-terminus. Another variant, designated Strep-tag II, which does not show this limitation, has been introduced [34]. This octapeptide (WSHPQFEK) possesses a binding affinity towards streptavidin and even a higher one towards recombinant core streptavidin [K_d ca. 10^{-6} M], named Strep-Tactin [42]. One advantage of the Strep-tag system is that elution of the bound Strep-tag fusion protein from the affinity matrix occurs in the native state and under very mild buffer conditions.

Here we report the feasibility and the compatibility of the Strep-tag affinity purification with our cell-free protein biosynthesis system because, in most cases, further analyses require a purified protein. In addition, the limitations from the use of the ultrafiltration membranes during product removal from a CFCF-reactor should be overcome by the affinity system described here.

Materials and Methods

Construction of plasmids

Standard methods for molecular biology were used [30]. The plasmids pHMFA and pHMFA+StII were prepared by Helmut Merk [23]. Both contain the sequence for the 5' untranslated region of phage T7 gene 10 followed by the coding sequence for FABP and the T7 transcription terminator 150 bp downstream of the coding sequence. The pHMFA+StII has in addition a 30 bp sequence coding for a linker and the Strep-tag II at the C-terminus of the protein. Another plasmid, named pFA+NstII, coding for Strep-tag II and a two amino acid linker at the N-terminus of the protein but otherwise identical to pHMFA, was constructed. Therefore, a PCR with pHMFA, T7-primer and reverse primer (5'-CACCATG-GCTTTTTCGAACTGCGGGTGGCTCCACATGGTATATCTCCTTCTTAAAG-3') was performed using Pfu DNA polymerase (Stratagene) according to the supplier's recommendation. The PCR product was digested with XbaI and NcoI and the resulting fragment was cloned into the adequate digested pHMFA vector.

The plasmid pCAT, coding for CAT and containing all elements necessary for efficient *in vitro* transcription/translation, was constructed in two steps. Firstly, the NcoI/BamHI-fragment of pCAT3 (Promega) was inserted into pET-3d. Secondly, the SphI/EcoRI-fragment of this modified pET-3d was cloned into the pUC 19 vector.

The plasmid pCAT served as template for the construction of PCR-products with a Strep-tag I and II at the end of the coding sequence. The forward primer was complementary to the T7 promoter and identical for both products. The reverse primers for introducing the Strep-tag I with linker and XbaI restriction site were ST1 (5'-GCTCGGCCGTCTAGATTAACCACCGAACTGCGGGTGACGCC-AAGCAGCGCTCGCCCCGCCCTGCCACTCATCGCAGTA-3') and ST2 (5'-GCT-CGGCCGTCTAGATTATTTTTCGAACTGCGGGTGGCTCCAAGCG-CTCGCCCCGCCCTGCCACTCATCGCAGTA-3'), which introduces the Strep-tag II with linker and XbaI restriction site. PCR was performed as described above. The PCR products were digested with XbaI and the XbaI/XbaI-fragments were subsequently cloned into the pCAT vector. The two additional plasmids coding for CAT with C-terminal Strep-tag I and II were termed pCAT+StI and pCAT+StII respectively.

Coupled *in vitro* transcription/translation

The coupled *in vitro* transcription/translation reaction is based on an *Escherichia coli* S30 lysate (strain D12) and was carried out as described [44] with some modifications. The lysate was completed with 500 U/ml T7 phage RNA polymerase (Stratagene) and 200 μM L-[14C]leucine (25 dpm/pmol; Amersham). Plasmids were used in concentrations of 0.5 to 2 nM. The reactions were incubated for 90 min at 37°C or for 120 min at 30°C. Control reactions were performed under identical conditions without plasmids.

Analysis of the synthesized protein

The incorporation of L-[^{14}C]leucine into the synthesized proteins was determined by liquid scintillation counting of the trichloroacetic acid-insoluble material as described [22]. The reaction products were also analyzed by denaturing poly-acrylamide gel electrophoresis (SDS-PAGE) [16] followed by autoradiography in a 'Phosphorimager' system (Molecular Dynamics).

CAT-Assay

The activity from *in vitro* synthesized CAT was detected with the FAST CAT® (deoxy) chloramphenicol acetyltransferase assay kit according to the manufacturer's protocol (Molecular Probes), with some modifications. The supernatant of a coupled transcription/translation reaction after centrifugation at 15,000 × g for 5 minutes was diluted 500-fold with a buffer (50 mM Tris-HCl, pH 7.8/ 2 mM DTT/ 0.03% BSA), and between 1 µl and 17 µl were used in a total volume of 24 µl (same buffer). For the enzymatic analysis, 4 µl of each solution, the FAST CAT substrate and the 9 mM acetyl CoA, were added. The reaction was stopped by extraction with 400 µl of ice-cold ethyl acetate. After a short centrifugation, the top 300 µl of ethyl acetate was transferred to a clean tube, the solvent evaporated, the dry sample dissolved in 20 µl of ethyl acetate and, finally, 3 µl of this solution was analyzed after thin layer chromatography in the 'Fluoroimager' system (Molecular Dynamics).

Strep-tag affinity purification

After *in vitro* protein biosynthesis

Purification of the Strep-tag fusion proteins was performed by affinity chromatography according to the manufacturer's protocol (IBA, Göttingen) except that the volume of the affinity column was reduced to 230 µl to purify 150 µl reaction mixture. The wash and elution volumes were 230 µl and 130 µl respectively. Reaction mixtures were briefly centrifuged after coupled transcription/translation and subjected to the column. The isolated fractions were analyzed by TCA precipitation and by autoradiography after SDS-PAGE as described.

Removal of fusion protein from a batch system during *in vitro* protein synthesis

After the affinity matrix (50 µl) was equilibrated with translation buffer (50 mM HEPES-KOH (pH 7.6), 70 mM KOAc, 30 mM NH$_4$Cl, 10 mM MgCl$_2$, 0.1 mM EDTA (pH 8.0), 0.002% NaN$_3$), the reaction mixture (150 µl) for the coupled transcription/translation was added. The coupled reaction was carried out with vigorous shaking, so that the matrix remained as a suspension. The matrix was collected by centrifugation for 1 minute at 220 × g after protein synthesis and between the purification steps. After removing the supernatant, the matrix was washed three times with 100 μl washing buffer (100 mM Tris-HCl (pH 8.0), 150 mM NaCl, 1 mM EDTA) followed by elution of the fusion protein with four times 100 μl elution buffer (= washing buffer with 2.5 mM desthiobiotin).

Removal of fusion protein
from a SFCF-reactor during synthesis

A coupled transcription/translation using a SFCF-reactor (Fig. 1) was performed for 20 h at 30°C. The volume of the reaction chamber was 750 μl and, in addition to the affinity column and the connecting hoses, the total volume of reaction mixture was 2150 μl. 6 ml of feeding buffer were used, consisting of translation buffer with 5 mM dithiothreitol, 4 mM MgCl$_2$, 0.02% NaN$_3$, 100 μM folinic acid, 400 μM L-[^{14}C]leucine (0.75 dpm/pmol, Amersham), 400 μM of each of the other 19 amino acids, 1 mM each of ATP and GTP, 0.5 mM each of CTP and UTP, 30 mM phosphoenolpyruvate (Boehringer Mannheim), 10 mM acetyl phosphate (Sigma). During protein synthesis, the reaction mixture was continuously pumped from the reaction chamber onto the affinity column, which was filled with 530 μl StrepTactin sepharose, and sent back into the reactor. To isolate the product, the

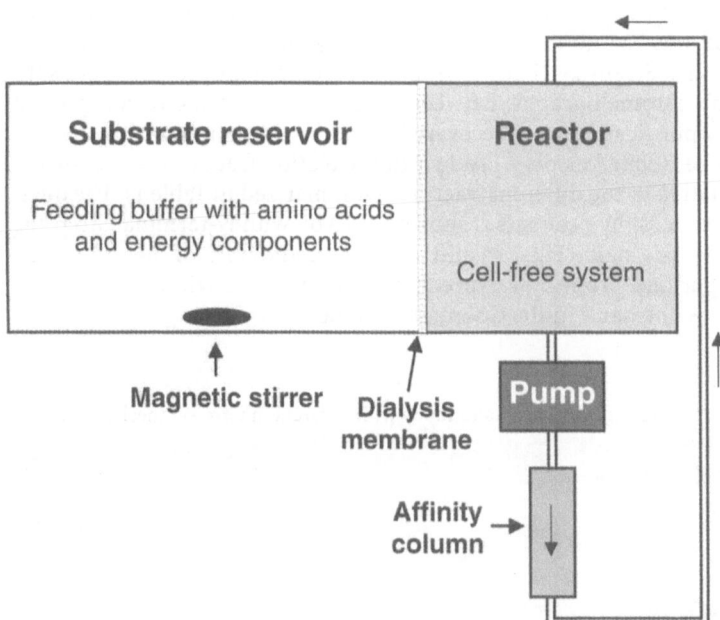

Fig. 1. Schematic drawing of the standard dialysis reactor used for product removal during protein synthesis

column was washed three times with 800 μl washing buffer, followed by elution of the fusion protein with six times 400 μl elution buffer. The isolated fractions were analyzed as described.

Results

Affinity purification of cell-free synthesized Strep-tag fusion proteins

First of all we have examined the quality and compatibility of Strep-tag purification with our cell-free protein synthesis system. To do this, we fused the two Strep-tag versions I and II to the C-terminus of the CAT-gene and Strep-tag II to the FABP-gene by PCR methods and cloned them into plasmids containing all the elements for an efficient *in vitro* transcription/translation. We chose these two genes because FABP is a well known standard in our laboratory and CAT reveals the influence of the Strep-tag on the activity of the fused protein on the basis of its enzymatic activity [35]. It was not known whether the additional 33 and 30 bp, encoding the Strep-tag I and II respectively, would influence the *in vitro* expression of the 'new genes', if the fused peptide would disturb the native structure of the proteins or if the tag would be accessible for affinity chromatography.

The newly constructed and *in vitro* synthesized fusion proteins showed no significant difference with regard to the amount of product when compared with that produced by the constructs without Strep-tag. The yields with CAT were 190 μg/ml, with the Strep-tag I 191 μg/ml and with the Strep-tag II 184 μg/ml. The amount of FABP was 228 μg/ml, with the C-terminal Strep-tag II 232 μg/ml and with the N-terminal Strep-tag II 201 μg/ml. The recombinant proteins were subjected to affinity chromatography. Between 70% and 87% of the fusion protein used for affinity purification were recovered from the column and between 60% and 82% could be isolated as pure product in the elution fractions, as calculated by TCA precipitation of the different fractions (summarized in Table 1). The quality of the chromatography products is shown for FABP with N-terminal Strep-tag II with the Coomassie stain (Fig. 2A) and as an autoradiogram of the protein gel (Fig. 2B). The purified product was predominantly isolated within one elution fraction visible as one band in the Coomassie stained gel.

Table 1. Results of Strep-tag affinity chromatography with different in vitro synthesized proteins

Protein	Strep-tag, position	synthesized protein	amount of eluted protein
CAT	–	190 μg/ml	2%
CAT	I, C-terminal	191 μg/ml	60%
CAT	II, C-terminal	184 μg/ml	72%
FABP	–	228 μg/ml	3%
FABP	II, C-terminal	232 μg/ml	82%
FABP	II, N-terminal	201 μg/ml	75%

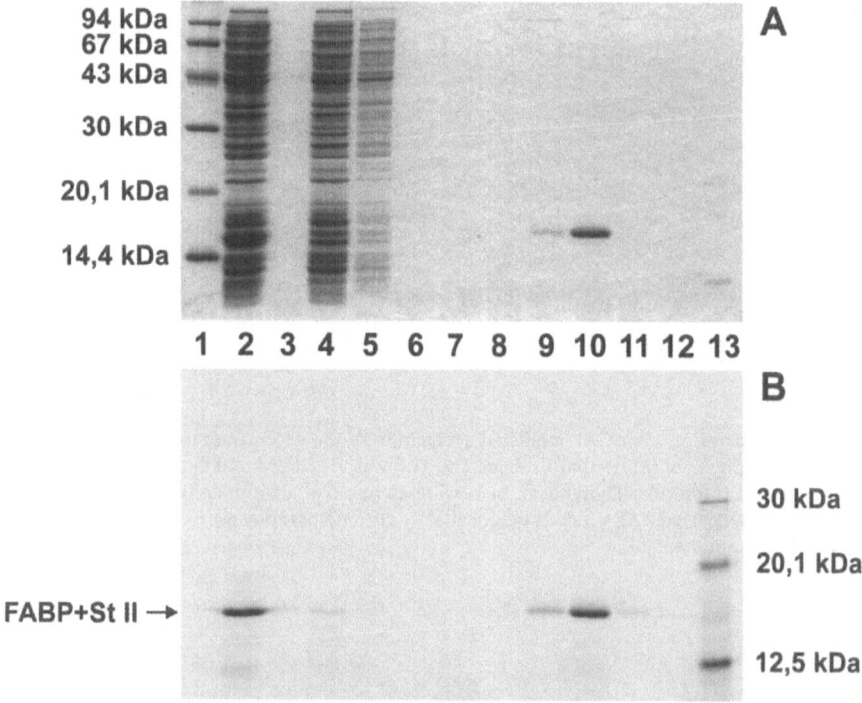

Fig. 2. Purification of FABP containing the Strep-tag II after cell-free synthesis using a Strep-Tactin affinity column. Comparable amounts of every isolated fraction were analyzed by SDS-PAGE. (A) Coomassie stain and (B) autoradiography of the radioactively labeled products. The samples in the numbered lanes are as follows: (1) molecular weight markers, (2) reaction mixture, (3) sample loading, (4–6) wash fractions 1–3, (7–12) elution fractions 1–6 and (13) [^{14}C] molecular weight standard (only partly visible in the Coomassie stain)

The influence of the Strep-tag to the enzymatic activity of the *in vitro* synthesized CAT [35] with and without Strep-tag, before and after affinity chromatography, was assayed using a fluorescent deoxychloramphenicol substrate (Molecular Probes). The observed activities were comparable or even higher than with a commercially available CAT (Sigma) (Fig. 3 and Table 2). We did observe an influence of the Strep-tag on the solubility of CAT but not for FABP. CAT is relatively insoluble in our system and its solubility is not improved by fusion with the Strep-tag I and II. The amount of soluble CAT was increased by two changes to the transcription/translation reaction. Firstly, the temperature was decreased from 37°C to 30°C and secondly, the concentration of polyethyleneglycol 2000 (PEG) was decreased from 4% to 3% (Fig. 4). These modified conditions led to an increased amount of soluble CAT with and without Strep-tag (Fig. 5).

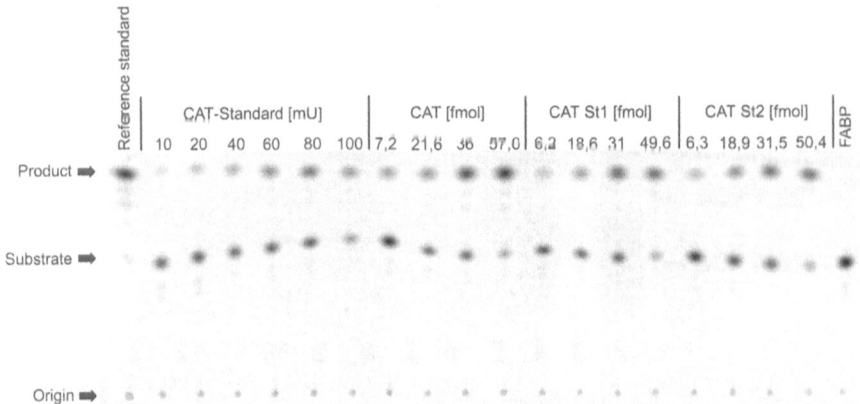

Fig. 3. TLC separation after CAT-mediated acetylation of the green fluorescent BODIPY FL 1-deoxychloramphenicol (Molecular Probes). The TLC was visualized and detected by the 'Fluorimager' system (Molecular Dynamics). Increasing amounts of a commercial CAT (Sigma) and the *in vitro* synthesized CAT versions before affinity chromatography were used to acetylate the modified chloramphenicol

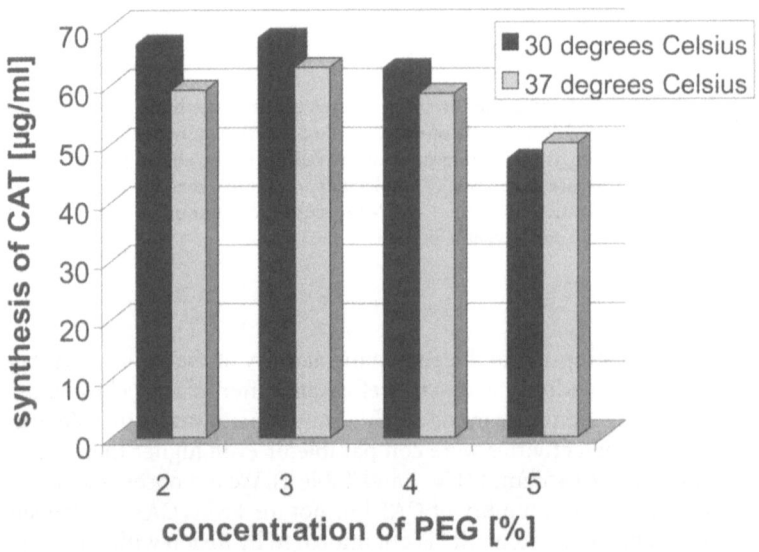

Fig. 4. Influence of the temperature and the PEG concentration on the amount of soluble CAT from an *in vitro* synthesis

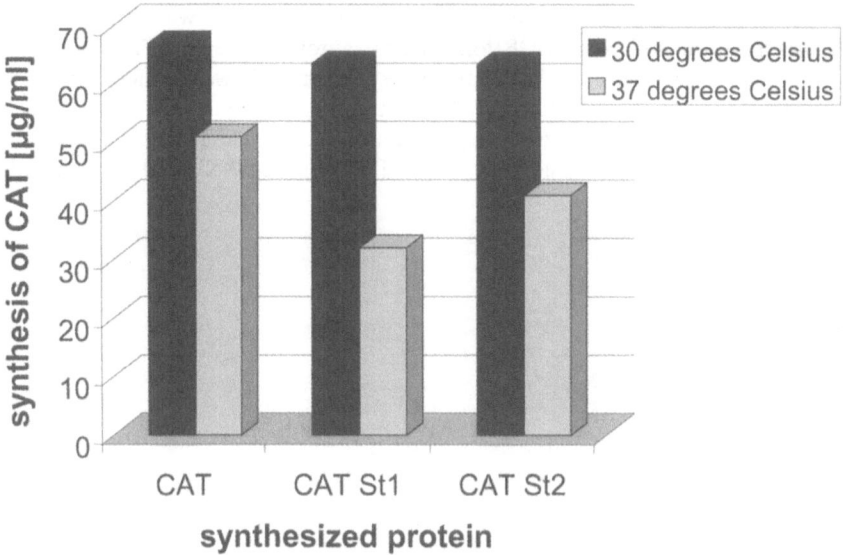

Fig. 5. Amount of soluble CAT with and without Strep-tag from an *in vitro* synthesis with a PEG concentration of 3%. The influence of the reaction temperature is shown

Table 2. Enzymatic activity of the *in vitro* synthesized CAT with and without Strep-tag, before and after affinity chromatography

Protein	assayed CAT activity before affinity chromatography	after affinity chromatography
commercial CAT	3 U/pmol	–
CAT	3.9 ± 0.7 U/pmol	4.8 ± 0.9 U/pmol*
CAT-Strep-tag I	3.4 ± 0.7 U/pmol	3.2 ± 0.6 U/pmol
CAT-Strep-tag II	2.9 ± 0.4 U/pmol	3.0 ± 0.3 U/pmol

* The enzyme was subjected to a streptavidin sepharose column and recovered after treatment with washing buffer.

Influence of affinity matrix on coupled transcription/translation

After the purification system had performed according to our expectations, we tried to isolate the Strep-tag fusion proteins during the process of protein synthesis. For that reason the influence of the StrepTactin sepharose (IBA, Göttingen) on the coupled transcription/translation reaction was examined. 20 µl StrepTactin sepharose were added to one out of two identical 60 µl coupled reaction mixtures, and a plasmid coding for FABP without Strep-tag was used to determine the total amount of synthesized protein after translation. The products were analyzed by TCA precipitation and SDS-PAGE followed by autoradiography (Fig. 6A). The amount of synthesized protein in the presence of the matrix was reduced by 6% ± 2% compared with the unchanged reaction. It is interesting to note that the by-products were decreased to an even larger extent (Fig. 6B). The rest of the reaction mixture with matrix was treated with 0.5 % SDS (30 min, 50 °C) and an identical

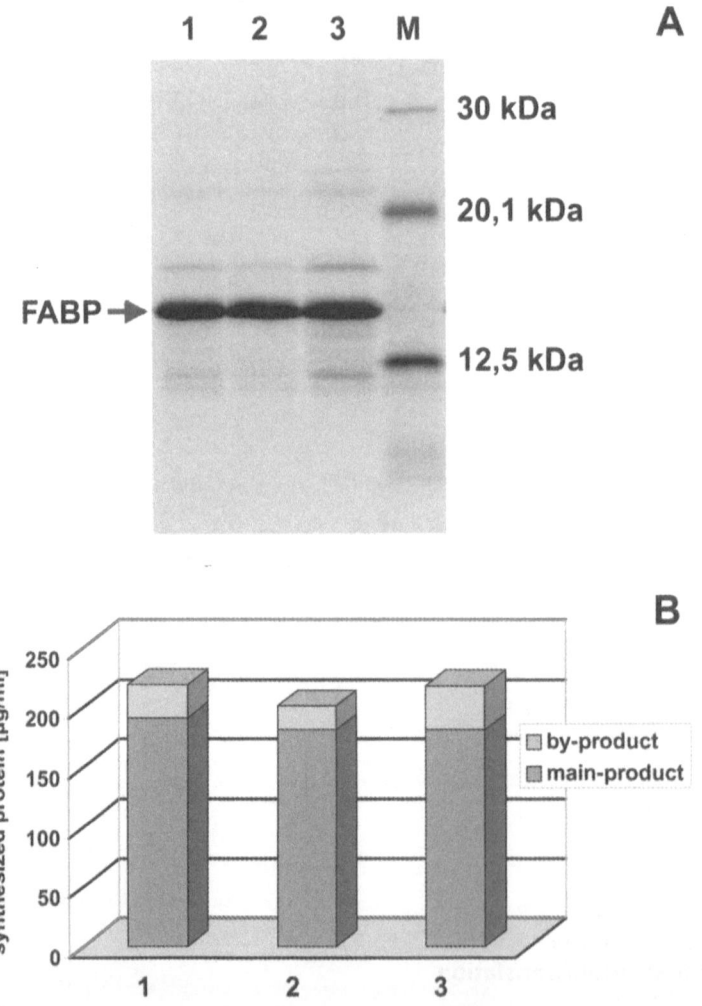

Fig. 6. Influence of StrepTactin sepharose on a coupled transcription/translation reaction of FABP without an affinity tag. One out of two identical 60 μl reactions was carried out in the presence of a 20 μl affinity matrix. After centrifugation, both supernatants were analyzed by TCA precipitation and SDS-PAGE. The rest of the reaction mixture with matrix was treated with 0.5% SDS, and comparable volumes were analyzed as described above. (A) Autoradiogram of the SDS-PAGE and (B) amount and distribution of the products. (1) standard reaction, (2) reaction with matrix, (3) reaction with matrix treated with SDS

volume was also analyzed by SDS-PAGE to determine whether some product was bound to the matrix. The autoradiogram revealed that the by-products were increased in this sample, but not the main product (Fig. 6A). The FABP itself seems to have no affinity for the matrix and the slightly reduced performance is probably a consequence of the matrix present in the system. The by-products are likely to be unfolded insoluble fractions of the protein with some affinity for the matrix.

Removal of CAT+StII during *in vitro* protein synthesis

The insignificant influence of the StrepTactin sepharose on the translation system gave us the opportunity of separating a protein with Strep-tag II during a coupled transcription/translation reaction. Thus, CAT with Strep-tag II was produced in the presence of StrepTactin sepharose and purified. About 82% of the synthesized product was bound to the matrix, of which about 87% could be isolated with reasonable purity. The chromatography results are shown in the Coomassie stain (Fig. 7A) and in the protein gel autoradiogram (Fig. 7B). The amount of eluted protein was comparable with purification via the column method. Although the overall synthesis is generally decreased in such a batch system, the amount of soluble CAT+Strep-tag II was slightly increased (data not shown.).

Removal of FABP+StII from a SFCF-reactor during protein synthesis

One advantage of a dialysis system is the longer reaction time of cell-free expression and the consequently higher yields. The removal of fusion proteins via StrepTactin sepharose from a SFCF-system (Fig. 1) was performed for 20 h, followed by washing the affinity column and eluting the bound fusion protein. 41% of the synthesized protein was found in the reaction mixture, 13% in the wash fractions and 46% could be eluted from the affinity column. The Coomassie stain (Fig. 8A) of the protein gel and the corresponding autoradiogram (Fig. 8B) demonstrate the results of the chromatography.

Discussion

The need to establish a one-step purification system for *in vitro* synthesized proteins is obvious. In general, short affinity peptides do not interfere with the biological function of the protein and therefore need not be removed via proteolysis [25]. This is a very important requirement, because the conditions during proteolysis and the protease itself have often a negative influence on the activity and the stability of the protein. In addition, the affinity chromatography should mimic physiological conditions as closely as possible so that the fusion protein can be obtained in the native state. The Strep-tag affinity peptide was tested with regard to these requirements to investigate if it could be a useful tool for protein purification and for our *in vitro* translation system. The newly constructed fusion proteins could be

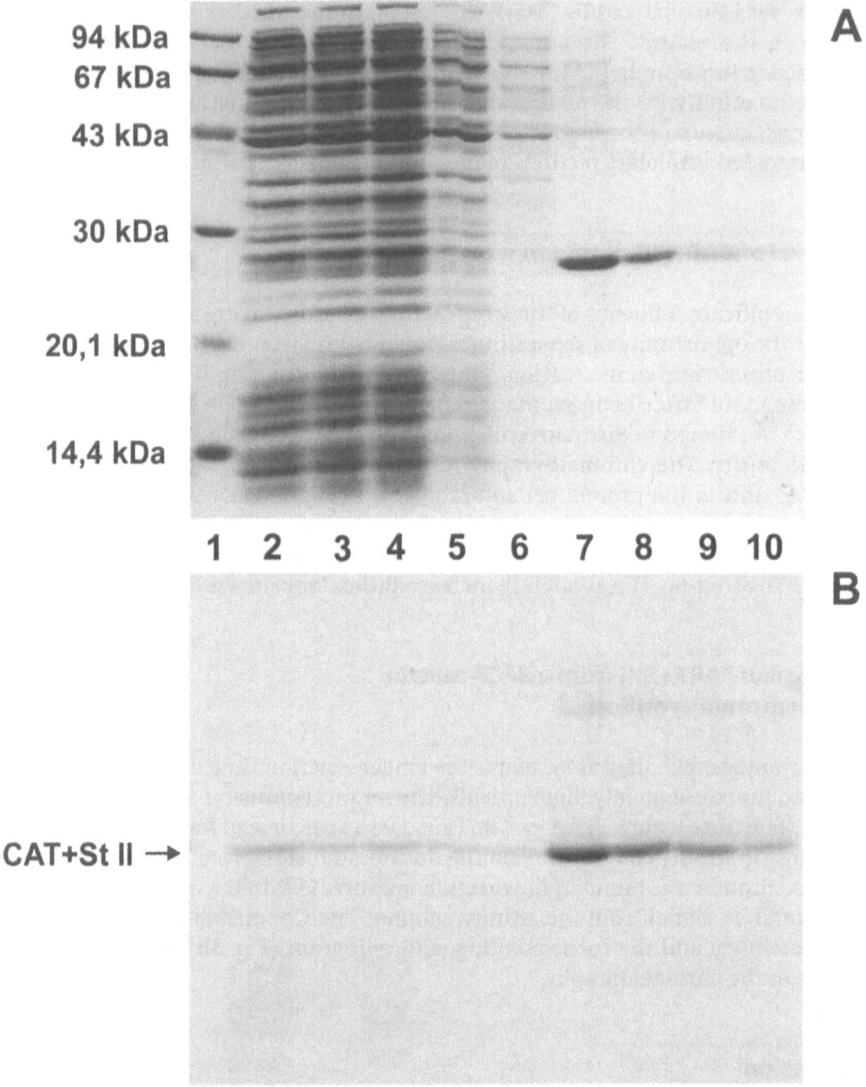

Fig. 7. Removal of CAT with Strep-tag II during *in vitro* protein synthesis in the batch mode via StrepTactin sepharose. A comparable amount of every isolated fraction except for the elution fractions was analyzed by SDS-PAGE. From the elution fractions the fourfold amount was separated to check the purity. The samples in the numbered lanes are as follows: (1) molecular weight marker, (2) the whole reaction after settling, (3) the reaction after centrifugation (5 minutes, 15000 × g), (4–6) wash fractions, (7–10) elution fractions. (A) Coomassie stain and (B) autoradiogram of the SDS-PAGE

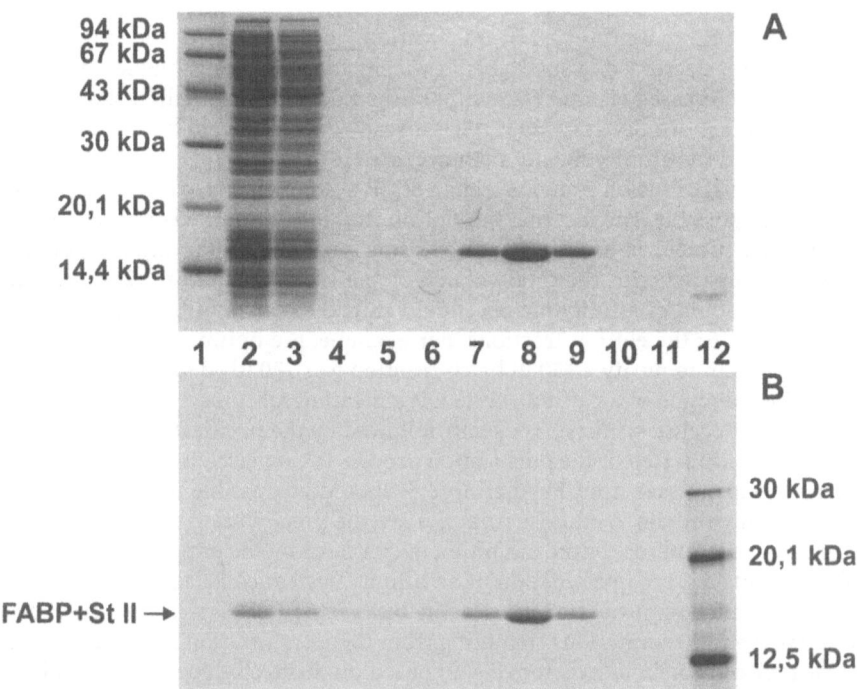

Fig. 8. Removal of FABP with Strep-tag II from a SFCF-reactor during protein synthesis using a StrepTactin sepharose column. 15 µl of every isolated fraction were analyzed by SDS-PAGE. The samples in the numbered lanes are as follows: (1) molecular weight marker, (2) reaction mixture, (3–5) wash fractions 1–3, (6–11) elution fractions 1–6 and (12) [^{14}C] molecular weight marker (only partly visible in the Coomassie stain). (A) Coomassie stain and (B) autoradiogram of the SDS-PAGE

synthesized properly in our cell-free system and their purification via Strep-tag was also achievable without any problems (Fig. 2). We conclude that the fused Strep-tag does not affect expression, and that it is available for affinity chromatography. The influence of the Strep-tag on enzymatic activity was tested with CAT fusion proteins. The observed activities were comparable with a commercial available CAT (Sigma), so that neither the Strep-tag nor the conditions of the chromatography seem to affect the biological function of the fused protein. We could see an influence of the Strep-tag on the solubility of CAT but not on that of FABP. CAT is partly insoluble in our system and that effect was increased by fusion with the Strep-tag. It is known that hydrophobic interactions at the C-terminus are essential for folding and stabilization of CAT [41]. One possible reason for the increase in misfolding would be the disruption of these interactions caused by the presence of the Strep-tag.

The performance of *in vitro* protein synthesis in the presence of StrepTactin sepharose was only slightly reduced (Fig. 6B) and it was even possible to bind fusion proteins to the affinity matrix during the reaction (Fig. 7). CAT with Strep-tag II was produced and purified in this way, with the amount of eluted protein comparable with that produced by purification via a column. Although the over-all synthesis is decreased in such a batch system, the amount of soluble CAT+St II was slightly increased (Lamla *et al.* unpublished data.). This method should also work with other affinity systems and has already been reported for dihydrofolate reductase and the affinity ligand methotrexate [19].

The removal of fusion proteins from a SFCF-system via StrepTactin sepharose (Fig. 8) was possible. The fact that the elution fractions contained only 46% of the synthesized protein is as a result of the small volume of the affinity column (530 μl) compared with the total volume of the reaction mixture (2150 μl). An increased amount of affinity matrix should shift the contribution of the synthe-sized protein to the elution fractions. For example, the optimized volume of an affinity column to purify a 150 μl batch reaction was 230 μl (data not shown).

The introduction of an affinity system is a meaningful complement to the dial-ysis system, because synthesis is usually followed by the purification of the prod-uct. Since the first step of the purification process takes place during protein syn-thesis, it helps to save time. Furthermore, it should be possible to avoid negative effects during protein synthesis; examples are the precipitation of the product at a certain concentration, or the inhibition of synthesis by the product itself. Beside these advantages, the introduction of an affinity system can also solve the prob-lem of product removal from a protein bioreactor (CFCF-reactor). This step would prevent the limitations stemming from the ultrafiltration membrane. Such a new type of protein bioreactor (Fig. 9) based on this technique still has a ultra-

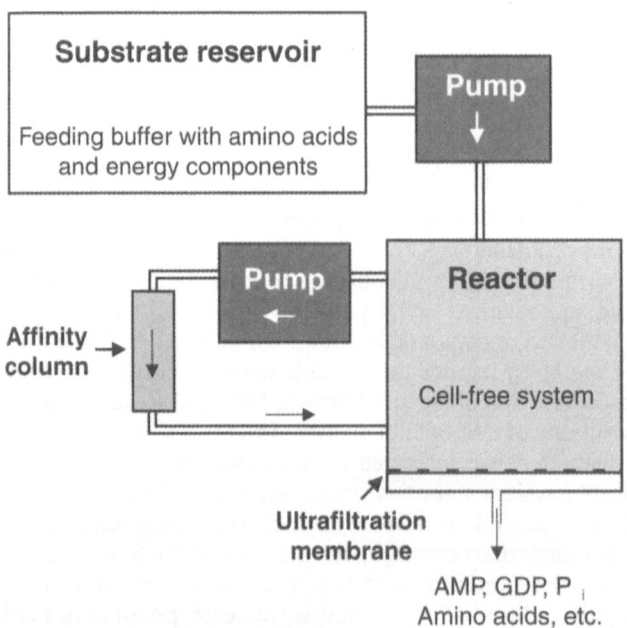

Fig. 9. Schematic drawing of a novel kind of continuous flow protein bioreactor using an affin-ity system for continuous product removal

filtration membrane, but in this case this is only to remove small molecular 'waste-products' and therefore a MW cut-off of ≤10 kDa is sufficient. We think that our new type of reactor will be much more efficient for the synthesis of proteins with a high molecular mass (≥50 kDa), since it is more difficult or even impossible to find membranes for bigger proteins which permit the passage of the protein through the membrane, while at the same time retaining the components of the protein-synthesizing machinery in the reactor.

References

1. Alimov AP, Khmelnitsky AY, Simonenko PN, Spirin AS and Chetverin AB (2000) Cell-free synthesis and affinity isolation of proteins on a nanomole scale. *BioTechniques* 28: 338–344
2. Baranov VI, Morozov IY, Ortlepp SA and Spirin AS (1989) Gene expression in a cell-free system on the preparative scale. *Gene* 84:463–466
3. Burks EA, Chen G, Georgiou G. and Iverson BL (1997) *In vitro* scanning saturation mutagenesis of an antibody binding pocket. *Proc. Natl. Acad. Sci. USA* 94:412–417
4. Cornish VW and Schultz PG (1995) Site-directed mutagenisis with an expanded genetic code. *Annu. Rev. Biophys. Biomol. Struct.* 24:435–462
5. Ellman JA, Volkman BF, Mendel D, Schultz PG and Wemmer DE (1992) Site-specific isotopic labeling of proteins for NMR studies. *J. Am. Chem. Soc.* 114:7959–7961
6. Hanes J and Plückthun A (1997) *In vitro* selection and evolution of functional proteins by using ribosome display. *Proc. Natl. Acad. Sci. USA* 94:4937–4942
7. Kawarasaki Y, Kawai T, Nakano H and Yamane T (1995) A long-lived batch reaction system of cell-free protein synthesis. *Anal. Biochem.* 226:320–324
8. Kigawa T and Yokoyama S (1991) A continuous cell-free protein synthesis system for coupled transcription-translation. *J. Biochem.* 110:166–168
9. Kigawa T, Muto Y and Yokoyama S (1995) Cell-free synthesis and amino acid-selective stable isotope labeling of proteins for NMR analysis. *J. Biomol. NMR* 6:129–134
10. Kigawa T, Yabuki T, Yoshida Y, Tsutsui M, Ito Y, Shibata T and Yokoyama S (1999) Cell-free production and stable-isotope labeling of milligram quantities of protein. *FEBS Letters* 442: 15–19
11. Kim DM, Kigawa T, Choi CY and Yokoyama S (1996) A highly efficient cell-free protein synthesis system from *Escherichia coli*. *Eur. J. Biochem.* 239:881–886
12. Kim DM and Choi CY (1996) A semicontinuous prokaryotic coupled transcription/ translation system using a dialysis membrane. *Biotechnol. Prog.* 12:645–649
13. Kim DM and Swartz JR (2000) Prolonging cell-free protein synthesis by selective reagent additions. *Biotechnol. Prog.* 16:385–390
14. Kudlicki W, Kramer G and Hardesty B (1992) High efficiency cell-free synthesis of proteins: refinement of the coupled transcription/translation system. *Anal. Biochem.* 206: 389–393
15. Lanar DE and Kain KC (1994) Expression-PCR (E-PCR): overview and applications. *PCR Methods Appl.* 4:92–96
16. Laemmli UK (1970) Cleavage of structural proteins during the assembly of the head of bacteriophage T4. *Nature* 227:680–685
17. Ludlam CFC, Sonar S, Lee CP, Coleman M, Herzfeld J, RajBhandary UL and Rothschild KJ (1995) Site-directed labeling and ATR-FTIR difference spectroscopy of bacteriorhodopsin: The peptide carbonyl group of Tyr 185 is structurally active during the bR → N transition. Biochemistry 34:2–6
18. Madin K, Sawasaki T, Ogasawara T and Endo Y (2000) A highly efficient and robust cell-free protein synthesis system prepared from wheat embryos: plants apparently contain a suicide system directed at ribosomes. *Proc. Natl. Acad. Sci. USA* 97:559–564

19. Marszal E and Scouten WH (1996) Dihydrofolate reductase synthesis in the presence of immobilized methotrexate. An approach to a continuous cell-free protein synthesis system. *J. Mol. Recognit.* 9:543–548

20. Martemyanov KA, Spirin AS and Gudkov AT (1997) Direct expression of PCR products in a cell-free transcription/translation system: synthesis of antibacterial peptide cecropin. *FEBS Lett.* 414:268–270

21. Mattheakis LC, Bhatt RR and Dower WJ (1994) An *in vitro* polysome display system for identifying ligands from very large peptide libraries. *Proc. Natl. Acad. Sci. USA* 91:9022–9026

22. Merk H, Stiege W, Tsumoto K, Kumagai I and Erdmann VA (1999) Cell-free expression of two single–chain monoclonal antibodies against lysozyme: Effect of domain arrangement on the expression. *J. Biochem.* 125:328–333

23. Merk H (2000) Steigerung der Effizienz des *Escherichia coli in vitro* Translationssystems durch Optimierung der Nukleinsäurekomponenten. PhD Thesis, Freie Universität Berlin

24. Nakano H, Shinbata T, Okumura R, Sekiguchi S, Fujishiro M and Yamane T (1999) Efficient coupled transcription/translation from PCR template by hollow-fiber membrane bioreactor. *Biotechnol. Bioeng.* 64:194–199

25. Nygren P-A, Stahl S and Uhlen M (1994) Engineering proteins to facilitate bioprocessing. *Trends Biotechnol.* 12:184–188

26. Park Y, Luo J, Schultz PG and Kirsch JF (1997) Noncoded amino acids replacement probes of the aspartate aminotransferase mechanism. *Biochemistry* 36:10517–10525

27. Patnaik R and Swartz JR (1998) *E.coli*-based in vitro transcription/translation: in vivo-specific synthesis rates and high yields in a batch system. *BioTechniques* 24:862–868

28. Pavlov MY, Freistroffer DV and Ehrenberg M (1997) Synthesis of region-labelled proteins for NMR studies by in vitro translation of column-coupled mRNAs. *Biochimie* 79:415–422

29. Roberts RW and Szostak JW (1997) RNA-peptide fusions for the *in vitro* selection of peptides and proteins. *Proc. Natl. Acad. Sci. USA* 94:12297–12302

30. Sambrook J, Fritsch EF and Maniatis T (1989) Molecular cloning: A laboratory manual, 2nd edn. Cold Spring Harbour Laboratory, Cold Spring Harbor, NY

31. Sassenfeld HM (1990) Engineering proteins for purification. *Trends Biotechnol.* 8:88–93

32. Schmidt TGM and Skerra A (1993) The random peptide library-assisted engineering of a C-terminal affinity peptide, useful for the detection and purification of a functional Ig Fv fragment. *Protein Eng.* 6:109–122

33. Schmidt TGM and Skerra A (1994) One-step affinity purification of bacterially produced proteins by means of the "Strep-tag" and immobilized recombinant core streptavidin. *J. Chromatog. Sect. A.* 676:337–345

34. Schmidt TGM, Koepke J, Frank R and Skerra A (1996) Molecular interaction between the Strep-tag affinity peptide and its cognate target, streptavidin. *J. Mol. Biol.* 255:753–766

35. Shaw WV and Leslie AGW (1991) Chloramphenicol acetyltransferase. *Annu. Rev. Biophys. Chem.* 20:363–386

36. Short GF, Golovine SY and Hecht SM (1999) Effects of release factor 1 on in vitro protein translation and the elaboration of proteins containing unnatural amino acids. *Biochemistry* 38:8808–8819

37. Sonar S, Lee CP, Coleman M, Patel N, Liu X, Marti T, Khorana HG, RajBhandary UL and Rothschild KJ (1994) Site-directed isotope labeling and FTIR spectroscopy of bacteriorhodopsin. *Nat. Struct. Biol.* 1:512–517

38. Spirin AS, Baranov VI, Ryabova LA, Ovodov SY and Alakhov YB (1988) A continuous cell-free translation system capable of producing polypeptides in high yield. *Science* 242: 1162–1164

39. Stiege W, Erdmann VA (1995) The potentials of the in vitro protein biosynthesis system. *J. Biotechnol.* 41:81–90

40. Switzer WM and Heneine W (1995) Rapid screening of open reading frames by protein synthesis with an in vitro transcription and translation assay. *Biotechniques* 18:244–248

41. Van der Schueren J, Robben J, Goossens K, Heremans K and Volckaert G (1996) Identification of local carboxy-terminal hydrophobic interactions essential for folding or stability of chloramphenicol acetyltransferase. *J. Mol. Biol.* 256:878–888

42. Voss S and Skerra A (1997) Mutagenesis of a flexible loop in streptavidin leads to higher affinity for the Strep-tag II peptide and improved performance in recombinant protein purification. *Protein Eng.* 10:975–982
43. Yabuki T, Kigawa T, Dohmae N, Takio K, Terada T, Ito Y, Laue ED, Cooper JA, Kainosho M and Yokoyama S (1998) Dual amino acid-selective and site-directed stable-isotope labeling of the human c-Ha-Ras protein by cell-free synthesis. *J. Biomol. NMR* 11:295–306
44. Zubay G (1973) In vitro synthesis of protein in microbial systems. *Annu. Rev. Genet.* 7:267–287

CHAPTER 3

Improved Composition and Energy Supply for Bacterial Batch Systems

3

DONG-MYUNG KIM, JAMES R. SWARTZ*

Introduction

Cell-free protein synthesis is an attractive alternative to the conventional technologies for protein production such as bacterial or yeast fermentation and mammalian or insect cell culture. In contrast to the *in vivo* gene expression methods where protein synthesis is carried out in the context of cell physiology and is surrounded by cell walls and membranes, cell-free protein synthesis provides a completely open system. This allows direct access to the reaction and direct control of reaction conditions. At the same time, most of the cellular functions other than protein synthesis need not be maintained in the cell-free system. Thus, it can be optimized with significantly wider latitude than for a living organism. For example, such reaction parameters as pH, redox potential, and ionic strength can be measured directly and changed without concern for harmful effects on the growth and viability of cells and with the assurance that these changes will directly impact the relevant reactions. In addition, the products of cell-free protein synthesis are less likely to affect continued productivity.

Although existing methods for cell-free protein synthesis are still expensive, based on the quantity of protein produced, there is good reason to believe that the system can be improved until its cost is lower than *in vivo* production methods for protein pharmaceuticals. Other potential advantages are lower capital requirements and reduced exposure of the product to modifying reactions during the expression period. These considerations suggest that cell-free synthesis has the potential to be the method of choice for large scale production of pharmaceutical proteins and long oligopeptides.

Another promising potential for cell-free synthesis can be found in its suitability for high throughput expression systems. The demand for such systems has become greater than ever as a result of the successful progress of the human genome project. It will be nearly impossible to keep pace with the exponentially growing amount of genetic information solely with current expression technologies. In addition to the extensive amount of time and labor required for cloning, transformation and cell culture, conventional *in vivo* expression systems cannot easily be set up for simultaneous expression of multiple genes. In contrast, cell-

* Department of Chemical Engineering, Stanford University, Stanford, CA 94305–5025

free protein synthesis, when conducted as a batch reaction, can easily be expanded into a multiplex format. For example, expression of different proteins can be carried out in a multi-well plate simply by adding different plasmids or PCR products to each well containing the mixture of cell-extract and other reagents required for protein synthesis. Furthermore, biological activities of synthesized proteins can often be determined directly in the same reaction plate. Cell-free systems also allow the direct isolation of products to shorten the time required for preparing purified proteins.

Limiting Factors in Cell-Free Protein Synthesis

The potential of cell-free synthesis system as a high throughput expression method can be realized only when it provides acceptable and consistent product yields. To date, batch cell-free systems have not produced proteins in significant amounts. This is mainly because of the short duration of the synthesis reaction. In general, cell-free protein synthesis continues for less than 30 minutes. Kim and Swartz discovered that in a conventional cell-free synthesis system derived from Escherichia coli, phosphoenol pyruvate (PEP), the secondary energy source for ATP regeneration, and several amino acids are rapidly degraded during the cell-free protein synthesis reaction (Kim and Swartz 1999, 2000a). The degradation of such compounds takes place even in the absence of protein synthesis and severely reduces the capacity for protein synthesis. However, the lost potency is completely recovered when the reaction mixture is supplied with additional PEP and amino acids, suggesting that the catalytic activity is relatively stable.

These results suggested that the duration of protein synthesis could be extended through the periodic supplementation of amino acids and PEP. Amino acid analysis of reaction supernatants from a reaction with PEP supplements revealed that the concentrations of alanine and aspartic acid/asparagine increase over the incubation periods while those of arginine, cysteine and tryptophan rapidly decrease. Only the three decreasing amino acids were required to extend protein synthesis. The benefit from repeated PEP addition, however, is limited by the accumulation of inorganic phosphate in the reaction mixture. The phosphate concentration increases in proportion to the amount of added PEP as a result of both coupled and uncoupled PEP hydrolysis. Inorganic phosphate concentrations above approximately 40 mM inhibit protein synthesis, apparently by complexing magnesium.

Besides PEP, such compounds as creatine phosphate (Anderson et al. 1983) and acetyl phosphate (Ryabova et al. 1995) have been used for the regeneration of ATP, in combination with the enzymes creatine kinase and acetate kinase respectively. As in the PEP/pyruvate kinase system, these enzymes catalyze the transfer of the high-energy phosphate bonds of the secondary energy sources to the ADP that is generated from ATP turnover. However, all of these conventional secondary energy sources are subject to uncoupled, enzymatic degradation (Kim and Swartz 2000a) by the phosphatase activities present in the cell-extract. Thus the above-mentioned problems of ATP depletion and phosphate accumulation are inevitable as long as we rely on the conventional ATP regeneration methods. To attain an extended reaction period in the batch system, therefore, a different approach for

ATP regeneration is needed. To avoid the accumulation of inorganic phosphate, we looked for a system that does not require a high concentration of compounds with high-energy phosphate bonds. In other words, we needed a system that could regenerate the high-energy phosphate donor *in situ* as well as the ATP.

Pyruvate as an alternative energy source

Use of pyruvate oxidase for in situ generation of secondary energy source

Pyruvate oxidase (E.C.1.2.3.3) plays an important role in the aerobic growth of lactobacteria by catalyzing the oxidative decarboxylation of pyruvate in several steps. Importantly, in the presence of the cofactors thiamine pyrophosphate (TPP) and FAD, this enzyme catalyzes the condensation of pyruvate and inorganic phosphate to generate acetyl phosphate. Since acetyl phosphate can be used to regenerate ATP by acetate kinase, pyruvate and pyruvate oxidase can be employed to supply ATP without the accumulation of inorganic phosphate. The inorganic phosphate produced either from protein synthesis or by degradation of acetyl phosphate is recycled to generate another molecule of acetyl phosphate. A simplified diagram for this strategy is given in Fig.1a. In addition, since both acetyl phosphate and ATP

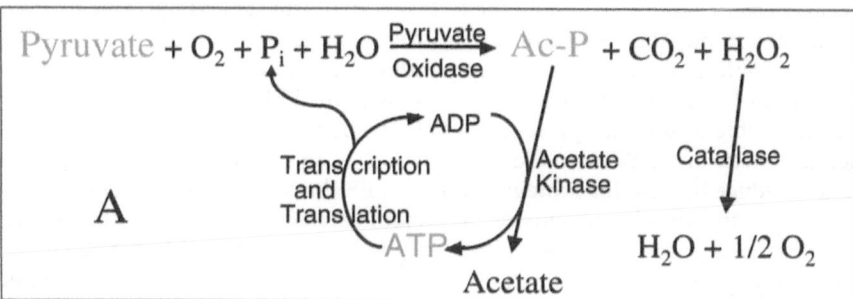

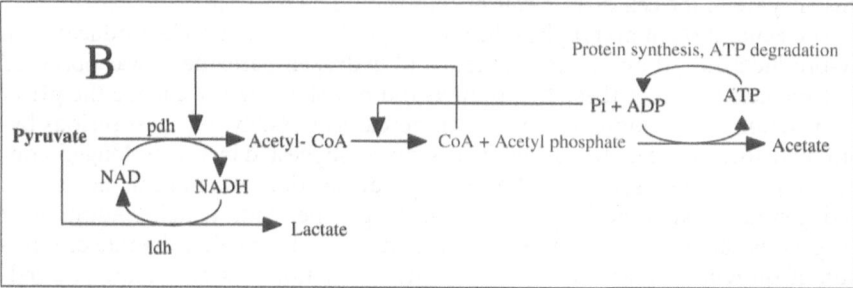

Fig. 1. Diagrams of the enzymatic reactions used for (A) regeneration of ATP from ADP using pyruvate and exogenous pyruvate oxidase and for (B) regeneration of ATP from ADP using pyruvate and endogenous enzymes

are continually generated and depleted, the peak concentration of both is likely to be kept low enough to discourage non-productive hydrolysis.

To examine this scheme under the reaction conditions for cell-free protein synthesis, the PEP and pyruvate kinase of the conventional batch reaction were replaced with a mixture of pyruvate, TPP, FAD and pyruvate oxidase. The measurements of TCA precipitatable radioactivity suggested that the new strategy for ATP regeneration does support protein synthesis (Kim and Swartz 1999). The amount of synthesized chloramphenicol acetyltransferase (CAT) increased with increasing concentrations of exogenous pyruvate oxidase and leveled off at 6.6 U/ml pyruvate oxidase. While relatively insensitive to the concentration of FAD, the yield of protein synthesis responds sharply to the concentrations of TPP and inorganic phosphate, giving maximal protein synthesis at 3.3 mM and 6.6 mM respectively. The yield of protein synthesis is almost negligible when pyruvate oxidase is removed from the reaction mixture, supporting the assumption that acetyl phosphate generated from pyruvate serves as the principal energy source.

Time course analyses of protein synthesis show that synthesis continues for up to 2 hours in the pyruvate system, whilst the reaction using PEP slows drastically after 20 minutes (Fig.2a). The time course of ATP concentration coincides nicely with the duration of protein synthesis. As shown in Fig.2b, the ATP concentration in the reaction mixture decreases rapidly in the absence of any secondary energy sources. Complete depletion of ATP is observed within 10 minutes. Although the addition of PEP can significantly increase ATP concentration, it is not sufficient to prevent an exponential decrease in ATP concentration. In contrast, although the average level is lower than in the PEP system, pyruvate and pyruvate oxidase maintain a stable ATP concentration for over 60 minutes.

As a result, despite the significantly reduced initial rate of protein synthesis (192 µg/ml-hr, compared with 300 µg/ml-hr with the PEP-driven system), the new system is able to produce protein at a volumetric yield higher than the conventional system using PEP. In addition, the use of pyruvate has the potential to greatly reduce the cost for protein synthesis. PEP, the conventional energy source, is so expensive that it accounts for approximately 80% of total reagent cost. In contrast, the cost for pyruvate is almost negligible compared with that of PEP or even acetyl phosphate and creatine phosphate. The current commercial cost of pyruvate oxidase offsets most of the pyruvate cost advantage, but expression of this enzyme in the cell extract source cells should be possible.

The generation of acetyl phosphate through this reaction cycle produces such by-products as carbon dioxide, acetate and hydrogen peroxide. It was observed that the generation of these by-products did not significantly change the pH of the reaction buffer. Moreover, hydrogen peroxide, possibly the most serious by-product, does not seem to poison the system, as suggested by the prolonged, constant rate of protein synthesis. Presumably, our reaction system contains enough endogenous catalase activity to avoid hydrogen peroxide toxicity. Addition of exogenous catalase to the reaction mixture did not affect either the rate or duration of protein synthesis. When the synthesis reaction was conducted in a fed-batch mode where the mixture of pyruvate and amino acids was added every hour, significant ATP levels were maintained and protein synthesis continued for over 7 hours to yield 0.7 mg/ml of CAT.

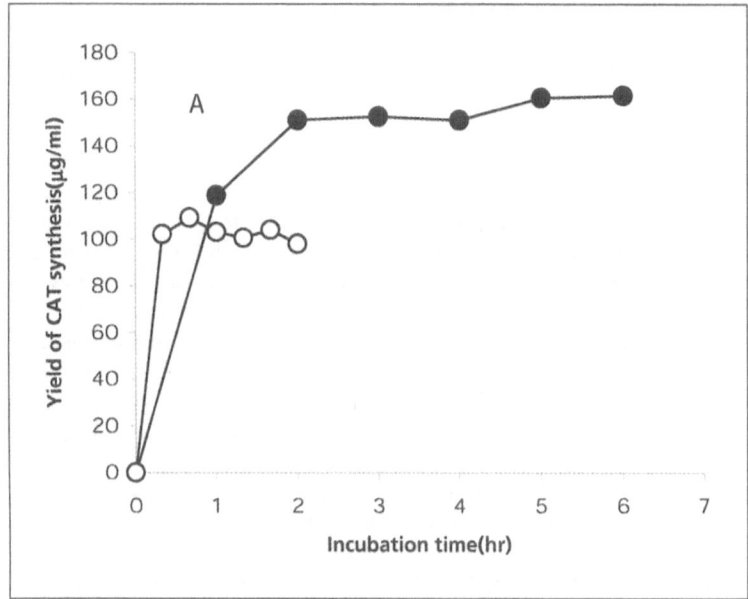

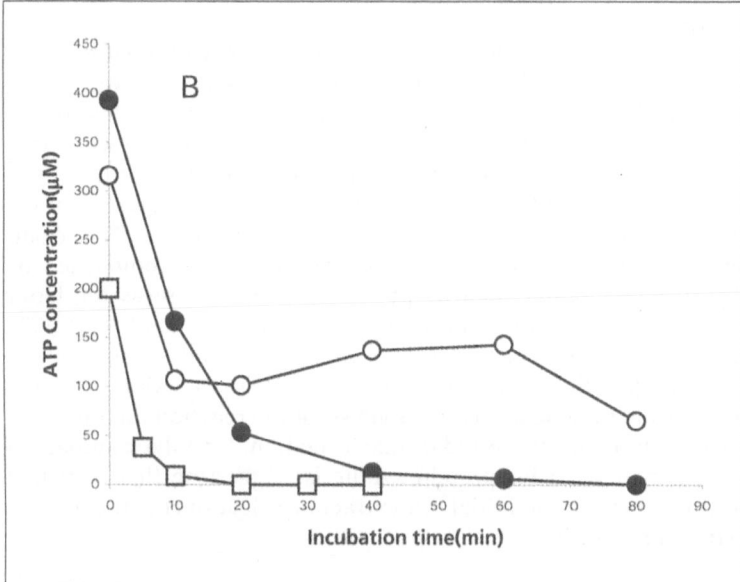

Fig. 2. Time course of CAT synthesis and ATP concentration in the conventional system (open circles) and in our new system (closed circles). 120 ml reaction mixtures of each system were prepared and incubated at 37°C. Panel A, 10 μl samples were withdrawn every 30 minutes and the [¹⁴C]leucine-labeled radioactivity of newly synthesized protein was measured; Panel B, 5 μl of reaction mixture was taken at the given time points, diluted with 495 μl of assay buffer and the ATP concentration was measured using a luciferase-based assay. Open squares represent the time course of ATP concentration without any energy sources

Utilization of pyruvate without exogenous enzymes

Even though the new system greatly extends the duration of protein synthesis, the current requirement for exogenous pyruvate oxidase (E.C.1.2.3.3) poses a significant limitation to the flexibility and cost of this synthesis system. More importantly, since this enzyme requires molecular oxygen for the oxidation of pyruvate into acetyl phosphate, the synthesis reaction cannot easily be scaled-up in a simple batch configuration because of oxygen transfer limitations. For example, when we simply increased the reaction volume from 15 µl to 120 µl in the same reaction tubes, the final volumetric yield of synthesis decreased more than 60%. To avoid these limitations, we sought to eliminate the requirements for exogenous enzyme and oxygen. For pyruvate to be used for ATP regeneration in the cell-free system, it first needs to be converted to acetyl phosphate. In *E.coli* cells, this conversion normally occurs via two reactions. First, pyruvate dehydrogenase catalyzes the condensation of Coenzyme A (CoA) and pyruvate to make acetyl-CoA in the presence of NAD as a cofactor. NAD is reduced to NADH during this reaction. Subsequently, acetyl-CoA is converted to acetyl phosphate by phosphotransacetylase (Fig.1b). On the other hand, pyruvate-formate lyase can also make acetyl-CoA from pyruvate, producing formate as the by-product. These enzymes were assumed to be in the cell-extract. We thus tested the cofactors NAD and CoA for their ability to stimulate the regeneration of ATP from pyruvate in support of cell-free protein synthesis.

33 mM sodium pyruvate was added to reaction mixtures with or without the cofactors. After a 2-hr incubation, significant synthesis was observed in the presence of NAD, with about 20% additional synthesis obtained by also adding CoA (Kim and Swartz 2001). The final yield of protein synthesis with both cofactors was about 70% of that from the reaction in which PEP was used. The use of pyruvate without the cofactors resulted in low protein synthesis, equivalent to the control reaction without any secondary energy source. This result indicates that the conversion of pyruvate to acetyl-CoA is accomplished by pyruvate dehydrogenase rather than by pyruvate-formate lyase as the latter enzyme does not require cofactors. Since only a catalytic amount of NAD (0.33 mM) is required, it seems obvious that NAD is also regenerated during the synthesis reaction. One possibility is that the reduced NAD is reoxidized by the conversion of pyruvate into lactate. HPLC analysis of the reaction mixture confirmed approximately equal acetate and lactate formation. A catalytic amount of CoA (0.27 mM) is beneficial for protein synthesis, suggesting that sufficient CoA is either not present in the initial cell extract or is lost during the dialysis step of cell extract preparation.

Stabilization of ATP with a metabolic inhibitor

Even though pyruvate is more stable than a conventional energy source such as PEP, pyruvate still undergoes metabolic conversion by the enzymes present in the cell-extract. Stabilization of pyruvate can be expected to further improve the productivity of cell-free protein synthesis. Among the metabolic reactions that consume pyruvate, we were particularly concerned about the conversion of pyruvate into PEP that is catalyzed by phosphoenol pyruvate synthetase (Pps). The reaction is driven by the conversion of ATP into AMP and could substantially reduce the energy supply during protein synthesis.

The Pps of *Escherichia coli* is strongly inhibited by oxalate (Narindrasorasak and Bridger 1978), which acts as a transition state analogue by mimicking the structure of enolpyruvate. When oxalate is added to a cell-free system using pyruvate as the energy source, it leads to improved protein synthesis. The final yield of CAT synthesis after a 3-hr incubation is increased by approximately 30% by the addition of 2.7 mM oxalate (Kim and Swartz 2000b). The time course of protein synthesis shows that the addition of oxalate extends protein synthesis without significantly affecting the initial rate. The result of ATP analysis during the reaction period indicates that the prolonged protein synthesis is as a result of an enhanced supply of ATP.

The inhibition of Pps by oxalate is also confirmed by amino acid analysis. As was reported previously (Kim and Swartz 2000a), aspartic acid and/or asparagine concentrations increase during cell-free protein synthesis. The addition of oxalate significantly reduces the rate of aspartic acid/asparagine generation during protein synthesis (Kim and Swartz 2000b). Concentrations of the other amino acids are not significantly affected. Since the precursor of aspartic acid/asparagine synthesis is oxalacetate that is produced from PEP, the slower generation of these amino acids supports the idea that oxalate is inhibiting Pps.

Utilization of pyruvate generated from the PEP: PANOx system

The reactions of ATP regeneration using pyruvate can also be used to improve the utilization efficiency of the conventional energy source, PEP. After being used for ATP regeneration or being degraded by phosphatase activities in the cell-extract, PEP produces an equimolar amount of pyruvate. If the cofactors NAD and CoA are present in the reaction mixture, at least half of the newly generated pyruvate is available for ATP regeneration (the other half is probably used for the regeneration of NAD). As a result, through the two-stage utilization of the energy source, the overall concentration of ATP is elevated and prolonged, and the productivity of protein synthesis is improved (Fig.3). Further improvements might be gained if the non-productive hydrolysis of PEP could be avoided, but at least the use of the resulting pyruvate allows some benefit to be salvaged.

The oxalate benefit is preserved when PEP is used in the presence of NAD and CoA. In addition, increasing the initial concentrations of amino acids further stimulates protein synthesis. When the ATP supply was increased by the addition of the cofactors and oxalate, both the rate and duration of protein synthesis were significantly improved by increasing the initial concentrations of amino acids from 0.5 to 2 mM. As a result, the final yield of CAT synthesis after a 1-hr incubation averages 350 μg/ml. Furthermore, periodic additions of fresh amino acids, PEP and magnesium allow the synthesis reaction to continue for over 3 hours, resulting in a final yield of 750 μg/ml (Fig.4). Synthesized CAT is visible as the

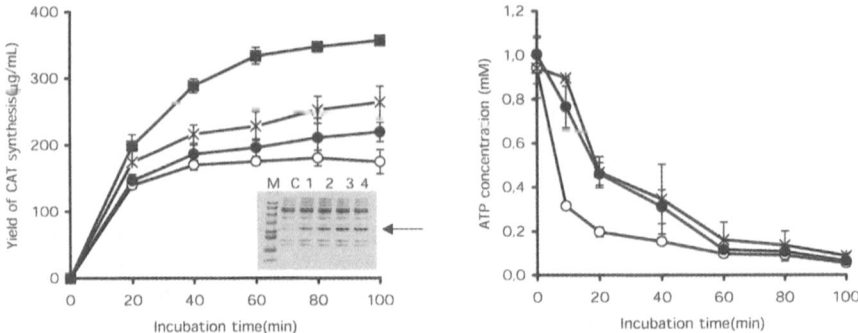

Fig. 3. Time course of protein synthesis and ATP concentration using endogenous enzymes. 120 μL standard reaction mixtures with 33 mM PEP were prepared and incubated in the presence of 0.33 mM NAD, 0.27 mM CoA and 2.7 mM sodium oxalate. At the given time points, 5 μl samples were taken and TCA-insoluble radioactivities were counted to measure protein synthesis (A). In order to measure the ATP concentrations (B), 10 μL samples were withdrawn, mixed with the same volume of 10% TCA solution, and centrifuged for 10 min. 10 μL of the supernatant was used for ATP analysis with a luciferase based assay. Open circles, control reaction with PEP only; filled circles, with NAD and CoA; asterisks, with NAD, CoA and oxalate; filled squares, with NAD, CoA, oxalate and 2 mM amino acids. At the end of each reaction, 5 μL samples were taken to run a 16 % SDS-PAGE gel. The gel was stained with Coomassie Blue following standard procedures (inset of panel A). Lanes: M, standard molecular weight markers; C, reaction without the template plasmid; 1, standard reaction; 2, reaction with 0.33 mM NAD and 0.27 mM CoA; 3, reaction with 0.33 mM NAD, 0.27 mM CoA, and 2.7 mM sodium oxalate; 4, reaction with 0.33 mM NAD, 0.27 mM CoA, 2.7 mM sodium oxalate and 2 mM amino acids

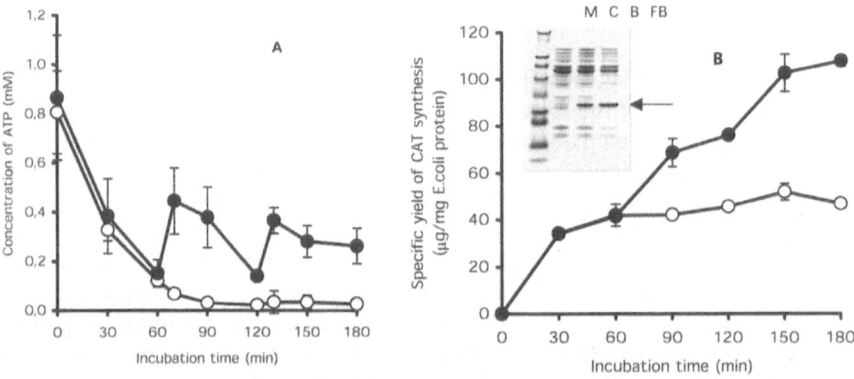

Fig. 4. Supplementation of PEP, amino acids and magnesium during protein synthesis. A synthesis reaction was carried out in the presence of 2 mM amino acids, 33 mM PEP, 0.33 mM NAD, 0.27 mM CoA and 2.7 mM sodium oxalate in a 120 μL volume. During the incubation period, the initial concentrations of PEP, amino acids and magnesium acetate were added to the reaction every hour. The same volumes of water were added to the control reaction. 5 μL samples were taken at the given time points to measure the concentration of ATP (panel A) and the yield of CAT synthesis (panel B). Open circles, batch reaction with water additions; filled circles, reaction with the additions. Samples taken at the end of each reaction were analyzed on a SDS-polyacrylamide gel followed by Coomassie Blue staining (inset of panel B). M, standard molecular weight markers; C, control reaction without template plasmid; B, batch reaction with water additions; FB, reaction with the additions of PEP, amino acids and magnesium acetate. The arrow indicates the expressed CAT

major band on a SDS-PAGE gel after Coomassie Blue staining. Approximately 60% of the expressed CAT is soluble and the measured specific activity is consistently higher than the published value of 125 units/mg (Shaw 1977). Since this fed-batch procedure requires only two additions, it is amenable to multiplex synthesis formats for high throughput screening.

Glucose-6-phosphate as an alternative secondary energy source

The successful use of pyruvate by endogenous enzymes suggests the possibility of using earlier glycolytic intermediates as ultimate energy sources. Since the current cell-extract contains most of the soluble *E.coli* enzymes, it is expected that pyruvate can be generated *in situ* from precursor compounds in the glycolytic pathway. Glucose-6-phosphate was examined first as it is the first intermediate of the glycolytic pathway. When 33 mM glucose-6-phosphate is used under the same reaction conditions as in the pyruvate/NAD system, it does support protein synthesis (Kim and Swartz 2001). In addition, although the initial rate is lower than with PEP, protein synthesis continues for over 2 hours. As a result, approximately 30% more CAT is produced at the end of incubation. This is most likely to be as a result of the remarkably extended maintenance of ATP concentrations. The time course of ATP concentration during protein synthesis with glucose-6-phosphate is characterized by a period of stable maintenance of ATP level followed by a slow decrease over the incubation period (Fig.5).

This seems to be as a result of the enhanced supply of ATP, which can be explained since glucose-6-phosphate offers a greater potential to regenerate ATP compared with PEP or pyruvate. Whilst pyruvate can regenerate, at best, only the equivalent number of ATP molecules and PEP only slightly more, 3 molecules of

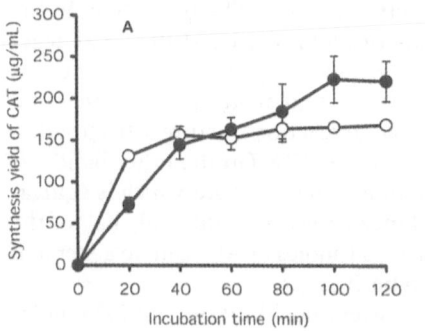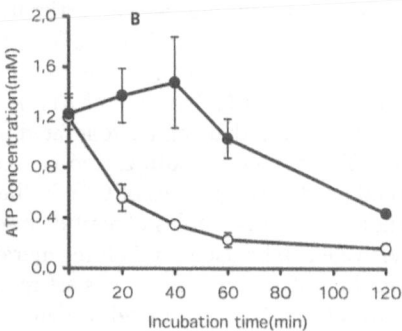

Fig. 5. Expression of CAT using glucose-6-phosphate as the secondary energy source. 33 mM glucose-6-phosphate, 0.33 mM NAD and 0.27 mM CoA were added to a 120 μL synthesis reaction. 5 μL and 10 μL samples were taken to determine protein synthesis and ATP concentration respectively and were assayed as in Figure 3. Open circles, conventional reaction using PEP; closed circles, reaction using glucose-6-phosphate as the energy source. Panel A, time course of CAT synthesis; panel B, time course of ATP concentration

ATP can be generated during the oxidation of glucose-6-phosphate into pyruvate. (The two molecules of pyruvate generated from the glycolytic pathway are probably required to regenerate the two NAD molecules that are reduced by glyceraldehyde-3-phosphate dehydrogenase.) Although the reason for the relatively slow initial rate remains unknown, the results strongly imply that we can use any of the glycolytic intermediates as a secondary energy source to support cell-free protein synthesis. Unlike the reactions using PEP or pyruvate, the addition of sodium oxalate does not improve the maintenance of ATP concentration.

The ATP regeneration with glucose-6-phosphate indicates that all of the glycolytic enzymes required to convert glucose-6-phosphate into pyruvate are active under the present reaction conditions. This provides flexibility in choosing a secondary energy source for protein synthesis. In theory, any of the glycolytic intermediates between glucose-6-phosphate and pyruvate can be used for ATP regeneration.

In addition to the postulated mechanisms described above, oxidative phosphorylation of ADP is another possible pathway for ATP regeneration in the new system. However, addition of oligomycin, an inhibitor of $F_1F_0ATPase$, has no effect on ATP regeneration or protein synthesis. In addition, a reaction under anaerobic conditions gives a similar yield of protein synthesis. Therefore, it is very likely that the ATP is regenerated through the pathways suggested above.

Estimation of ATP utilization efficiency

Because each mRNA is used multiple times, we estimate that the translational demand for ATP dominates. During protein synthesis, the EF-Tu cycle consumes 2 GTP equivalents per peptide bond (Ehrenberg et al. 1990) and EF-G requires another molecule of GTP for the translocation of ribosome. In addition, since a molecule of ATP is hydrolyzed to AMP during the aminoacylation of tRNAs, we assume that 5 molecules of ATP are required to add an amino acid residue. Based on these assumptions, we can estimate the efficiencies of ATP utilization. In the standard reaction using PEP, the final amount of synthesized CAT in a 15 μL reaction is 99.3 pmole, which represents 2.2 x 10–8 moles of peptide bonds. Thus we can assume that 0.11 μmole of ATP is used for protein synthesis. Since the total amount of ATP that can be generated in the same reaction mixture is 0.5 μmoles, the efficiency of ATP utilization is estimated to be 22%. On the other hand, we could produce 66 pmoles of CAT in the reaction using pyruvate, which is equivalent to 1.50 x 10–8 moles of peptide bonds. Since we estimate that only half of the pyruvate can be used for ATP regeneration, 0.25 μmoles of ATP can be generated and the utilization efficiency is estimated to be 30%.

By adding NAD, CoA, and oxalate to the conventional PEP system; 182 pmoles of CAT is synthesized in a 15 μL reaction. Since the estimated total amount of available ATP is 0.75 μmoles (0.5 μmoles from PEP and 0.25 μmoles from the pyruvate produced from PEP), the efficiency of ATP utilization is estimated to be 27.0%. However, when the initial concentrations of amino acids are raised to 2 mM, the amount of produced CAT increases to 313 pmoles and the efficiency of ATP utilization reaches 47.4%.

In contrast, the efficiency of ATP utilization with glucose-6-phosphate is only 8.8% (note that 1 molecule of glucose-6-phosphate can generate 3 molecules of ATP). Thus only a small fraction of the potential ATP pool is used for protein synthesis. This suggests either that the majority of the regenerated ATP is degraded by ATPase activities present in the current cell extract, that side reactions are degrading glycolytic intermediates, or that other factors become limiting.

Future Opportunities

Our eventual target is to use glucose as the secondary energy source. The use of glucose would provide a cell-free system that is highly competitive with traditional technologies for protein expression in terms of economic efficiency. Our initial results suggest such a possibility. Even though the yield of synthesis was relatively low (approximately 80 μg/ml), the use of glucose along with hexokinase did support protein synthesis. However, the results with glucose were inconsistent as well as low, and we are currently working to improve this potentially attractive system. Furthermore, even though ATP is not regenerated by oxidative phosphorylation in our system, we do not exclude the possibility that the cell-extract contains active membrane vesicles. If we can use these vesicles, oxidative phosphorylation would offer an extremely efficient method of supplying ATP for protein synthesis.

In addition, as was reported previously (Kim and Swartz 1999), the presence of non-specific phosphatase activities in the cell-extract may also be degrading the PEP and glucose-6-phosphate. Even though repeated addition of the secondary energy source maintains the ATP supply, accumulated inorganic phosphate eventually inhibits protein synthesis. This should be less severe with glucose-6-phosphate than with PEP, but is still a concern. In both cases, the identification and removal of the phosphatase activities should improve system performance. These can be removed by inactivating the corresponding chromosomal gene, or, if the enzymatic activity is essential for growth, we can genetically mark the enzyme with an affinity tag for removal during cell-extract preparation. The same approach is being taken to avoid the problem of amino acid degradation (Kim and Swartz 2000a). Through such a "genetic optimization" of the E.coli strain combined with the improved ATP regeneration systems, we expect to develop a highly efficient batch cell-free protein synthesis system.

Finally, although this work focused on the regeneration of ATP from ADP, it should be remembered that many reactions generate AMP and that the regeneration of all the nucleotide triphosphates is required for the coupled, transcription/translation system, especially GTP. As we continue to improve the ATP supply, we may encounter other limitations in phosphate transfer reactions (for example, the conversion of AMP to ADP by adenylate kinase) that hamper system performance. Clearly, there are many questions still to address and many opportunities to improve the system. It is this potential that makes the cell-free system so exciting.

References

Anderson CW, Straus JW, Dudock BS (1983) Preparation of a cell-free protein-synthesizing system from wheat germ. Methods Enzymol 101:635–644.

Ehrenberg M, Rojas AM, Weiser J, Kurland GG (1990) How many EF-Tu molecules participate in aminoacyl-tRNA binding and peptide bond formation in *Escherichia coli* translation? J Mol Biol 211:739–749.

Kim D-M and Swartz JR (1999) Prolonging cell-free protein synthesis with a novel ATP regeneration system. Biotechnol Bioeng 66:180–188.

Kim D-M and Swartz JR (2000a) Prolonging cell-free protein synthesis by selective reagent additions. Biotechnol Prog 16:385–390.

Kim D-M and Swartz JR (2000b) Oxalate improves protein synthesis by enhancing ATP supply in a cell-free system derived from *Escherichia coli*. Biotechnol Lett 22:1537–1542.

Kim D-M and Swartz JR (2001) Regeneration of adenosine triphosphate from glycolytic intermediates for cell-free protein synthesis. Biotechnol Bioeng 74:309–316.

Narindrasorasak S and Bridger WA (1978) Probes of the structure of phosphoenolpyruvate synthetase: effects of a transition state analogue on enzyme conformation. Can J Biochem 56:816–819.

Ryabova LA, Vinokurov LM, Shekhovtsova E.A, Alakhov YB, Spirin AS (1995) Acetyl phosphate as an energy source for bacterial cell-free translation systems. Anal Biochem 226:184–186.

Shaw WV (1975) Chloramphenicol acetyltransferase from chloramphenicol-resistant bacteria. Methods in Enzymol 43:737–755.

Chapter 4

PURE System

4

New Cell-Free Translation System Reconstructed with Purified Components

Takuya Ueda[1], Akio Inoue[1], Midori Kaida[1],
Ryoko Baba[1], Yoshihiro Shimizu[2]

An *Escherichia coli* cell-free translation system was developed using thirty-one purified soluble factors overexpressed in *E. coli*. The system named PURE system is capable of producing more than 0.1 mg protein in 1 ml reaction within one hour and has increased durability compared with conventional cell-extract in a batch system. The PURE system produced active enzymes without the aid of molecular chaperones. The PURE system that is reinforced with trans-translation system shows increased productivity and addition of human placental RNase inhibitor also resulted in prolonged production time.

Introduction

Development of highly productive and easily controllable cell-free translation system will undoubtedly provide a platform for biotechnology in the postgenome era. A. Spirin and his coworkers firstly demonstrated that cell-extract is able to produce proteins in high yield using a continuous flow system in 1988 (Spirin *et al.* 1988). Subsequent improvement of cell-free translation for protein production has been based mostly on cell-extracts derived from *E. coli* or wheat germ. Recently, Y. Endo's group succeeded in preparing wheat-germ extract with remarkable production (Madin *et al.* 2000). Despite their productivity, the exploitation of cell-extracts will potentially result in similar problems encountered in overexpression *in vivo*, such as inclusion bodies, protease attack, and so on.

In order to evade these problems, we have addressed the construction of cell-free translation systems with a new concept. The system, which we designated the PURE (protein synthesis using recombinant elements) system, is composed of purified soluble factors involved in the initiation, elongation, termination and aminoacylation processes (Shimizu et al. 2000). More than twenty years ago, Weissbach's group reconstructed the translation system using factors purified from wild type *E. coli*

[1] Department of Integrated Biosciences, Graduate School of Frontier Sciences,
[2] Department of Chemistry and Biotechnology, Graduate School of Engineering,
The University of Tokyo, Bldg. FSB-4015-1-5 Kashiwanoha,
Kashiwa, Chiba Prefecture 277-8562, Japan
e-mail: ueda@k.u-tokyo.ac.jp, telephone:+81(0)471-36-3641, fax:+81(0)471-36-3642

and observed scarce activity of peptide synthesis (Kung *et al.* 1977). However, for-
tuitously, their attempts showed that ribosome recycling factor is essential for the
transition process from termination to initiation (Kung *et al.* 1975). In these two
decades, the explosion of gene manipulation technology has enabled us to easily
overexpress and purify these components on a milligram scale over a period of one
week. On the basis of this progress, we challenged ourselves to prepare all the soluble factors of homogeneity as active forms with a peptide-tag. Eventually, we have
established the PURE system, capable of producing submilligrams of protein in 1
ml reaction. Here we describe the characterization and improvement of the PURE
system and discuss its feasibility in cell-free technology.

Construction of the PURE system

Components of the PURE system

The bacterial protein biosynthesis is governed by ten soluble protein factors: IF1,
IF2, IF3, EF-Tu, EF-G, EF-Ts, RF1, RF2, RF3 and RRF. All genes encoding these factors were cloned into an expression vector after the amplification by the PCR
method from *E. coli* genomic DNA. The resultant genes were expressed as His-tag
fusion forms in *E. coli* cells and carefully purified by Ni-column. In *E. coli,* twenty aminoacyl-tRNA synthetases are indispensable for tRNA activation, and
methionyl-tRNA transformylase is involved in the formylation of methionine
attached with initiator tRNA. These enzymes were also expressed and purified in
a similar manner.

It should be emphasized that substance inhibiting protein biosynthesis must be
carefully eliminated during purification of these factors. Insufficient purification
causes contamination of ATPase activity, which subsequently reduces the efficiency of peptide synthesis by abortive consumption of ATP in the PURE system. The
elution profile of seryl-tRNA synthetase from Ni-column is shown in Fig. 1. The
seryl-tRNA synthetase was eluted by the imidazole gradient, and the fractions only
containing a trace amount of impurities were carefully collected. The stepwise elution is irrelevant for ensuring quality in the PURE system. By observing the homogeneity of thirty-one factors by SDS-PAGE, it appeared that all the samples were
more than 95% pure. The measurement of the activities of all the factors and
aminoacyl-tRNA synthetases indicated that all the enzymes were purified as active
forms. The attachment of the His-tag peptide caused the loss of activity in the case
of some factors, which was circumvented by altering the attaching terminus.

The composition and productivity of the PURE system

The reconstruction of a genuine cell-free transcription/translation-coupled system has been carried out by utilizing the purified components mentioned above
and T7 RNA polymerase, which was also purified as a His-tag fusion form, as for
the other protein components. Reaction mixtures were basically prepared in the
polymix buffer as reported (Jelenc and Kurland 1979; Wagner *et al.* 1982) with a

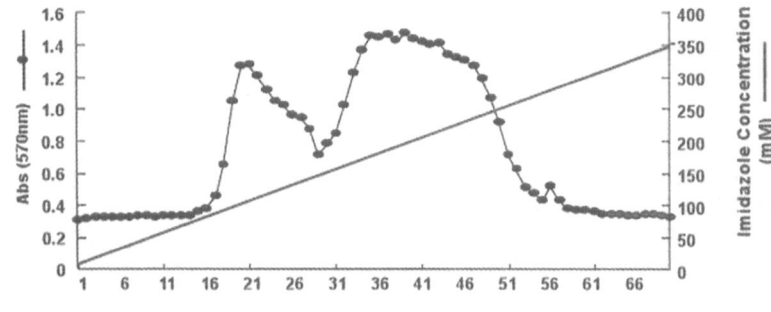

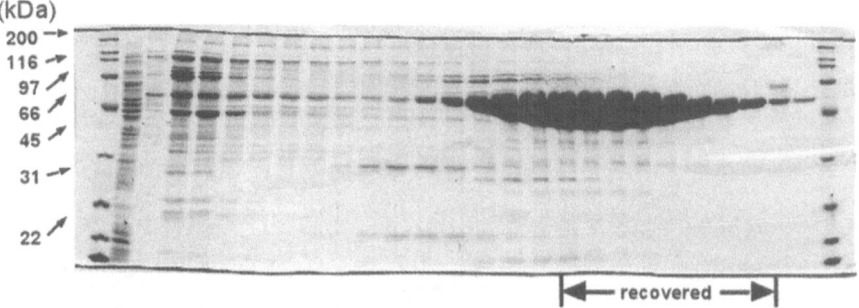

Fig. 1. The elution profile of seryl-tRNA synthetase overexpressed as His-tagged protein in *E. coli*. The gene corresponding to seryl-tRNA synthetase was amplified by PCR and cloned into pET21a vector. The enzyme was expressed in *E. coli* BL21/DE3 strain in the presence of 0.1 mM IPTG for 4 hours. The cell lysate was prepared by sonication, followed by centrifugation at 100,000 g for one hour, and was applied onto Ni^2-column. His-tagged enzyme was eluted with a liner gradient from 10 mM to 400 mM imidazole. Protein concentrations were determined by the Bio-Rad protein assay kit. The fractions were pooled after dialysis and stored at –80°C.

slight modification as shown in Table 1. The PURE system is able to produce proteins without the energy recycling system because abortive consumption of ATP does not take place as a result of low contamination of ATPase activity, in contrast to the systems that use crude cell extract. However, to maximize productivity the system was subsidized by an energy recycling system composed from creatine kinase, myokinase and nucleoside diphosphate kinase. In the presence of these enzymes, the ADP, AMP and GDP that resulted from the translation reaction are efficiently converted to their triphosphate form using the chemical energy of the creatine phosphate supplied (this is further discussed below). In addition, inorganic pyrophosphatase was supplied to hydrolyze pyrophosphate, which may reduce productivity through chelation with metal ions.

Several proteins, such as DHFR, λ lysozyme, GFP, GST, and T7 *gene*10, have been synthesized from the plasmid encoded genes inserted between the T7 promoter and T7 transcriptional terminator. By SDS–PAGE analysis of [^{35}S]-methionine labeled products, it appeared that the transcription/translation-coupled PURE system is capable of synthesizing the proper molecular weights. In the

Table 1. The reaction condition of the PURE system. Specific activities of ARS and MTF were measured using radioactive amino acids. One unit of activity was defined as the amount of enzyme that catalyzes the formation of 1 pmol of aminoacyl-tRNA in 1 min. The amount of each translation factor was optimized by dose-dependent translation assay.

[Buffer]
5 mM potassium phosphate, pH 7.3
95 mM potassium glutamate
9 mM magnesium acetate
5 mM ammonium chloride
0.5 mM calcium chloride
1 mM spermidine
8 mM putrescine
1 mM DTT

[Energy sources]
2 mM ATP, GTP
1 mM CTP, UTP
10 mM creatine phosphate

[Others]
0.1 mM 20 amino acids
10 µg/ml 10-formyl-5,6,7,8-tetrahydrophilic acid
56 A_{260}/ml tRNA$_{mix}$ (Roche, Mannheim, Germany)
0.24 µM ribosome

[Translation factors]
2.7 µM IF1
0.40 µM IF2
1.5 µM IF3
0.26 µM EF-G
0.92 µM EF-Tu
0.66 µM EF-Ts
0.25 µM RF1
0.17 µM RF3
0.50 µM RRF

[ARS and MTF]
1900 U/ml AlaRS (27 U/µg)
2500 U/ml ArgRS (1300 U/µg)
20 mg/ml AsnRS (N.D.)
2500 U/ml AspRS (310 U/µg)
630 U/ml CysRS (500 U/µg)
1300 U/ml GlnRS (230 U/µg)
1900 U/ml GluRS (150 U/µg)
5000 U/ml GlyRS (520 U/µg)
630 U/ml HisRS (1600 U/µg)
2500 U/ml IleRS (63 U/µg)
3800 U/ml LeuRS (940 U/µg)
3800 U/ml LysRS (580 U/µg)
6300 U/ml MetRS (3000 U/µg)
1300 U/ml PheRS (15 U/µg)
1300 U/ml ProRS (120 U/µg)
1900 U/ml SerRS (1000 U/µg)
1300 U/ml ThrRS (200 U/µg)
630 U/ml TrpRS (600 U/µg)
630 U/ml TyrRS (1800 U/µg)
3100 U/ml ValRS (1700 U/µl)
4500 U/ml MTF (230 U/µl)

[Other enzymes]
4.0 µg/ml creatine kinase (Roche, Mannheim, Germany)
3.0 µg/ml myokinase (Sigma)
1.1 µg/ml nucleoside-diphosphate kinase
2.0 units/ml pyrophosphatase (Sigma)
10 µg/ml T7 RNA polymerase

semi-optimized system using ribosomes at a ten-fold concentration and increasing several translation factors, the amount of synthesized DHFR, λ lysozyme, GFP, GST, and T7 *gene*10 in the 1 ml reaction over a period of 1 h are 160 µg, 170 µg, 120 µg, 11 µg, 100 µg in turn. More than 100 µg of each peptide was synthesized in each 1 ml reaction, except for GST. Although we currently have no good explanation for the unexpectedly low productivity of GST, it can be concluded that all the components in the PURE system are sufficient for translation.

Protein folding in PURE system

In cells, newly synthesized proteins are folded into proper three-dimensional structures. It has been proposed that several molecular chaperones, such as GroEL, GroES, DnaK, DnaJ, and GrpE participate in the correct folding of nascent peptides (Hartl *et al.* 1994). The PURE system inevitably lacks these folding related components. To evaluate their importance, we compared the specific activity of DHFR synthesized by the PURE and S30 systems and found no differences. The molecular chaperones described above have been supplied to the PURE system translating DHFR and GFP (Fig. 2), but we observed no enhancement in the specific activity of the product by these factors. These results are consistent with the finding that chaperone-deficient cell extract is able to produce DHFR with the same specific activity as wild-type extract (Kudlicki *et al.* 1994). Thus, it is likely that these molecular chaperones are not essential for co-translational folding of proteins. In

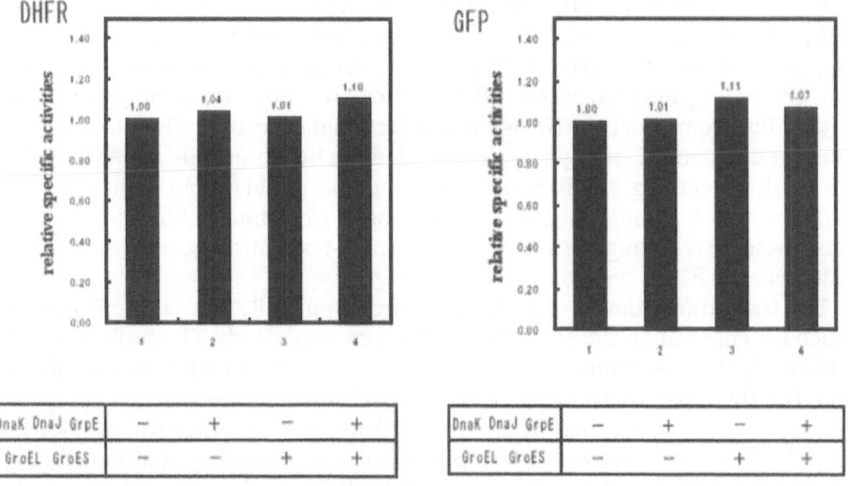

DnaK DnaJ GrpE	−	+	−	+
GroEL GroES	−	−	+	+

DnaK DnaJ GrpE	−	+	−	+
GroEL GroES	−	−	+	+

Fig. 2. The effects of DnaK, DnaJ and GrpE, GroEL and GroES on specific activities of DHFR synthesized by the PURE system. The concentrations of these molecular chaperones in the reaction mixtures described in Table I, were 10 µg/ml. The DHFR activity was determined according to the literature (Poe *et al.* 1972) and fluorescence of GFP was measured by spectrofluorometer 850 (Hitachi). The amounts of proteins were calculated by the incorporation of [35S]-methionine. Details of preparation of molecular chaperones will be published elsewhere.

contrast, activities of mammalian mitochondrial rhodanese and eukaryotic enzymes are dependent on the presence of molecular chaperones in a cell-free translation system (Kudlicki *et al.* 1994). As the proteins examined here might be folded into the native structure because of their simple tertiary structure, it is necessary to examine and check a number of proteins derived from various origins.

Improvement of the PURE system using a trans-translation system

Unfortunately a minimal amount of nuclease activity has been detected in the PURE system. By checking all the components, we found that nucleases existed in ribosome and enzyme samples. Although we are currently attempting to eliminate nucleases from these samples by careful purification procedures, we tried the following experiment to improve the PURE system in another way.

First, the degradation of mRNA in the mRNA-directed PURE system in the presence or absence of HPRI (RNase inhibitor derived from human placenta) was analyzed. In the presence of HPRI, 80% of mRNA still remained at the intact length after 3 h incubation, while the absence of HPRI resulted in degradation of 70 % of mRNA in the PURE system (data not shown). In the presence of HPRI, we observed an increase in protein synthesis of the mRNA-directed PURE system. Previously, we described that most of the DHFR mRNA in crude cell extract derived from *E. coli* is digested within half an hour. Thus, we believe that the PURE system, even in its present state, is an improved device for protein production, with regard to the stability of template nucleic acid.

We also examined the effect of HPRI on the transcription/translation coupled PURE system and found an increase in productivity. As mRNA was continuously supplied by the transcription reaction in the coupled system, we suspected that deceleration of protein synthesis after one hour was not a result of a shortage of mRNA, but from short mRNA without a termination codon. The possibility of shortage of chemical energy was clearly ruled out by monitoring the ATP level in the PURE system (Fig. 3a). To examine our hypothesis that mRNA without the termination codon through nuclease attack causes ribosome stalling, which hampers ribosome recycling, we introduced the trans-translation system (Karzai *et al.* 2000) into the PURE system.

The trans-translation system is composed from tmRNA and SmpB protein, which are encoded on the ssrA and smpB genes respectively. The tmRNA, kindly provided by Prof. A. Muto, Hirosaki University, was a native RNA with modification. The SmpB protein was first expressed as a fusion form in *E.coli* and prepared in a native form by cutting off the his-tag using protease after the purification (Fig. 3b). The activity of the ssrA-SmpB system was checked by addition of a tag-peptide encoded on tmRNA using truncated mRNA. As shown in Fig. 3c, supplement of ssrA-SmpB to the PURE system elongated durability of peptide synthesis. Thus, break-off of ribosome stalling caused by shortened mRNA may be effective in increasing productivity of protein synthesis. Furthermore, the combination of HPRI and ssrA-SmpB had a synergistic effect on the mRNA-directed PURE system. As shown in Fig. 3c, the PURE system with the trans-translation system and

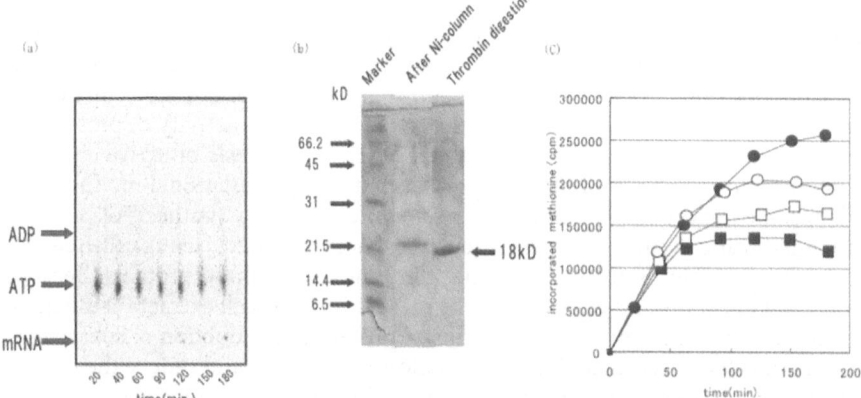

Fig. 3. Effects of trans-translation on the PURE system. (a) the ATP level in the PURE system. The PURE system was carried out in the presence of [^{32}P]-αATP and aliquot was spotted on POLY-GRAM CEL300 PEI (Macherey-Nagel) and developed. (b) SDS-PAGE of purified smpB protein. Overexpressed SmpB protein was purified with the his-tag peptide, which was removed by thrombin digestion. (c) The PURE system with HPRI and/or trans-translation system. DHFR was synthesized in the PURE system in the absence (square), or in the presence of 0.4 μM tmRNA and 10 mg/ml smpB protein (circle). HPRI was added at the concentration of 2.2 units/μl (closed circle, open square). The amount of DHFR was calculated by incorporation of [^{35}S]-methionine.

HPRI is capable of synthesizing peptide for more than three hours, while the PURE system without these supporting components stopped after around one hour. Because HPRI inhibits only pyrimidine-specific nucleases, other nucleases might generate truncated mRNA causing ribosome stalling. The trans-translation system might play a role in the recruitment of ribosomes, resulting in enhanced activity of protein production.

We consider that the reinforced system demonstrated here is very important in mass production of protein, because preparation of a sufficient amount of nuclease-free ribosomes is a difficult task in the construction of the PURE system.

Conclusion

Here we describe the characterization and the improvement of the PURE system. In the semi-optimized system, the productivity is similar to cell extract prepared from *E. coli*. Because of incomplete optimization of reaction conditions so far, we believe that there is further potential in the PURE system that will be elicited in the near future. One of the great advantages of the PURE system is that concentrations of individual components can be manipulated. Through this approach, we are attempting to develop a system capable of producing protein on a milligram scale per 1 ml reaction in the batch system.

Furthermore, the introduction of various subsystems into the PURE system is highly advantageous. Because of the ease of control of events in the system, we can design an appropriate cell-free translation system corresponding to the target proteins. We have demonstrated the PURE system to work with the trans-translation system or with molecular chaperones. In the former subsystem, we succeeded in recycling ribosome in the system. Although so far supplement of molecular chaperones has not been shown to be effective in increasing the activity of protein, we believe this reinforcement will bring about efficient synthesis of active enzymes with unstable structures. Studies on proteins with posttranslational modification will become a more important part of proteomics research. Synthesis of proteins with modifications such as glycosylation, phosphorylation, etc., using cell-free system will be highly useful in evaluating the function of these proteins. Because more than one thousand components are present in crude cell extracts, it might be difficult to control the final mature forms of proteins. Introduction of subsystems into crude cell extract might lead to complicated reactions and reduce the usefulness. In contrast, manipulation of all the events in the PURE system even with subsystems for modification processes is relatively easy. Thus, the PURE system provides a fundamental system with which to generate modified proteins.

References

Hartl FU, Hlodan R and Langer T (1994) Molecular chaperones in protein folding: the art of avoiding sticky situations. *Trends in Biochemical Sciences* 19:20–5.

Jelenc PC and Kurland CG (1979) Nucleoside triphosphate regeneration decreases the frequency of translation errors. *Proceedings of the National Academy of Sciences of the United States of America*, 76, 3174–8.

Karzai AW, Roche ED and Sauer RT (2000) The SsrA-SmpB system for protein tagging, directed degradation and ribosome rescue. *Nature Structural Biology* 7:449–55.

Kudlicki W, Mouat M, Walterscheid JP, Kramer G and Hardesty B (1994) Development of a chaperone-deficient system by fractionation of a prokaryotic coupled transcription/translation system. *Analytical Biochemistry* 217:12–9.

Kung H, Spears C and Weissbach H (1975) Purification and properties of a soluble factor required for the deoxyribonucleic acid-directed in vitro synthesis of beta-galactosidase. *Journal of Biological Chemistry* 250:1556–62.

Kung HF, Redfield B, Treadwell BV, Eskin B, Spears C and Weissbach H (1977) DNA-directed in vitro synthesis of beta-galactosidase. Studies with purified factors. *Journal of Biological Chemistry* 252:6889–94.

Madin K, Sawasaki T, Ogasawara T and Endo Y (2000) A highly efficient and robust cell-free protein synthesis system prepared from wheat embryos: plants apparently contain a suicide system directed at ribosomes. *Proceedings of the National Academy of Sciences of the United States of America* 97:559–64.

Poe M, Greenfield NJ, Hirshfield JM, Williams MN and Hoogsteen K (1972) Dihydrofolate reductase. Purification and characterization of the enzyme from an amethopterin-resistant mutant of Escherichia coli. *Biochemistry* 11:1023–30.

Shimizu Y, Inoue A, Tomari Y, Suzuki T, Yokogawa T, Nishikawa K and Ueda T (2001) Cell-free translation reconstituted with purified components. Nature Biotechnology 19:751–5

Spirin AS, Baranov VI, Ryabova LA, Ovodov SY, Alakhov YB (1988) A continuous cell-free translation system capable of producing polypeptides in high yield. Science. 242:1162–1164.

Wagner EG, Jelenc PC, Ehrenberg M and Kurland CG (1982) Rate of elongation of polyphenylalanine in vitro. *European Journal of Biochemistry* 122:193–7.

All journal titles here are not abbreviated

CHAPTER 5

Direct Expression of PCR Products in Cell-Free Translation Systems   5

ANATOLY T. GUDKOV, KIRILL A. MARTEMYANOV[1]

Abstract. An effective methodology is described for the direct expression of PCR-generated DNA fragments in cell-free transcription/translation systems without cloning DNA into plasmids. The methodology is accomplished here for the synthesis of the active antibacterial peptide cecropin using the synthetic coding DNA sequence. The proposed approach can be generally used for the cell-free synthesis of polypeptides and proteins that are unstable in living cells, or proteins that are strongly cytotoxic. Moreover, PCR-generated copies and/or multi-copies of genes and their fragments from genomic libraries (or genomic DNA) can be expressed directly in a transcription/translation system and then tested, without the subcloning of DNA.

Introduction

Early cell-free translation systems were based on DNA-free cytoplasmic extracts of bacterial or eukaryotic cells programmed by endogenous mRNA molecules. In these systems, the ribosomes continued polypeptide synthesis encoded by mRNA available in the extract. Later, the cell-free translation systems were improved, and the use of exogenous messages, being either natural mRNA or synthetic polyribonucleotides, became possible (Nirenberg and Matthaei 1961).

The next step in the development of cell-free translation systems was achieved when exogenous DNA instead of mRNA was introduced into an RNA polymerase-containing cell extract, and thus mRNA synthesis took place directly in the cell-free system mixture (DeVries and Zubay 1967). In the case of bacterial systems, a proper DNA construct can be added to the DNA-free cell extract, and the endogenous RNA polymerase in the extract synthesizes the corresponding mRNA *in situ*. The eukaryotic extracts for cell-free translation, however, are prepared from the cytoplasmic fraction, so that they lack the endogenous RNA polymerase activity. This limitation can be overcome by the addition of an exogenous RNA polymerase, typically bacteriophage T7 or SP6 RNA polymerase; in this case DNA

---

[1] Institute of Protein Research, Russian Academy of Sciences,
142290 Pushchino, Moscow Region, Russia
e-mail: gudkov@vega.protres.ru, Phone/Fax: 7 (095) 924-0493

constructs must be supplemented with T7 or SP6 promoters in order to produce mRNA *in situ*.

Translation in cell-free systems is commonly used in studies that investigate protein biosynthesis mechanisms over a long period of time. During the last decade, cell-free protein synthesis systems were also proposed for biotechnological applications (for reviews, see Spirin 1992). In most cases reported up until recently, the genes for mRNA production *in vitro* were isolated from living cells, modified if necessary, inserted into plasmids, cloned by the routine *in vivo* procedure and isolated in the form of a plasmid suitable for expression. A fully cell-free version of the transcription/translation cell-free system, where living cells do not participate in any steps in the process, was proposed several years ago (Spirin 1992). This approach includes chemical synthesis of a gene and its PCR amplification, with subsequent direct expression of the PCR product in a cell-free transcription/translation system. A practical demonstration of this strategy is given below, using the synthesis of cecropin, an antibacterial peptide of 31 amino acid residues in length (Martemyanov et al. 1997) as an example.

Construction of DNA for expression in a cell-free system

The fused gene expression technique is often used to synthesize foreign or unstable proteins to avoid problems with stability, solubility, folding and yield of polypeptide products. In particular, genes encoding for relatively short polypeptides can be poorly expressed because of the proteolytic vulnerability and instability of products, whereas their fusion products may be stable (Piers et al. 1993; Martemyanov et al. 1996). PCR is a convenient methodology for producing fused genes (Higuchi et al. 1988). Several PCR techniques have been used to obtain duplicated genes (Yon and Fried 1989) and multiple repeats of a coding sequence, especially in the cases of DNA sequences encoding for oligopeptides (Hemat and McEntee, 1994; Nakajima and Yaoita 1997). A similar procedure to obtain the multimeric (up to seven repeats) gene encoding for antibacterial polypeptide cecropin has also been developed within our group (Martemyanov and Gudkov 1997). In this case, the coding sequence was deduced from the amino acid sequence of cecropin (SWLSKTAKKLENSAKKRISEGIAIAIQGGPR) and synthesized chemically (Martemyanov et al. 1996).

A general scheme of the DNA construction is depicted in Fig. 1. Oligonucleotide primers (Pr1 and Pr2) with overhanging 5'-ends, complementary to the opposite sites of the DNA, are necessary to ligate the DNA fragments by PCR. For identical duplication of the coding sequence, the 5'-end of Pr1 is complementary to the 3'-encoding region upstream of the termination codon, and contains an ATG triplet, while its 3'-end is complementary to the 5'-end of the gene (Fig. 1). The 5'-end of Pr2 is complementary to the 5'-end of the coding sequence and contains no stop (TAA) codon. Using primers Pr1 and Pr2, a tandemly repeated cecropin gene without a termination codon within the linear DNA fragment was obtained.

Megaprimer (163 nt) containing regulatory elements for efficient expression of genes (T7 promoter, *s10* leader and Shine-Dalgarno sequences, as well as the additional 45 nucleotides upstream of the T7 promoter) and the 3'-end complemen-

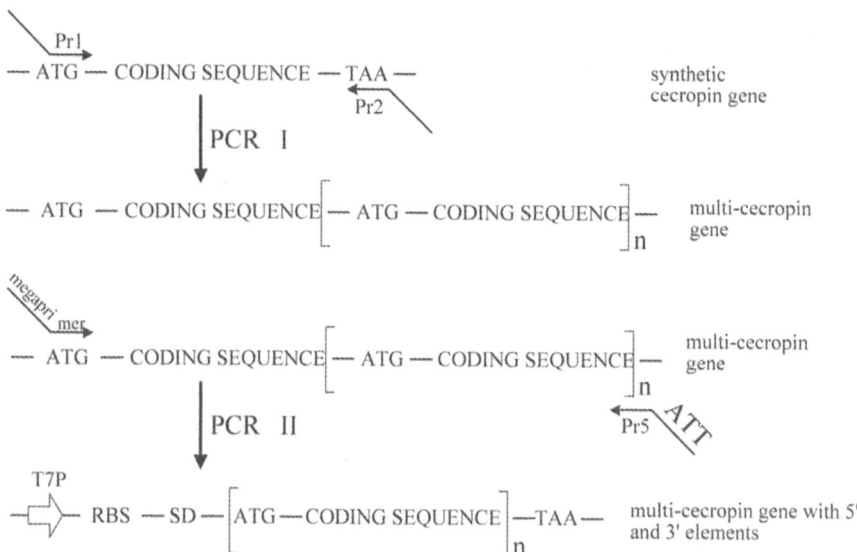

Fig. 1. General scheme of DNA construction for its cell-free expression (see text, section 2)

tary to the beginning of the cecropin gene, was obtained by PCR using the usual primers Pr3 and Pr 4 from plasmid pET21d(+) (Novagen) carrying the synthetic cecropin gene. The final construct was obtained with megaprimer and 3'-flanking primer Pr5 containing the TAA stop codon in the second PCR run (Martemyanov et al. 1997).

Direct expression of PCR products in a cell-free system

Unpurified PCR mixture was added to the E. coli S30 extract supplemented with T7 RNA polymerase for direct expression of the DNA coding sequence in a cell-free transcription/translation system. The cecropin yield was about 3 nmol/ml (Fig. 2). For comparison, the DNA encoding for dimeric cecropin cloned in plasmid and expressed under the same cell-free conditions produced 2.1 nmol/ml (Fig. 2). Gel electrophoresis proved the formation of a product with the expected molecular weight (Fig. 3). Antibacterial activity was tested by the growth zone inhibition assay using E. coli strain D21 (Fig. 4) after cleavage with BrCN (cecropin oligomers were not active in this assay) (Martemyanov et al. 1997).

The yield of cecropin in the E. coli cell-free transcription/translation system from the monomeric synthetic gene cloned in plasmid was 800 pmol/ml. The expression of the DNA fragment coding for dimeric cecropin molecules cloned in plasmid gave 2100 pmol/ml of cecropin monomer. Direct PCR product expression produced up to 3050 pmol/ml of cecropin calculated per monomer. As the genetic elements at the 5'- and 3'-ends were the same in all the constructs used, the obvious reasons for the best results in the case of the direct expression of PCR

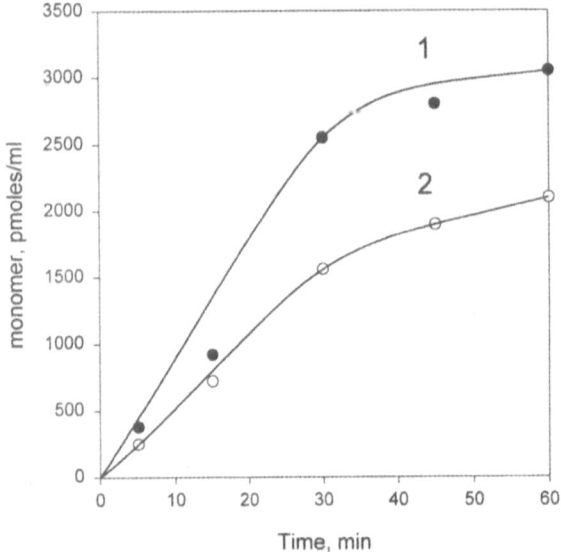

Fig. 2. Synthesis of cecropin in cell-free transcription/translation systems: (1) direct expression of PCR product (multimeric cecropin genes), (2) expression of plasmid with dimeric cecropin gene. Experiments on the *in vitro* expression were carried out in the transcription/translation system based on the *E.coli* MRE 600 S30 extract supplemented with the T7 phage RNA polymerase. The reaction mixture contained 50 units RNAsin, 100 units T7 polymerase, 20 µM [^{14}C]Leu (282 Ci/mole) (Amersham), 1/10 volume S30 extract, 100 µM of each amino acid and 1 mM of each NTP. A 20 µl aliquot of the PCR reaction mixture corresponding to 0.5–1 µg of the DNA synthesized was added to 50 µl reaction volume. In the control expression experiment, 1 µg of plasmid with the dimeric cecropin gene was used in the same translation mixture. 10 µl aliquots were taken at consecutive time points for the estimation of [^{14}C]-Leu incorporation into hot TCA insoluble material. Synthesized products were also analyzed by electrophoresis followed by autoradiography (Martemyanov et al. 1997)

Fig. 3. Gel electrophoresis of the cell-free transcription/translation products: autoradiography of the polypeptides synthesized (1) in the plasmid-directed system (dimeric cecropin), and (2) in the PCR product-directed system after CNBr cleavage (monomeric cecropin)

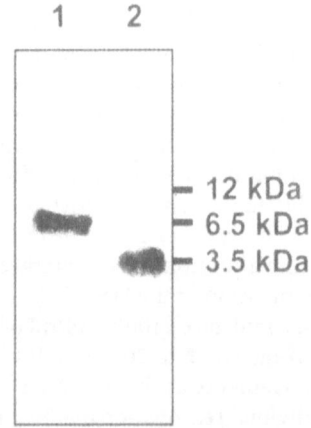

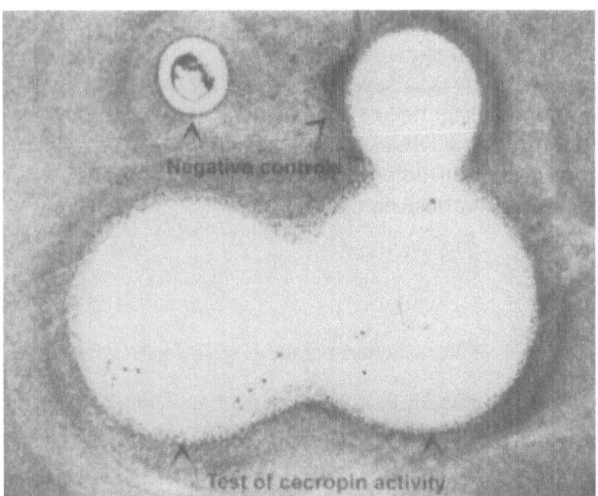

Fig. 4. Assay of cecropin activity. Translation mixtures after CNBr cleavage of the polypeptide product were tested. Antibacterial activity was tested by the growth zone inhibition assay using *E.coli* strain D21. Multimeric cecropin products were cleaved with BrCN to produce active monomeric cecropin. For this purpose, 200 μl of the translation mixture were clarified by centrifugation, dialyzed against 5 mM NaH_2PO_4/Na_2HPO_4 buffer (pH 7.4), dried and dissolved in 70% CH_3COOH. After the addition of 0.5 mg BrCN the mixture was incubated overnight in the dark. The mixture was then diluted 10fold by distilled water, dried and dissolved in 20 μl water. 10 μl aliquots were applied to the bacterial lawn. As a negative control, the same procedure was applied except that no cecropin-encoding DNA was added to the translation system

products could be the stability of the multimeric gene, the use of multimeric mRNA and/or the multimeric polypeptide product synthesized in the cell-free system. This finding correlates with the fact that cecropin fused to protein A could be successfully synthesized *in vivo* (Martemyanov et al. 1996). We believe that our PCR constructs could be further improved by addition of a stable structural element at the 3'-end of the gene to impede the 3'-exonucleolytic degradation. Such elements can be introduced with the primer in the last round of PCR.

Brief discussion remarks

The proposed approach can generally be used for the cell-free synthesis of polypeptides and proteins that are unstable in living cells, or proteins that are strongly cytotoxic. Moreover, PCR-generated copies and/or multi-copies of genes and their fragments from genomic libraries (or genomic DNA) can be directly expressed in a transcription/translation system and then tested, without subcloning of DNA procedures. In particular, *in vitro* expression of DNA fragments with open reading frames for the purpose of protein identification, or test expression of some genes before construction of special vectors could be useful applications of the methodology reported here.

The results described above were obtained in the batch format of the transcription/translation cell-free system. It is expected that the yield of polypeptides synthesized in this way could be increased several times if the continuous flow cell-free (CFCF) or the continuous-exchange cell-free (CECF) expression system was used (Spirin et al. 1988). Multimeric DNA sequences encoding for short cytotoxic polypeptides can be also useful for the production of an active peptide after chemical or enzymatic cleavage of synthesized product.

References

DeVries JK, Zubay G (1967) The synthesis of the α-fragment of the enzyme β-galactosidate.Proc Natl Acad Sci USA 57: 1011–1012

Hemat F, McEntee K (1994) A rapid and efficient PCR-based method for synthesizing high-molecular-weight multimers of oligonucleotides. Biochem Biophys Res Commun. 205: 475–481

Higuchi R, Krummel B, Saiki RK (1988) A general method of in vitro preparation and specific mutagenesis of DNA fragments: study of protein and DNA interactions. Nucleic Acids Res. 16: 7351–67

Martemyanov KA, Gudkov AT (1997) An efficient PCR method for producing gene tandem repeats. Doklady Biochemistry. 357: 158–160

Martemyanov KA, Spirin AS, Gudkov AT (1996) Synthesis, cloning, and expression of genes for antibacterial peptides: cecropin, magainin and bombinin. Biotechnology Lett. 18: 1357–1362

Martemyanov KA, Spirin AS, Gudkov AT (1997) Direct expression of PCR products in a cell-free transcription/translation system: Synthesis of antibacterial peptide cecropin. FEBS Lett. 414: 268–270

Nakajima K, Yaoita Y (1997) Construction of multiple-epitope tag sequence by PCR for sensitive Western blot analysis. Nucleic Acids Res. 25: 2231–2232

Nirenberg MW, Matthaei JH (1961) The dependence of cell-free synthesis in E. coli upon naturally occurring or synthetic polynucleotides. Proc Natl Acad Sci USA 47: 1588–1602

Piers KL, Brown MH, Hanckock RE (1993) Recombinant DNA procedures for producing small antimicrobial cationic peptides in bacteria. Gene. 134: 7–13

Spirin AS (1992) Cell-Free Protein Synthesis Bioreactor. In: Todd P, Sikdar SK, Bier M (eds) Frontiers of Bioprocessing II. American Chemical Society, Washington, DC, pp. 31–43

Spirin AS, Baranov VI, Ryabova LA, Ovodov SY, Alakhov YB (1988) A continuous cell-free translation system capable of producing polypeptides in high yield. Science. 242: 1162–1164

Yon J, Fried M (1989) Precise gene fusion by PCR. Nucleic Acids Res. 17:4895

CHAPTER 6

Yeast Cell-Free Translation Systems 6

MICHAEL ALTMANN, HANS TRACHSEL[1]

Summary

Translation in eukaryotes is well conserved among higher eukaryotes, such as mammals, and in lower eukaryotes, such as the yeast *Saccharomyces cerevisiae*. Since yeast cells are easy to handle and amenable to very powerful genetic and biochemical analysis, the yeast system represents an attractive model system for studies on the mechanism and regulation of eukaryotic translation. The first cell-free systems capable of initiating translation on exogenous mRNA were prepared from the yeast *Saccharomyces cerevisiae* in the late 1970's. These extracts show cap- and poly(A)-dependent translation, whereby the cap structure and the poly(A) tail on the mRNA lead to synergistic stimulation. Yeast extracts can be made dependent on exogenous translation factors by inactivation of the endogenous factor activity through various means, including downregulation of factor synthesis, expression of conditional lethal factor or knock-out mutants and inhibition of factor activity by antibodies. Such systems have been used to extend and deepen our knowledge of the mechanism of translation initiation, protein targeting and translation termination.

Introduction

The mechanism and regulation of translation, including translation initiation, is well conserved between the yeast *Saccharomyces cerevisiae* and mammalian cells. This is supported by the existence of structural and functional similarities between mammalian and yeast translation factors, mainly translation initiation factors (eIFs), and by the ability of some mammalian translation initiation factors to substitute for the corresponding yeast initiation factors in yeast cells *in vivo*. Accordingly, the yeast system is widely used as a model system for studies of eukaryotic translation, taking advantage of the well developed genetic, molecular genetic and biochemical techniques available for this organism. In the past, this has contributed considerably to our knowledge of eukaryotic translation initiation (for a review, see McCarthy 1998), protein targeting and translation termina-

[1] Institute of Biochemistry and Molecular Biology,
University of Berne, Bühlstrasse 28, 3012 Berne, Switzerland

tion. Cell-free translation systems are important experimental tools which allow the investigator to manipulate translation by changing buffer components, translation factor and mRNA concentrations more easily than in *in vivo* systems. Therefore, it was an important achievement when conditions for the preparation of yeast extracts able to initiate translation on exogenous mRNA were first reported in 1979 (Gasior et al. 1979). Since then, cell-free translation systems derived from many yeast strains have been used in quite a few laboratories to study eukaryotic translation, mainly the function of translation initiation factors.

Preparation of yeast cell-free translation systems

Early attempts to produce yeast *Saccharomyces cerevisiae* extracts active in translation resulted in extracts competent for elongation but unable to translate exogenous mRNA (reviewed in Tuite and Plesset 1986). In 1979 (Gasior et al. 1979) and in 1980 (Tuite et al. 1980), procedures for the preparation of yeast extracts that are able to initiate upon and translate exogenous mRNA efficiently were published. Modifications of lysate were reported and the individual steps of the procedures were later discussed in detail (Tuite and Plesset 1986).

Apart from *Saccharomyces cerevisiae*, mRNA-dependent cell-free translation systems have also been developed from the human pathogenic fungus *Candida albicans*, using the glass bead disruption method (see below). After micrococcal nuclease treatment, this lysate was able to translate exogenous mRNAs (Colthurst et al. 1991). Furthermore, the preparation of a translation system from heat shock treated *Aspergillus nidulans* was reported, which is capable of synthesing proteins efficiently and faithfully from natural and *in vitro* transcribed eukaryotic mRNAs (Devchand et al. 1988).

Procedures

Several methods for the preparation of yeast cell-free translation systems have been described, differing mainly in the ways in which cells are lysed. In our preferred procedure, cells are lysed by homogenization after spheroblasting (Gasior et al. 1979, see below), whereas other procedures use either glass beads (Hofbauer et al. 1982) or a French Press (Kreutzfeldt and Lochmann 1983, Rothblatt and Meyer 1986) to break the cells. The method we routinely use for yeast extract preparation in our laboratory is a modified version of the procedure described by Gasior et al. (Gasior et al. 1979, Altmann and Trachsel 1997). The system is made dependent on exogenous mRNA by treatment with micrococcal nuclease (Pelham and Jackson 1976). Below, we present the protocol for extract preparation and cell-free translation (Altmann and Trachsel 1997):

Cell growth and harvest:
● One liter of yeast culture is grown overnight in rich medium (e.g. YPD: 1% yeast extract, 2% peptone, 2% glucose) at 25°-30°C (25°C for temperature-sensitive strains) to a final A_{600} of ~1.2–1.5.

- Cells are pelleted by centrifugation at 3,000x g for 5 min and washed once by resuspension in 50 ml H_2O and centrifugation. The yield is ~4–5 grams of wet cells per liter of culture.

Spheroblasting:
- Cells are resuspended in 40 ml of 1M sorbitol, 2 mM EDTA, 14 mM ß-mercaptoethanol and incubated for 30 min at room temperature with occasional shaking.
- Cells are collected by centrifugation at 3,000x g and resuspended in 40 ml of 1M sorbitol (ß-mercaptoethanol is omitted in order not to accelerate the zymolyase treatment to uncontrollable levels). 60 µl Zymolyase 20'000 (50 mg/ml stock; stored at -40°C in 1M sorbitol, 20 mM Tris-HCl, pH 7.5, 50% glycerol) is added and cells are incubated at 25°C with gentle shaking. Spheroblasting is monitored by measuring the A_{600}-decrease of the spheroblast suspension. This is done by resuspension of spheroblasts in H_2O (e.g. resuspend 10 µl cells in 1 ml H_2O). As a control, spheroblasts are resuspended in 1M sorbitol. When the A_{600} drops to 50% of the value before zymolyase addition (normally after 30 to 45 min) the treatment is stopped by centrifugation of the spheroblasts at 3,000x g for 5 min. Note that the amount of zymolyase required to obtain optimal spheroblasting may vary from strain to strain and may be higher when the cells are grown to higher densities.

Regeneration:
- The spheroblasts are resuspended in 40 ml YPD:1M sorbitol (1:1), incubated for 1 h at 25°C with gentle shaking and then centrifuged.

Preparation of the extract:
- All of the subsequent steps are performed at 4°C. The spheroblasts are resuspended in 8–10 ml of buffer A (30 mM HEPES-KOH, pH 7.4, 100 mM K(OAc), 2 mM Mg(OAc)$_2$, 2 mM DTT) and homogenized with 30 strokes in a Dounce homogenizer with a tight-fitting pistil (type A).
- The homogenate is centrifuged for 15 min at 40,000x g, the supernatant transferred to a new tube and recentrifuged for 15 min at 110,000x g. The lipid layer on the surface is removed by suction. The concentration of the extract is determined by measuring the A_{260}. It should be ~100–200 units/ml. If desired, extracts can be frozen at -80°C at this stage and kept for months.
- To remove low molecular weight components, e.g. amino acids, about 1.5 ml of extract is loaded on a 20 ml Sephadex G25 (fine) column equilibrated with buffer A. Fractions of 1 ml are collected, the A_{260} measured and the main fractions pooled. Usually, 2 ml of extract with ~60–100 A_{260} /ml are obtained. Extracts of lower concentration show reduced translational activity.
- The extract is frozen in aliquots of 75–150 µl and kept in liquid nitrogen for prolonged storage (years) without loss of activity.

In vitro translation assay:
We routinely treat extracts with micrococcal nuclease to remove endogenous mRNA and to render the extract dependent on added mRNA (Pelham and Jack-

son 1976). This treatment allows monitoring of *de novo* initiation of protein synthesis as opposed to ribosome run-off, e.g. ribosomes terminating the synthesis of nascent polypeptide chains whose synthesis was initiated *in vivo*.

Nuclease treatment:
- Incubate the following components for 10 min at 20°C:
  150 μl extract
  2.5 μl creatine phosphokinase (5 mg/ml)
  4.0 μl micrococcal nuclease (3 units/μl)
  3.0 μl CaCl$_2$ (50 mM)
- Stop nuclease treatment by adding 3.0 μl EGTA (100 mM) and put the extract on ice.

Translation mix:
Add to the digested extract (on ice):
25 μl 10x cocktail (see below),
5 μl creatine phosphate (0.6 M)
2.5 μl Mg(OAc)$_2$ (100 mM)

At this stage the extract may be preincubated for 5–10 min at 20°-25°C. This leads to ribosome run-off on mRNA fragments remaining after micrococcal nuclease treatment and reduces background incorporation.
Add 3 μl [35S]-methionine (15 μCi/μl), final volume: 200 μl translation mix
Use 12 μl translation mix per assay and add mRNA, factors or H$_2$O to a final volume of 15 μl per assay.
Translation cocktail (10x):
Mix:
200 μl HEPES-KOH, pH 7.4 (1 M)
40 μl Mg(OAc)$_2$ (0.5 M)
760 μl KCl (2 M)
200 μl GTP (0.04 M)
200 μl ATP (0.1 M)
400 μl 19 amino acid mix (2.5 mM each, minus methionine)
200 μl H$_2$O

Parameters which may require testing:
We use total yeast RNA as mRNA to optimize translation in extracts derived from different strains. In our experience, the optimal concentrations of Mg^{2+}, GTP and extract in the final incubation vary somewhat from extract to extract and should be optimized for each strain.

Reaction temperature:
Yeast extracts translate mRNA best at temperatures between 20°C and 25°C. Higher temperatures activate an inhibitor that causes reversible inhibition of translation during short incubations (up to 15 min) and irreversible inhibition during prolonged incubation (15 min and more, see Mandel and Trachsel 1989). Extracts may therefore be preincubated for 5–10 min at temperatures between 30°C to

37°C before adding the radioactively labelled amino acid and continuing incubation at 20°C to 25°C. This two-step incubation, though often not required, may be used to deplete extracts from temperature-sensitive strains of endogenous factor activity.

Measurement of methionine incorporation:
- 3 µl aliquots of reaction mixture at desired time points after the start of the incubation (e.g. 10 min and 30 min) are spotted on filter paper discs to stop the reaction.
- Discs are then soaked in 5% trichloroacetic acid (TCA) to fix the protein. They can be kept in this solution until all samples have been collected.
- The discs are transferred to fresh 5% TCA and boiled for 3 min to deacylate charged met-tRNA, which will otherwise increase the background incorporation.
- The discs are rinsed with cold 5% TCA, followed by H_2O, ethanol and acetone.
- Finally, the discs are air-dried and counted in a scintillation counter. Alternatively, 3-5 µl aliquots of reaction mixtures may be analyzed by SDS polyacrylamide gel electrophoresis and autoradiography.

The results of a typical *in vitro* translation experiment are shown in Fig. 1. The extract used for this experiment was made from cells carrying mutant eIF4E (cap-binding protein). Therefore, the extract had to be supplemented with exogenous eIF4E to obtain wild-type translational activity.

Experimental work using yeast extracts

Extracts derived from wild-type and mutant yeast strains have been used to study the function of translation initiation factors in the various steps of translation initiation, the functional interaction of the poly(A) tail with the cap structure in translation, as well as translation termination and protein targeting (import into ER vesicles or mitochondria). Examples illustrating the use of yeast extracts are presented below.

Studies on translation initiation

Initiation of translation is a multi-step biochemical pathway leading to binding of a ribosome at the initiator AUG codon on an mRNA and able to translate the open reading frame into a polypeptide chain. This is achieved either by the cap-dependent or by the internal initiation pathway (for reviews, see Pain 1996; McCarthy 1998; Hershey and Merrick 2000). While we know rather little about internal initiation of translation in *Saccharomyces cerevisiae*, the cap-dependent initiation pathway has been well documented and is (with a few exceptions) well conserved between yeast and higher eukaryotes (McCarthy 1998). In this pathway, the initiation factor eIF2 forms a ternary complex with initiator Met-tRNA$_i$ and GTP and associates with eIF1, eIF3 and eIF5 to form a multi-factor complex

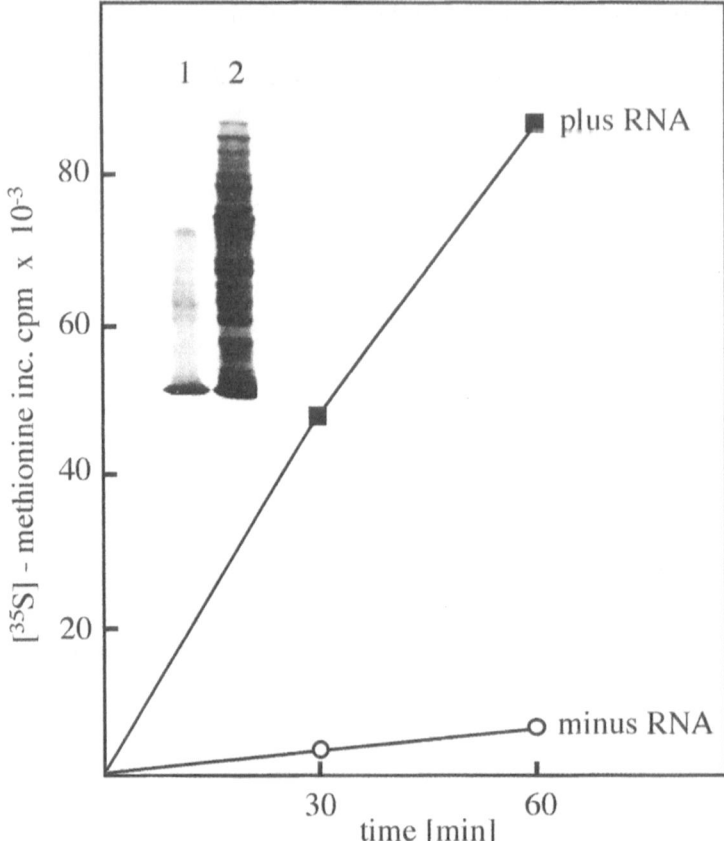

Fig. 1. Cell-free translation in yeast *Sacharomyces cerevisiae* extract. The extract was prepared as described above from strain ts 4–2 (Altmann et al. 1989), digested with micrococcal nuclease and 20 μl reaction mixtures supplied with 10 μg of total yeast RNA as mRNA source and 50 ng purified eIF4E. Reaction mixtures were incubated at 23°C and 4 μl aliquots taken for analysis of (^{35}S)-methionine incorporation as described above. The inset (autoradiogram) shows 5 μl aliquots of reaction mixtures incubated for 60 min without total yeast RNA (lane 1) or with total yeast RNA (lane 2), fractionated by SDS gel electrophoresis and analyzed by autoradiography

(Asano et al. 2000) which binds to the 40S ribosomal subunit. The resulting 43S pre-initiation complex then binds at or close to the cap structure m^{7}GpppX at the 5' end of the mRNA. This step is mediated by the initiation factors eIF4E (cap-binding protein) and eIF4G which recruit the factors eIF4A and eIF4B, which in turn facilitate ribosome binding by unwinding RNA secondary structure in the 5'-untranslated region of the mRNA. The bound 43S pre-initiation complex then moves along the mRNA in the 5' to 3' direction searching for the initiator AUG. At the AUG codon, hydrolysis of GTP bound to eIF2 is triggered by eIF5, initiation factors are released and the 60S ribosomal subunit joins the 40S ribosomal subunit to form an 80S ribosome which is competent to translate the open reading frame of the mRNA.

Factor-dependent extracts

To study the function of individual initiation factors in translation, extracts dependent on exogenous factors for translational activity were produced. Yeast extracts dependent on eIF3 (Prt1- and Rpg1-subunit), eIF4E, eIF4A, Ded1 (a second RNA helicase involved in translation), eIF4B, eIF4G, eIF5 and poly(A)-binding protein (PABP) were developed (a) by preparation of extracts from yeast strains carrying a temperature-sensitive mutation in the gene encoding the factor (eIF3: Danaie et al. 1995; eIF4E: Altmann et al. 1989; eIF4G: Neff and Sachs 1999, Dominguez et al. 2001; eIF5: Maiti et al. 2000), (b) by preparation of extracts carrying the gene encoding an initiation factor under the control of the galactose-regulated *GAL1* promoter and shut-down of factor synthesis by growing the cells for a limited time on glucose-containing medium prior to cell lysis (eIF4A: Blum et al. 1989), (c) by preparation of extracts from cells deleted for a non-essential factor gene (eIF4B: Altmann et al. 1995) and by inactivating or removing a factor with antibodies (PABP: Tarun and Sachs 1995).

An example of an initiation factor-dependent yeast extract (eIF4A-dependent extract) is shown in Fig. 2.

Interestingly, the eIF4A dependence of the extract derived from PL49 cells is complete (Fig. 2), even though the cells were still growing (albeit slowly) on glucose before harvest. An apparently stronger dependence on an initiation factor *in vitro* compared with *in vivo* was also seen with an eIF4B (Tif3)-dependent extract. The reason for this is most likely to be the higher dilution of translation components in the extract, since we observed that in eIF4B (Tif3)-dependent extract this effect depends on the concentration of the yeast extract and is most prominent in dilute extracts (unpublished results). A difference in susceptibility between intact cells and extracts was also observed for methylation of 18S ribosomal RNA, where a deficiency in methylation in a mutant strain (dim1–2) does not change the growth rate, whereas translation in extracts from the mutant is strongly impaired (Lafontaine et al. 1998).

Role of the poly(A) tail in initiation of translation

Translation in yeast extracts, like translation in living cells, is cap- and poly(A)-dependent. The presence of both a cap structure and a poly(A) tail leads to synergistic stimulation in the range of 3–10-fold. This was first shown in yeast extracts derived from cells deficient for the double-stranded RNA L-A (Iizuka et al. 1994; Iizuka et al. 1997) and later found to hold true for other yeast extracts too (Tarun and Sachs 1995). The poly(A) tail stimulates 43S preinitiation complex binding to the initiator codon on mRNA in a reaction which is dependent on poly(A)-binding protein (Pabp) but independent of the cap structure and the cap-binding protein eIF4E (Tarun and Sachs 1995). This was shown in extracts

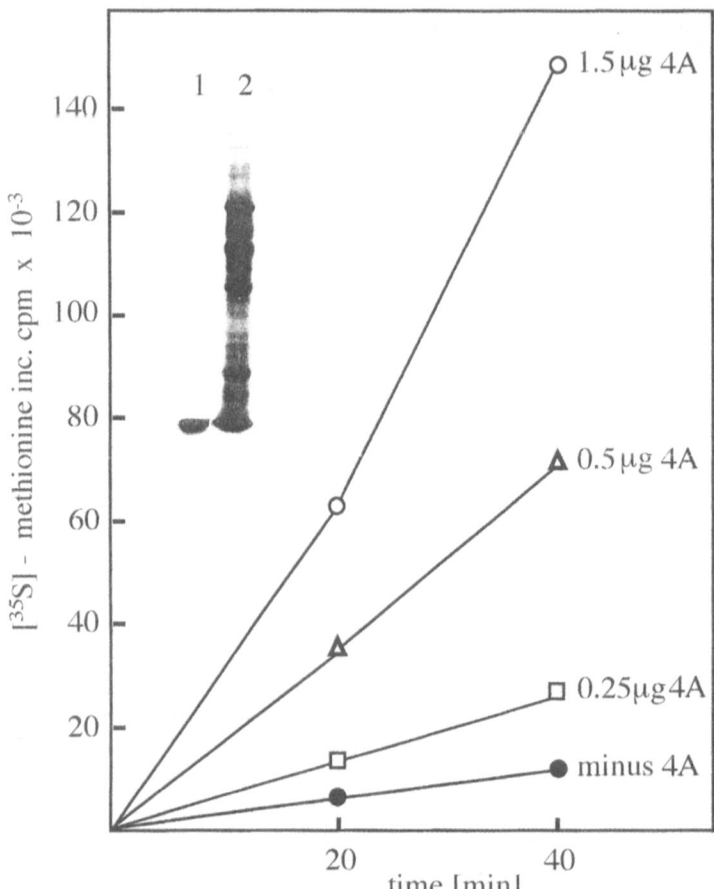

Fig. 2. Translation in initiation factor-dependent *Saccharomyces cerevisiae*. Cells of the yeast strain PL49 (with the only *TIF1* gene encoding eIF4A on a plasmid under the control of the *GAL1* promoter, Blum et al. 1989) were transferred from galactose-containing to glucose-containing medium and extract prepared as described above. The extract was digested with micrococcal nuclease and 20 μl reaction mixtures supplied with 10 μg of total yeast RNA as mRNA source and the indicated amount of purified eIF4A. Reaction mixtures were incubated at 25°C and 4 μl aliquots taken for analysis of (^{35}S)-methionine incorporation as described above. The inset (autoradiogram) shows 5 μl aliquots of reaction mixtures incubated for 60 min without exogenous eIF4A (lane 1) or with 1.5 μg of exogenous eIF4A (lane 2), fractionated by SDS gel electrophoresis and analyzed by autoradiography

dependent on either exogenous PABP or eIF4E. Positioning of the ribosome at an initiator codon by the poly(A)-PABP complex is even possible when the 5'-untranslated region of the mRNA is physically blocked for scanning ribosomes (by oligonucleotides or protein bound to the RNA), indicating that the poly(A)-PABP complex promotes internal binding of pre-initiation complexes (Preiss and Hentze 1998; for a review, see Preiss et al. 1998).

Initiation factor function and factor-factor interactions

Experiments with factor-dependent extracts were used to assay for factor activity during protein purification, to test potential translation initiation factors for stimulation of translation *in vitro*, and to test the functional importance of factor-factor interactions.

Factor eIF3:
The initiation factor eIF3 is a multi-subunit protein complex involved in several steps of the initiation pathway (for reviews, see Pain 1996; McCarthy 1998; Hershey and Merrick 2000). Using stimulation of translation in extracts dependent on Prt1 (Danaie et al. 1995) and Rpg1 (Valasek et al. 1998), at that time putative subunits of eIF3, as an assay during protein purification procedures, a protein complex containing Prt1, Rpg1 and three additional subunits was isolated and shown to represent the core of yeast eIF3.

Factors eIF4A, Ded1 and eIF4B:
When the genes *TIF1*, *DED1* and *TIF3* were cloned, they were found to encode proteins with similarities to mammalian eIF4A (Tif1, Ded1) and eIF4B (Tif3), initiation factors involved in mRNA binding to 43S preinitiation complexes. Definitive proof for their involvement in translation was derived from experiments with extracts devoid of either Tif1 (Blum et al.1989), Ded1 (Chuang et al. 1997) or Tif3 activity (Altmann et al. 1993; Altmann et al. 1995). These extracts were either prepared from cells carrying their only copy of the *TIF1* gene under the *GAL1* promoter on a plasmid and grown for several hours on glucose (Tif1-dependent extract), by immunodepletion of Ded1 (Ded1-dependent extract) or from cells lacking the *TIF3* gene (Tif3-dependent extract). Note that the *TIF3* gene is not essential for growth of yeast cells but renders cells growing slightly slower than wild-type and the derived extracts strongly Tif3-dependent.

Factor eIF4E:
The initiation factor eIF4E binds to the m^7GpppX cap structure of mRNA and to initiation factor eIF4G. Since eIF4G recruits further initiation factors (including the RNA helicase eIF4A) to the 5' end of mRNA and these factors catalyze the unwinding of RNA secondary structure, eIF4E is particularily important for translation of mRNAs with RNA secondary structure in the 5'-untranslated region. This is supported by experiments with an eIF4E-dependent extract (Altmann et al. 1989) that showed that translation of mRNA with an unstructured 5'-untranslated region is little affected by low levels of eIF4E activity (Altmann et al. 1990a) and that low levels of eIF4E activity favor internal initiation of translation (Altmann et al. 1990b).

Factor eIF4G:
The initiation factor eIF4G assembles other factors involved in mRNA binding into a multi-protein complex (see above). Using extracts derived from cells carrying a temperature-sensitive allele encoding eIF4G, the interaction between eIF4G and eIF4A was shown to be essential for translation (Neff and Sachs 1999, Dominguez et al. 2001).

The yeast eIF4G orthologs, Tif4631 and Tif4632, share a conserved PABP-binding site. Translation experiments in extracts derived from cells carrying a mutated *TIF4631* or *TIF4632* gene encoding eIF4G unable to interact with PABP showed that the PABP-eIF4G interaction is required for stimulation of translation by the poly(A) tail and for the synergism between the cap structure and the poly(A) tail *in vitro* (Tarun et al. 1997).

Interestingly, defects in the PABP-eIF4G interaction only showed an effect on growth of yeast cells when other initiation factors were mutated. This is another case where the *in* vitro system seems to be more sensitive to changes of factor activity than the *in vivo* system.

Factor eIF5:
Initiation factor eIF5 triggers the hydrolysis of GTP bound to eIF2 when the initiation complex reaches the initiator AUG on the mRNA. Through its C-terminal domain eIF5 interacts with the α-subunit of eIF2. *In vivo* studies and experiments with extracts derived from cells carrying a temperature-sensitive allele of the *TIF5* gene showed that this interaction is essential for translation (Das and Maitra 2000) and that the N-terminal domain stimulates GTP hydrolysis by the β-subunit of eIF2 (Das et al. 2001).

Translation termination

When the ribosome arrives at one of the three stop codons, the release factors (RF) eRF1 and eRF3 bind to the ribosome and release the polypeptide chain. Yeast wild-type eRF3 can aggregate (PSI+ phenotype), leading to changes in stop codon recognition. The effect of mutations in eRF3 on the efficiency of *in vitro* nonsense suppression was tested in extracts derived from wild-type and mutant yeast strains. The efficiency of all three types of yeast tRNA-mediated nonsense suppressor (ochre, amber and UGA) was found to be higher in cell-free systems prepared from wild-type strains than in extracts prepared from mutant strains (Tuite et al. 1983), indicating that aggregation of eRF3 reduces its activity in stop codon recognition.

Protein targeting

Yeast extracts are able to produce proteins which can be imported either into the endoplasmatic reticulum or into mitochondria. A cell-free yeast system supplied with yeast microsomes was shown to allow the translation, translocation and glycosylation of secreted proteins, specifically yeast mating factor alpha and invertase. Glycosylated alpha-factor precursor was found to be sequestered within the membrane vesicles. Similar results, including signal sequence cleavage, were observed for invertase (Rothblatt and Meyer 1986).

Concluding remark

Yeast extracts have proven to be powerful tools for studies of eukaryotic translation. They even reveal subtle changes in protein factor activity and factor-factor
interactions that are difficult to detect in yeast *in vivo*, as seen in some experiments with initiation factor-dependent extracts. This may also hold true for factor modifications (e.g. phosphorylation), making yeast extracts an important
experimental tool in the future.

References

Altmann M, Sonenberg N, Trachsel H (1989) Translation in *Saccharomyces cerevisiae*: initiation
factor 4E-dependent cell-free system. Mol. Cell. Biol. 9: 4467–4472.

Altmann M, Blum S, Wilson TMA, Trachsel H (1990a) The 5'-leader sequence of tobacco mosaic virus RNA mediates initiation factor-4E-independent, but still initiation-factor-4A-
dependent translation in yeast extracts. Gene 91: 127–129.

Altmann M, Blum S, Pelletier J, Sonenberg N, Wilson TMA, Trachsel H (1990b) Translation initiation factor-dependent extract from *Saccharomyces cerevisiae*. Biochim. Biophys. Acta
1050: 155–159.

Altmann M, Muller PP, Wittmer B, Ruchti F, Lanker S, Trachsel H (1993) A *Saccharomyces cerevisiae* homologue of mammalian translation initiation factor 4B contributes to RNA helicase
activity. EMBO J. 12: 3997–4003.

Altmann M, Wittmer B, Methot N, Sonenberg N, Trachsel H (1995) The *Saccharomyces* cerevisiae translation initiation factor Tif3 and its mammalian homologue, eIF-4B, have RNA
annealing activity. EMBO J. 14: 3820–3827.

Altmann M, Trachsel H (1997) Translation initiation factor-dependent extracts from yeast *Saccharomyces cerevisiae*. Methods: A Companion to Methods in Enzymology 11:343–352.

Asano K, Clayton J, Shalev A, Hinnebusch AG (2000) A multifactor complex of eukaryotic initiation factors, eIF1, eIF2, eIF3, eIF5, and initiator tRNA(Met) is an important translation initiation intermediate *in vivo*. Genes Dev. 14: 2534–2546.

Blum S, Mueller M, Schmid SR, Linder P, Trachsel H (1989) Translation in *Saccharomyces cerevisiae*: initiation factor 4A-dependent cell-free system. Proc. Natl. Acad. Sci. USA 86: 6043–6046.

Chuang RY, Weaver PL Liu Z, Chang TH (1997) Requirement of the DEAD-box protein Ded1p
for messenger RNA translation. Science 275: 1468–1471.

Colthurst DR, Chalk P, Hayes M, Tuite MF (1991) Efficient translation of synthetic and natural
mRNAs in an mRNA-dependent cell-free system from the dimorphic fungus *Candida albicans*. J. Gen. Microbiol. 137: 851–857.

Danaie P, Wittmer B, Altmann M, Trachsel H (1995) Isolation of a protein complex containing
translation initiation factor Prt1 from *Saccharomyces cerevisiae*. J. Biol. Chem. 270: 4288–4292.

Das S, Maitra U (2000) Mutational analysis of mammalian translation initiation factor 5 (eIF5):
role of interaction between the beta subunit of eIF2 and eIF5 in eIF5 function *in* vitro and *in
vivo*. Mol. Cell. Biol. 20: 3942–3950.

Das S, Ghosh R, Maitra U (2001) Eukaryotic translation initiation factor 5 functions as a
GTPase-activating protein. J. Biol. Chem. 276: 6720–6726.

Devchand M, Gwynne D, Buxton FP, Davies RW (1988) An efficient cell-free translation system
from *Aspergillus nidulans* and *in vitro* translocation of prepro-alpha-factor across *Aspergillus*
microsomes. Curr. Genet. 14: 561–566.

Dominguez D, Kislig E, Altmann M, Trachsel H (2001) Structural and functional similarities
between the central eukaryotic initiation factor (eIF)4A-binding domain of mammalian
eIF4G and the eIF4A-binding domain of yeast eIF4G. Biochem. J. 355: 223–230.

Gasior E, Herrera F, Sadnik I, McLaughlin CS, Moldave K (1979) The preparation and characterization of a cell-free system from *Saccharomyces cerevisiae* that translates natural messenger
ribonucleic acid. J. Biol. Chem. 254: 3965–3969.

Hershey JWB, Merrick, WC (2000) The pathway and mechanism of initiation of protein synthesis, p. 33–88. In N. Sonenberg, J.W.B. Hershey and M.B. Mathews (ed), Translational control of gene expression. Cold Spring Harbor Laboratory Press, Cold Spring Harbor, N.Y.

Hofbauer R, Fessl F, Hamilton B, Ruis H (1982) Preparation of a mRNA-dependent cell-free translation system from whole cells of Saccharomyces cerevisiae. Eur. J. Biochem. 122: 199–203.

Iizuka N, Najita L, Franzusoff A, Sarnow P (1994) Cap-dependent and cap-independent translation by internal initiation of mRNAs in cell-free extracts prepared from Saccharomyces cerevisiae. Mol. Cell. Biol. 14: 7322–7330.

Iizuka N, Sarnow P (1997) Translation-competent extracts from Saccharomyces cerevisiae: effects of L-A RNA, 5' cap, and 3' poly(A) tail on translational efficiency of mRNAs. Methods 11: 353–360

Kreutzfeldt C, Lochmann ER (1983) Preparation of a cell-free extract from yeast that is active in protein synthesis. FEMS Microbiol. Letts. 16: 179–182.

Lafontaine DL, Preiss T, Tollervey D (1998) Yeast 18S rRNA dimethylase Dim1p: a quality control mechanism in ribosome synthesis? Mol. Cell. Biol. 18: 2360–2370.

Mandel T, Trachsel H (1989) Yeast Saccharomyces cerevisiae cell-free translation: the inhibition of translation by high temperature is reversible. Biochim. Biophys. Acta 1007: 80–83.

Maiti T, Das S, Maitra (2000) Isolation and functional characterization of a temperature-sensitive mutant of the yeast Saccharomyces cerevisiae in translation initiation factor eIF5: an eIF5-dependent cell-free translation system. Gene 244: 109–118.

McCarthy JE (1998) Posttranscriptional control of gene expression in yeast. Microbiol. Mol. Biol. Rev. 62: 1492–1553.

Neff CL, Sachs AB (1999) Eukaryotic translation initiation factors 4G and 4A from Saccharomyces cerevisiae interact physically and functionally. Mol. Cell. Biol. 19: 5557–5564.

Pain VM (1996) Initiation of protein synthesis in eukaryotic cells. Eur. J. Biochem. 236:747–771.

Pelham HR, Jackson RJ (1976) An efficient mRNA-dependent translation system from reticulocyte lysates. Eur. J. Biochem. 67: 247–256.

Preiss T, Hentze MW (1998) Dual function of the messenger RNA cap structure in poly(A)-tail-promoted translation in yeast. Nature 392: 516–520

Preiss T, Muckenthaler M, Hentze MW (1998) Poly(A)-tail-promoted translation in yeast: implications for translational control. RNA 4: 1321–1331.

Rothblatt JA, Meyer DI (1986) Secretion in yeast: reconstitution of the translocation and glycosylation of alpha-factor and invertase in a homologous cell-free system. Cell 44:619–628.

Tarun SZ, Sachs AB (1995) A common function for mRNA 5' and 3' ends in translation initiation in yeast. Genes Dev. 9: 2997–3007.

Tarun SZ, Wells SE, Deardorff JA, Sachs AB (1997) Translation initiation factor eIF4G mediates in vitro poly(A) tail-dependent translation. Proc. Natl. Acad. Sci. USA 94: 9046–9051.

Tuite MF, Plesset J, Moldave K, McLaughlin CS (1980) Faithful and efficient translation of homologous and heterologous mRNAs in an mRNA-dependent cell-freee system from Saccharomyces cerevisiae. J. Biol. Chem. 255: 8761–8766.

Tuite MF, Cox BS, McLaughlin CS (1983) In vitro nonsense suppression in [psi+] and [psi-] cell-free lysates of Saccharomyces cerevisiae. Proc. Natl. Acad. Sci. USA 80:2824–2828.

Tuite MF, Plesset J (1986) mRNA-dependent yeast cell-free translation systems: theory and practice. Yeast 2: 35–52.

Valasek L, Trachsel H, Hasek J, Ruis H (1998) Rpg1, the Saccharomyces cerevisiae homologue of the largest subunit of mammalian translation initiation factor 3, is required for translational activity. J. Biol. Chem. 273: 21253–21260.

Poly(A)-Dependent Cell-Free Translation Systems from Animal Cells

GIOVANNA BERGAMINI, FÁTIMA GEBAUER*

Introduction

Initial studies of translation in animal cell extracts date back to the sixties, at the time when evidence was accumulating for the existence of a special class of RNA molecule: messenger RNA. Extracts derived from rabbit reticulocytes, Krebs II ascites cells and HeLa cells were instrumental in the identification of messenger RNA as the ultimate carrier of information to the protein synthesis machinery (Labrie 1969). Subsequently, in vitro systems became powerful tools for elucidating the mechanisms of polypeptide chain synthesis. Thus, the following two decades witnessed the rapid development of cell-free translation systems from a wide variety of animal origins (Villa-Komaroff et al. 1974). Studies using these in vitro systems led to findings such as the significance of the cap structure for translation (Edery et al. 1984), the mechanism of translation initiation and start site selection (Kozak 1983), and the effects of stress conditions on translation (Morley et al. 1985). Rabbit reticulocyte lysates (RRL) were optimized to produce a highly efficient system and became widely used (Pehlam and Jackson 1976; Jackson and Hunt 1983). However, this system showed properties rarely observed in intact cells, such as the frequent translation initiation at internal sites within the mRNA.

Although messenger RNA (mRNA) was known to contain a polyadenylate tail, it was not until the early eighties that this structure was implicated in translation (e.g. Jacobson and Favreau 1983). The poly(A) tail acts synergistically with the cap structure to stimulate translation of the mRNA in vivo (Gallie 1991). This important feature of translation was not recapitulated in RRL, precluding the use of this system for studying the critical contribution of the poly(A) tail. The synergism was, however, reproduced in yeast extracts (Iizuka et al. 1994). Using yeast lysates it could be shown that the poly(A) tail was able to recruit 40S ribosomal subunits (Tarun and Sachs 1995) and that the cap, in addition to recruiting 40S

* Corresponding author:
European Molecular Biology Laboratory, Gene Expression Programme
Meyerhofstrasse 1, D-69117 Heidelberg, Germany
Phone: 49-6221-387502, FAX: 49-6221-387518
e-mail: gebauer@embl-heidelberg.de

ribosomal subunits itself, tethers those recruited by the poly(A) to the 5' end of the mRNA (Preiss and Hentze 1998). These functional studies imply a physical contact between the 5' and 3' ends of the mRNA. The interaction between factors that recognize the cap and the poly(A) tail has been proposed as a molecular explanation for the synergy. Indeed, an interaction between one of the compo-nents of the cap binding complex, eIF4G, and poly(A)-binding protein (PABP) has been proven (Tarun and Sachs 1996; Wells et al. 1998; Imataka et al. 1998).

A number of poly(A)-dependent cell-free translation systems from insect and mammalian origin have recently been described, opening up the possibility of studing the contribution of the poly(A) tail in higher eukaryotes. In this chapter we will describe these systems in detail and will review their current uses.

Poly(A)-dependent cell extracts: protocols and properties

Several in vitro translation systems that recapitulate the synergism between the cap and the poly(A) tail have been reported in the last few years. These include crude extracts derived from *Drosophila melanogaster* embryos and ovaries (Gebauer et al. 1999; Castagnetti et al. 2000) and HeLa cells (Bergamini et al. 2000), as well as processed, micrococcal nuclease-treated extracts derived from *Drosophila* (Lie and Macdonald 2000) and rabbit reticulocytes (Borman et al. 2000; Michel et al. 2000). In this section we will describe the protocols we have used to obtain crude translation extracts and will compare their properties with those of other published systems.

Extract preparation

Lysates from HeLa cells and *Drosophila* embryos can be prepared in large scale. HeLa cells are typically grown in suspension, collected by centrifugation and washed with PBS before disruption. *Drosophila* embryos laid in agar plates are collected by differential filtration through sieves of decreasing pore size. The out-ermost layer of these embryos, the chorion, must be removed in an aqueous buffer. Ethanol-based buffers for dechorionation should be avoided. In contrast to HeLa cells or *Drosophila* embryos, attempts to obtain active preparations of *Drosophila* ovaries on a large scale have failed. However, translating ovary extracts can be prepared from a limited number of ovaries obtained by manual dissection of female flies. HeLa cells, ovaries or dechorionated embryos are subsequently disrupted in a hypotonic HEPES-based buffer (Figure 1). After disruption, the homogenate is centrifuged to eliminate the cell nuclei and debris. A wide range of centrifugation conditions have been used for this purpose (see Figure 1 legend for details). The resulting supernatant, or aqueous interphase if using *Drosophila* material, is either frozen in liquid nitrogen or used for translation straight away. Crude *Drosophila* embryo and HeLa cell extracts preserve their activity after mul-tiple freeze/thaw cycles. However, this is not the case for the crude ovary extracts, which should be used immediately after preparation.

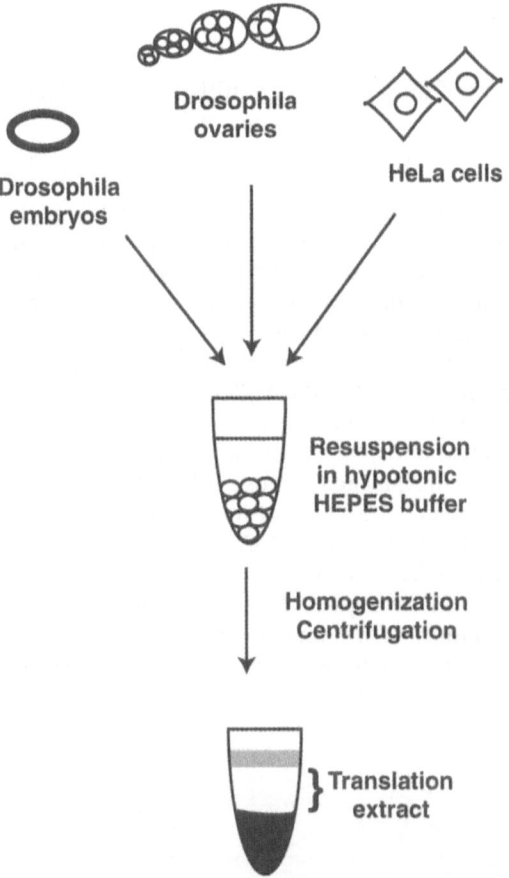

Fig. 1. Preparation of poly(A)-dependent translation extracts from *Drosophila* and HeLa cells. The common steps in obtaining translation extracts are shown. *Drosophila* embryos were collected in a pile of sieves, washed twice with 10 volumes of EW buffer (0.7% NaCl, 0.04% Triton X-100) and dechorionated in EW buffer supplemented with 3% sodium hypochlorite for 3 min at 25°C with vigorous agitation. Dechorionated embryos were first washed extensively with water, and then twice with 5 volumes of DE buffer (10 mM HEPES pH 7.4, 5 mM DTT). The embryo suspension was homogenized in one volume of DEI (DE buffer plus protease inhibitors) at 4°C using a Potter-Elvehjem homogenizer. The homogenate was spun by centrifugation at 40000 g for 20 min at 4°C. The aqueous interphase was removed, glycerol was added to a final concentration of 10% and the extract was frozen in liquid nitrogen and stored at a final concentration of –70°C. *Drosophila* ovaries were obtained by manual dissection of female flies in cold PBS. Ovaries were washed first with 12 volumes of a (1:1) mix of phosphate buffered saline (PBS):DEI, and then with 12 volumes of DEI. Excess buffer was removed and the ovaries were homogenized directly in the eppendorf tube using a plastic pestle. The homogenate was centrifuged at 14000 rpm for 10 min at 4°C, and the supernatant was recovered and used directly for translation. HeLa cells were grown in suspension in 2–6 l of Jocklik's medium, collected by centrifugation at 700 g for 15 min and washed 3 times with cold PBS. Pelleted cells were resuspended in one volume of ice-cold DEI buffer supplemented with 10 mM KOAc and 0.5 mM Mg(OAc)$_2$. After 5 min on ice, cells were homogenized using a Dounce homogenizer and centrifuged at 13000 g for 5 min at 4°C. The supernatant was frozen in liquid nitrogen and stored at –70°C

In vitro translation

The in vitro translation assay consists of the addition of a mRNA to a translation mix that includes the cell extract and several other components, followed by the incubation of the reaction mixture at an appropriate temperature. A convenient protocol to set up the reaction is described in Table 1. The assay contains 40% cell extract. Lower amounts of extract decrease the translation efficiency, while higher amounts do not result in enhanced translation. The optimal concentration of Mg^{++} and K^+ ions depends on the particular mRNA tested and should be optimized for each transcript. For example, higher salt concentrations are necessary for efficient translation of IRES-containing mRNAs in HeLa extracts than for capped mRNAs (Bergamini et al. 2000). In general, translation in HeLa extracts occurs at a higher Mg^{++} concentration (2.5-4 mM $Mg(OAc)_2$) compared to *Drosophila* extracts (0.2-0.6 mM $Mg(OAc)_2$). The translation reaction requires energy which is provided by the creatine phosphate-creatine phosphokinase ATP regenerating system, although addition of ATP and GTP also enhances translation efficiency. A HEPES-based buffer is preferred because of its low pK_a and lack of sensitivity to temperature changes between 0° C and 37° C.

The amount of protein translated reaches a plateau after incubation for 60 to 90 minutes, depending on the mRNA. To distinguish the translation products of the exogenously added mRNAs from those of endogenous transcripts present in the crude extracts, quantitative analysis such immunoprecipitation, Western

Table 1. In vitro translation assay

Reagent	Volume [a]	
	HeLa ext.[b]	Dros. ext.[b]
Cell extract	4 µl	4 µl
1 M HEPES pH7.4	0.16 µl	0.24 µl
2 M KOAc [c]	0.25 µl	0.15 µl
125 mM $Mg(OAc)_2$ [c]	0.2 µl	0.03 µl
2 mM Amino Acids	0.5 µl	0.3 µl
5 mM Spermidine [c]	0.2 µl	0.2 µl
100 mM ATP	0.08 µl	–
10 mM GTP	0.1 µl	–
1 M Creatine phosphate	0.2 µl	0.2 µl
10 mg/ml Creatine phosphokinase	0.04 µl	0.08 µl
4mg/ml Calf liver tRNA	0.2 µl	0.25 µl
100 ng/µl mRNA template [c]	0.1 µl	0.3 µl

[a] The indicated values are optimal for the translation of a capped and polyadenylated luciferase mRNA in a 10 µl reaction.
[b] Translation is typically performed by incubating for 60 minutes at 37° C for HeLa extracts or for 90 min at 25° C for Drosophila extracts.
[c] The amount of these components should be optimized for each mRNA.

blotting or a direct measurement of the biochemical activities of the translated products.

Cap-poly(A) synergy

The cell-free systems described above show a strong cap and poly(A)-dependence. Figure 2 shows the translation profiles of CAT mRNAs containing no end modifications (-), a canonical m^7GpppG cap (c), a poly(A) tail of 98 residues (a) or both a cap and a poly(A) (c-a) tail in a *Drosophila* embryo extract. Translation of the mRNA lacking any of the terminal structures is barely detectable, while that of the capped or polyadenylated mRNA is considerably higher. In addition, the combined effects of the cap and the poly(A) tail are synergistic because translation of the mRNA containing both modifications is severalfold higher than that of the mRNAs containing only one. The differences in translational efficiency are not related to mRNA stability. While uncapped mRNAs show a reduced stability in the extract, the stability of the capped mRNAs is similar whether they contain a poly(A) tail or not (Figure 2, lower panel). In addition, artificial stabilization of the messages by the use of an ApppG cap does not result in increased translation (Gebauer et al. 1999; Castagnetti et al. 2000; Bergamini et al. 2000). The synergy, defined as the translation efficiency of the c-a mRNA divided by the sum of the individual efficiencies of the c and a mRNAs, varies between 5- and 15-fold depending on the mRNA template and the batch of extract. This synergism is observed under a wide range of salt conditions and is independent of the translational efficiency of the mRNA. As previously observed in the yeast system (Preiss et al. 1998), the degree of synergy increases with the length of the poly(A) tail, the minimal size range being 30 to 50 adenosines (Gebauer et al. 1999; Bergamini et al. 2000).

The *Drosophila* ovary and embryo extracts reported by Lie and Macdonald (2000) display similar properties to those described above. These authors, however, further processed the extracts by including a step of gel filtration to remove low molecular weight components, and a subsequent step of micrococcal nuclease treatment to eliminate endogenous mRNAs. Preiss and Hentze (1998) showed that treatment of yeast lysates with micrococcal nuclease abrogated the translational synergism between the cap and the poly(A) tail of an exogenously added mRNA. This observation suggested that the poly(A) tail conferred a selective advantage for translation of capped mRNAs only under competitive conditions. However, as mentioned above, a similar treatment of the *Drosophila* extracts did not affect the stimulatory effect of the poly(A) tail on translation (Lie and Macdonald 2000; F. Gebauer, T. Preiss and M. W. Hentze, unpublished observations), pointing to differences in the properties of translation extracts between yeast and higher eukaryotes.

Another poly(A)-dependent in vitro system was obtained by ultracentrifugation of commercially available nuclease-treated RRL to partially deplete them of ribosome-associated factors (Borman et al. 2000; Michel et al. 2000). In standard nuclease-treated RRL, as well as in untreated lysates, positive effects of capping and polyadenylation are additive. The depletion employed by Michel et al. selectively reduced translation of non-polyadenylated, capped mRNAs, resulting in an

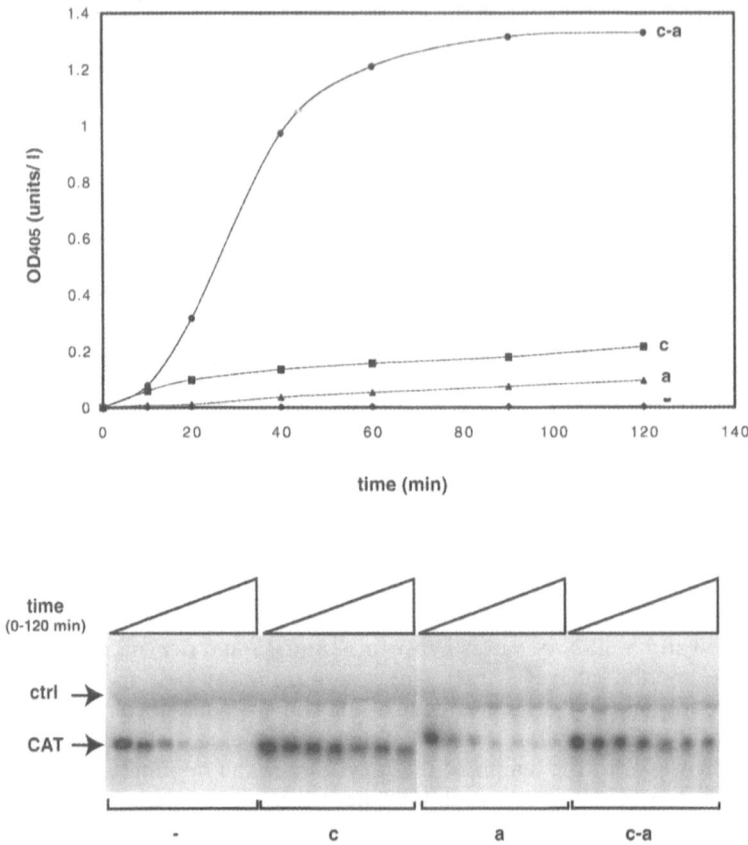

Fig. 2. Translation synergy between the cap and the poly(A) tail in *Drosophila* embryo extracts. Upper panel, translation time-course of trace-labelled CAT mRNAs containing the following end modifications: c, m⁷GpppG cap (squares); a, 98-residue poly(A) tail (triangles); c-a, cap and poly(A) tail (circles); -, no end modification (diamonds). Aliquots were taken at the indicated times and the amount of CAT protein was measured by CAT-ELISA. Lower panel, physical stabilities of the CAT RNAs. Total RNA was extracted and separated in a 1% denaturing agarose gel. Trace-labelled Luc mRNA (ctrl) was added as an extraction control. The gel was dried and exposed in the phosphorimager

observable cap-poly(A) synergy. Important for the poly(A)-dependence of this system was the alteration of the stoichiometry of the components of the translation machinery. Interestingly, specific disruption of the eIF4G-PABP interaction by a viral peptide known to bind to eIF4G at the PABP-binding site abrogated the poly(A) effect. This observation points to the relevance of this interaction for the cap-poly(A) synergy in RRL.

Finally, the HeLa cell-derived in vitro system proved useful to test the effect of the poly(A) tail on IRES-driven translation (Bergamini et al. 2000). In fact, translation from the three classes of picornavirus IRESes has been shown to be stimu-

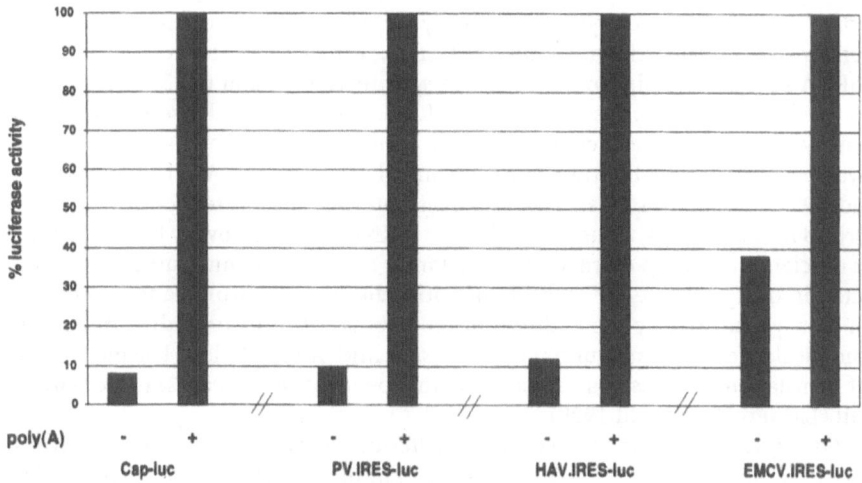

Fig. 3. IRES-driven translation is stimulated by the poly(A) tail. The indicated mRNAs were tested in the HeLa cell-free system. For each set of mRNAs, translation efficiency was expressed as the percentage of that of the polyadenylated form. Similar to translation of the capped mRNA, translation from picornavirus IRESes types I and III, PV and HAV IRESes was stimulated by the poly(A) tail by approximately 10 fold. In the case of the EMCV IRES, stimulation of translation by the poly(A) tail was less than 3 fold

lated by the poly(A) tail (Figure 3). In particular, translation from poliovirus (PV) and hepatitis A virus (HAV) IRESes, type I and III respectively, was enhanced by the presence of the poly(A) tail to a similar extent as that observed for capped mRNAs. On the other hand, EMCV (type II) IRES activity was increased by the poly(A) tail to a lesser extent. These significant differences may be the result of alternative modes of translation initiation from IRES type II with respect to IRESes types I and III. Nevertheless, a detailed understanding of the mechanism by which the poly(A) tail promotes IRES-dependent translation requires further experimentation.

Applications

Poly(A)-dependent *Drosophila* ovary and embryo extracts have been used to study a variety of translational control events. In *Drosophila*, translational regulation is often coupled to mRNA localization, and mutations in cis- and trans-acting elements frequently lead to pleiotropic phenotypic effects. The use of the in vitro systems has allowed for the study of translation separate from other biological processes. Most of the examples studied in vitro include the translational regulation of maternal mRNAs that are required for the proper establishment of the anteroposterior axis in the *Drosophila* embryo (reviewed in Lipshitz and Smibert 2000). One of the earliest events in axis formation is the localization-dependent translation of *oskar* mRNA, a mechanism whereby *oskar* mRNA localized at the posterior pole of the oocyte is selectively translated while the unlocalized RNA remains in a translationally repressed state. Translational repression of *oskar* mRNA requires the binding of Bruno to specific elements in the 3' UTR referred to as the Bruno Response Elements (BREs). The combined use of ovary and embryo extracts, which contain or lack Bruno respectively, has opened new avenues for studying the translational control of *oskar* mRNA (Lie and Macdonald 1999; Castagnetti et al. 2000). Consistent with the presence or absence of Bruno in the extracts, translation of *oskar* mRNA is poor in ovary lysates, while it is efficient in embryo extracts. Translation of a BRE-containing mRNA is stimulated in ovary extracts by addition of antibodies against Bruno, and is inhibited in embryo extracts by the addition of recombinant Bruno, providing direct biochemical evidence that Bruno represses *oskar* mRNA translation. The mechanism of translational repression seems to be independent of the cap and the poly(A) tail (Lie and Macdonald 1999).

Oskar stimulates the translation of posteriorly localized *nanos* mRNA. Translation of the remaining 96% unlocalized *nanos* message, however, is repressed by Smaug. This protein binds to discrete sequences in the 3' UTR of *nanos* mRNA called the Smaug Response Elements (SREs). SRE-dependent regulation can be recapitulated in extracts from early embryos (Smibert et al. 1999). Depletion of Smaug from these extracts abrogates translational repression of SRE-containing RNAs. Interestingly, translationally repressed *nanos* mRNA is associated with polysomes, and repression does not seem to impose stable arrest of the elongating ribosomes, suggesting a novel mechanism for translational repression of this message (Clark et al. 2000).

Poly(A)-dependent cell-free systems not only have proven useful in the study of the translational regulation of maternal mRNAs, but also in that of messages transcribed from the zygotic nuclei. Such is the case of the translational regulation of *male-specific-lethal* 2 (*msl*-2) mRNA, which encodes a product essential for dosage compensation. Translational inhibition of *msl*-2 mRNA by the female-specific RNA-binding protein Sex-lethal (SXL) ensures the viability of female flies by maintaining dosage compensation in a repressed state (Bashaw and Baker 1997; Kelley et al 1997). *Drosophila* embryo extracts, unlike other commercially available translation systems, are able to recapitulate proper inhibition of *msl*-2 mRNA translation by SXL. Translational repression is independent of the poly(A) tail and requires the functional cooperation between SXL-binding sites located in both UTRs of the transcript, suggesting a novel mode of regulation (Gebauer et al. 1999).

The development of mammalian in vitro systems that faithfully reproduce cap and poly(A) tail dependence now makes it possible to functionally test components of the translation machinery. An example is the recent report of a novel partner of PABP, Paip2, which decreases the affinity of PABP for polyadenylated RNA and disrupts the repeating structure of poly(A) ribonucleoprotein (Khaleghpour et al. 2001). Paip2 preferentially inhibits translation of polyadenylated mRNAs, either containing a cap or an IRES element, in Krebs-2 cell extracts that exhibit poly(A)-dependence.

A better understanding of the process of proteins synthesis in animal cells should be forthcoming from biochemical analysis using the recently developed in vitro systems.

Concluding remarks

In vitro translation systems from animal· cells have been instrumental in deciphering the mechanisms of the translation reaction in eukaryotes. However, many questions remain to be answered, especially regarding the mechanisms of translation initiation in its cap-dependent or cap-independent modes, and the crosstalk between the mRNA ends for translation. Investigating how regulators such as mRNA-binding proteins impinge on translation can help to dissect these mechanisms. The recent development of cell-free systems that reproduce translation regulatory events and show strong cap and poly(A)-dependence will certainly be instrumental to understand the biochemistry of the translation process.

References

Bashaw G, Baker B (1997) The regulation of Drosophila msl-2 gene reveals a function for Sex-lethal in translational control. Cell 89: 789–798.

Bergamini G, Preiss T, Hentze MW (2000) Picornavirus IRESes and the poly(A) tail jointly promote cap-independent translation in a mammalian cell-free system. RNA 6: 1781–1790.

Borman AM, Michel YM, Kean KM (2000) Biochemical characterisation of cap-poly(A) synergy in rabbit reticulocyte lysates: the eIF4G-PABP interaction increases the functional affinity of eIF4E for the capped mRNA 5'-end. Nuc Acids Res 28: 4068–4075.

Castagnetti S, Hentze MW, Ephrussi A, Gebauer F (2000) Control of oskar mRNA translation by Bruno in a novel cell-free system from Drosophila ovaries. Development 127: 1063–1068.

Clark IE, Wyckoff D, Gavis ER (2000) Synthesis of the posterior determinant Nanos is spatially restricted by a novel cotranslational regulatory mechanism. Curr Biol 10: 1311–1314.

Edery I, Lee KAW, Sonenberg N (1984) Functional characterization of eukaryotic mRNA cap binding protein complex: effects on translation of capped and naturally uncapped RNAs. Biochemistry 23: 2456–2462.

Gallie DR (1991) The cap and poly(A) tail function synergistically to regulate mRNA translational efficiency. Genes Dev 5: 2108–2116.

Gebauer F, Corona DFV, Preiss T, Becker PB, Hentze MW (1999) Translational control of dosage compensation in Drosophila by Sex-lethal: cooperative silencing via the 5' and 3' UTRs of msl-2 mRNA is independent of the poly(A) tail. EMBO J 18: 6146–6154.

Iizuka N, Najita L, Franzusoff A, Sarnow P (1994) Cap-dependent and cap-independent translation by internal initiation of mRNAs in cell extracts prepared from Saccharomyces cerevisiae. Mol Cell Biol 14: 7322–7330.

Imataka H, Gradi A, Sonenberg N (1998) A newly identified N-terminal aminoacid sequence of human eJF4G binds poly(A)-binding protein and functions in poly(A)-dependent translation. EMBOJ 17:7480–7489.

Jackson RJ, Hunt T (1983) Preparation and use of nuclease-treated rabbit reticulocyte lysates for the translation of eukaryotic messenger RNA. Methods Enzymol 96: 50–74.

Jacobson A, Favreau M (1983) Possible involvement of poly(A) in protein synthesis. Nuc Acids Res 11: 6353–6368.

Kelley R, Wang J, Bell L, Kuroda M (1997) Sex-lethal controls dosage compensation in Drosophila by a non-splicing mechanism. Nature 387: 195–199.

Khaleghpour K, Svitkin YV, Craig AW, DeMaria CT, Deo RC, Burley SK, Sonenberg N (2001) Translational repression by a novel partner of human poly(A) binding protein, Paip2. Mol Cell 7: 205–216.

Kozak M (1983) Comparison of initiation of protein synthesis in procaryotes, eucaryotes, and organelles. Microbiol Rev 47: 1–45.

Labrie F (1969) Isolation of an RNA with the properties of hemoglobin messenger. Nature 221: 1217–1222.

Lie YS, Macdonald PM (1999) Translational regulation of oskar mRNA occurs independently of the cap and the poly(A) tail in Drosophila ovarian extracts. Development 126: 4989–4996.

Lie YS, Macdonald PM (2000) In vitro translation extracts prepared from Drosophila ovaries and embryos. Biochem Biophys Res Com 270: 473–481.

Lipshitz HD, Smibert CA (2000) Mechanisms of RNA localization and translational regulation. Curr Opin Genet Dev 10: 476–488.

Michel YM, Poncet D, Piron M, Kean KM, Borman AM (2000) Cap-poly(A) synergy in mammalian cell-free extracts. J Biol Chem 275: 32268–32276.

Morley SJ, Buhl WJ, Jackson RJ (1985) A rabbit reticulocyte factor which stimulates protein synthesis in several mammalian cell-free systems. Biochim Biophys Acta 825: 57–69.

Pelham HRB, Jackson RJ (1976) An efficient mRNA-dependent translation system from reticulocyte lysates. Eur J Biochem 67: 247–256.

Preiss T, Hentze MW (1998) Dual function of the messenger RNA cap structure in poly(A)-tail-promoted translation in yeast. Nature 392: 516–520.

Preiss T, Muckenthaler M, Hentze MW (1998) Poly(A)-tail-promoted translation in yeast: implications for translational control. RNA 4: 1321–1331.

Smibert CA, Lie YS, Shillinglaw W, Henzel WJ, Macdonald PM (1999) Smaug, a novel and conserved protein, contributes to repression of nanos mRNA translation in vitro. RNA 5: 1535–1547.

Tarun SZ Jr, Sachs AB (1995) A common function for mRNA 5' and 3' ends in translation initiation in yeast. Genes Dev 9: 2997–3007.

Tarun SZ Jr, Sachs AB (1996) Association of the yeast poly(A) tail binding protein with translation initiation factor eIF-4G. EMBO J 15: 7168–7177.

Villa-Komaroff L, McDowell M, Baltimore D, Lodish HF (1974) Translation of reovirus mRNA, and bacteriophage Q β RNA in cell-free extracts of mammalian cells. Methods Enzymol 30: 709–723

Wells SE, Hillner PE, Vale RD, Sachs AB (1998) Circularization of mRNA by eukaryotic translation initiation factors. Mol Cell 2: 135–140.

PART III

Continuous-Flow and Continuous-Exchange   III
Cell-Free Translation Systems

CHAPTER 8

Continuous-Flow and Continuous-Exchange   **8**
Cell-Free Translation Systems and Reactors

VLADIMIR A. SHIROKOV, PETER N. SIMONENKO,
SERGEY V. BIRYUKOV, ALEXANDER S. SPIRIN*

Introduction

Progress in biotechnology generates a growing demand for convenient and productive technologies of gene expression on preparative scale. *In vivo* systems for expression of foreign genes in bacterial or eukaryotic cells are efficient and widely used. However, difficulties associated with cytotoxicity, proteolytic degradation or improper folding and aggregation of synthesized proteins are often encountered *in vivo*. The *in vivo* expression systems are restricted by mechanisms of cell control and allow limited opportunities for solving these problems. On the other hand, gene expression *in vitro*, in cell-free translation or transcription-translation systems, is an alternative that allows full control and high flexibility of conditions.

Cell-free systems for *in vitro* protein synthesis have been widely used during the last four decades for elucidation of molecular mechanisms of translation, identification of open reading frames, and so on (see review by Spirin, this volume). The *in vitro* protein-synthesizing systems, as performed in a test-tube or another container with a fixed volume (batch mode), served well for most analytical purposes. However, short lifetimes and low productivities of the batch systems have hampered their application for the synthesis of preparative amounts of protein. Novel continuous-duty cell-free translation and transcription-translation systems introduced some time ago (Spirin et al. 1988; Baranov et al. 1989) have opened up new prospects for *in vitro* protein synthesis. The principle of systems of this type was the continuous addition of reaction substrates (amino acids and nucleoside triphosphates) concurrently with continuous removal of reaction products through a porous membrane during incubation. The continuous systems provided a prolonged synthesis of proteins, as compared with the standard (batch) cell-free format, and correspondingly a higher yield of proteins synthesized. Continuous cell-free systems are now becoming a technique for practical biotechnology.

* Institute of Protein Research, Russian Academy of Sciences,
  142290 Pushchino, Moscow Region, Russia
  Tel./Fax: 007(095)924-0493, e-mail: spirin@vega.protres.ru

Continuous Cell-Free Methodology

In a cell-free translation system performed in the fixed volume of a test-tube (batch format) the reaction conditions change as a result of the consumption of substrates and the accumulation of products. Translation stops as soon as any essential substrate is exhausted, or any product or by-product reaches the inhibiting concentration, usually within 20 to 60 minutes of incubation. The principle of continuous cell-free systems is the maintenance of more or less constant reaction conditions by means of the persistent supply of consumable substrates for translation or transcription-translation and the removal of products and by-products from the reaction mixture. To achieve that, a porous (semi-permeable) barrier is used, which retains the high-molecular-weight components of the protein-synthesizing machinery (ribosomes, mRNA, ARSases, etc.) within a defined reaction compartment and separates it from another compartment containing a feeding solution with the resources of the low molecular weight components (substrates) of the reaction. In the so-called *continuous-flow cell-free, or CFCF systems* (Spirin et al. 1988; Baranov et al. 1989; Spirin 1991), the substrates are continuously supplied and the products are removed by the active (forced) flow of the feeding solution across an ultrafiltration membrane (or another type porous barrier). In this case, all of the products, including the protein synthesized, if the pore size is large enough, are continuously removed from the reaction compartment (Fig. 1A). In the dialysis or "semi-continuous" versions of the continuous systems, designated as *continuous-exchange cell-free, or CECF systems,* passive (diffusional) exchange

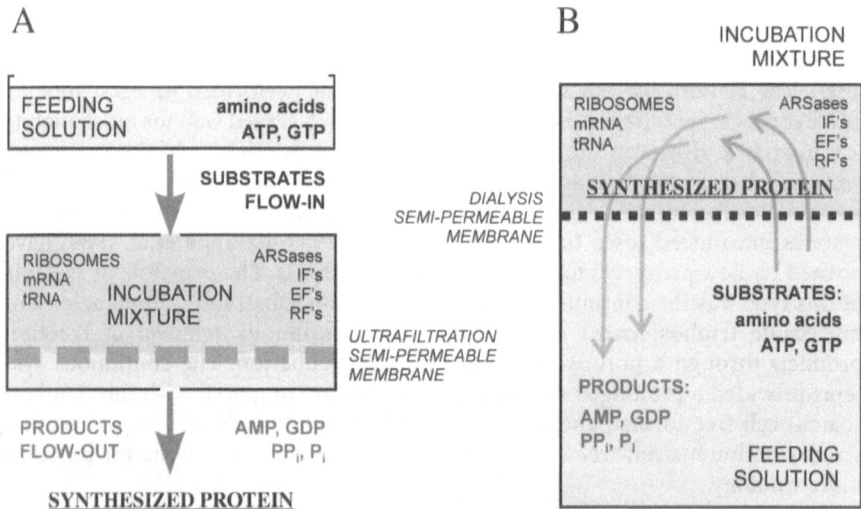

Fig. 1. Principal schemes for the continuous-flow (A) and continuous-exchange (B) cell-free translation systems. In addition to the key components indicated in the drawing, the reaction mixture and feeding solution contain the nucleoside triphosphate regeneration system, magnesium and potassium ions, buffer substances, antioxidant and other necessary components. In the transcription-translation systems, RNA polymerase and DNA template are also present in the reaction mixture, and UTP and CTP are added as transcription reaction substrates

of substrates and products takes place through the porous barrier (Alakhov et al. 1995; Kim and Choi 1996; Davis et al. 1996). When the dialysis membrane is used, all the protein synthesized is accumulating in the reaction compartment, whereas the low-molecular-weight substrates and products are in exchange with the feeding solution (Fig. 1B).

Continuous-Flow Cell-free Systems

In the first experiments investigating these systems, a standard Amicon 8 MC micro-ultrafiltration cell was used as a reactor for establishing the CFCF translation system (Spirin et al. 1988; Spirin 1991). Typically, the volume of the reaction mixture was 1 ml, the feeding solution was pumped in with a flow rate of 1 ml per hour, and the reaction products, including the protein synthesized, were removed at the same rate through an ultrafiltration membrane (either PM-30, or XM-50, or YM-100, or XM-300, depending on the size of the protein molecules to be synthesized). Both bacterial and eukaryotic (rabbit reticulocyte or wheat germ) extracts were used. Various proteins were synthesized under these conditions over 20 to 50 hours with more or less constant rate. The productivity reached 100 to 300 copies of polypeptide per mRNA molecule, corresponding to nanomoles of protein per ml, whereas the control batch systems under the same conditions produced no more than 1 to 3 copies of polypeptide per mRNA.

In subsequent experiments carried out, coupled and combined transcription-translation systems (see Spirin, this volume), as well as the coupled RNA replication-translation system (Morozov et al. 1993; Ryabova et al. 1994), were also made in the CFCF format. The introduction of the RNA-synthesizing machinery into the CFCF system provided the sustenance of mRNA concentration during the long run time. The use of crude bacterial extracts possessing essential RNAase activities became possible in coupled and combined transcription-translation CFCF systems with endogenous or phage DNA-dependent RNA polymerases respectively (Baranov et al. 1989; Kigawa and Yokoyama 1991; Kudlicki et al. 1992). Furthermore, it was found possible to introduce the phage (T7 or SP6) RNA polymerases into a eukaryotic extract and thus combine the transcription and translation processes in the same reaction mixture of the CFCF system based on wheat germ extract or rabbit reticulocyte lysate (Spirin 1991; Baranov and Spirin 1993; Nishimura et al. 1993, 1995). The mRNA replication by Qβ phage RNA-dependent RNA polymerase (Qβ RNA replicase) was also used in the CFCF systems (Ryabova et al. 1994). In this case, to make mRNA replicable by Qβ RNA replicase, the coding mRNA sequence was inserted into a special Qβ-replicable vector, the so-called RQ RNA (see Chetverin and Spirin 1995, for review). The protocols for translation, coupled transcription-translation, combined transcription-translation and coupled replication-translation CFCF systems using bacterial and eukaryotic extracts are given in Baranov and Spirin (1993) and Ryabova et al. (1998).

One of the more recent CFCF experiments is demonstrated in Fig. 2. The synthesis of green fluorescent protein (GFP) and chloramphenicol acetyltransferase (CAT) in a bacterial combined transcription-translation CFCF system has been performed in joint work between the Institute of Protein Research group, Pushchi-

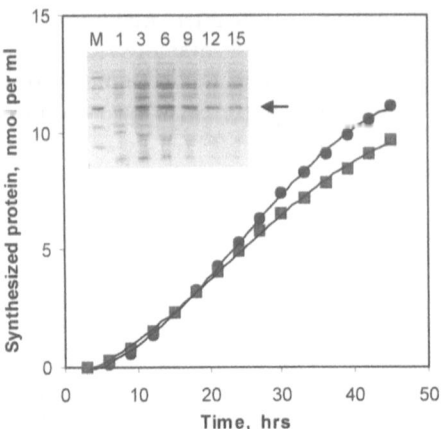

Fig. 2. Syntheses of green fluorescent protein (GFP, 26.9 kD), ●, and chloramphenicol acetyl-transferase (CAT, 23.6 kD), ■, in a continuous-flow transcription-translation cell-free system with bacterial ($E.\ coli$) S30 extract. The syntheses were performed in a Y-flow reactor (see Fig. 4) with YM-10 and XM-50 ultrafiltration membranes in 300 µl reaction volume. Reaction mixture contained 40 OU_{260}/ml $E.\ coli$ S30 extract (33% v/v), 2000U/ml T7 RNA polymerase, 100 U/ml RNase inhibitor (HPRI), 18 µg/ml plasmid with T7 gene10 ribosome binding site and GFP or CAT gene respectively, 0.48 mg/ml total $E.\ coli$ tRNA, 0.25 mM each amino acid, 10 µg/ml folinic acid, 1 mM ATP, 0.8 mM GTP, 0.6 mM each UTP and CTP, 30 mM acetyl phosphate in 100 mM HEPES-KOH buffer pH 7.6 with 14 mM $Mg(OAc)_2$, 226 mM KOAc, 3 mM NaN_3, 2 mM DTT, 1 mM EDTA, 4% glycerol. Feeding solution contained the same components except S30 extract, plasmid, RNA polymerase and RNase inhibitor. The reactor was kept at 30°C and feeding solution supplied at flow rate of 0.6 ml/h. The protein product was removed in the outflow through the XM-50 membrane at 0.09 ml/h and collected in fractions, 0.28 ml each. Concentration of GFP in fractions was determined by fluorescence measurement (395 nm/509 nm) relative to the reference GFP. CAT was assayed by functional test. The amount of the functional protein synthesized per ml of the reaction mixture is plotted against the incubation time. Inset: SDS-PAGE analysis of GFP in effluent fractions. Lane M: molecular mass markers (14.3: 20.1; 26.9; 39.2; 55.6 kDa); lanes 1, 3, 6, 9, 13, 15: 50 µl fractions Nos. 1, 3, 6, 9, 13 and 15 respectively. Gel was stained by Coomassie G-250. Arrow indicates position of GFP band

no, Russia, and the Roche Molecular Biochemicals group, Penzberg, Germany. The CFCF system kept its protein-synthesizing activity over two days. The yield of the functional protein was more than 10 nmol (about 300 µg) per ml of the reaction mixture, this being 12 to 15 higher than that of the parallel batch reactions (0.6 and 0.8 nmol/ml). The synthesized protein is seen as the major band in SDS-PAGE analysis of the outflow fractions. In this and all the previous CFCF experiments, however, the classic cell-free system protocols were used. Recently, a number of improvements in performing the cell-free systems have been made, significantly enhancing their rate and productivity.

The experiments with the CFCF systems have been reproduced by several other groups. As a whole, bacteriophage MS2 coat protein (Spirin et al. 1988), brome mosaic virus coat protein (Spirin et al. 1988), calcitonin polypeptide (Spirin 1991), globin (Ryabova et al. 1989), functionally active dihydrofolate reductase (DHFR) (Baranov et al. 1989; Spirin 1991; Endo et al. 1992; Kudlicki et al. 1992), chloramphenicol acetyltransferase (CAT) (Spirin 1991; Kigawa and Yokoyama 1991; Ryabova et al. 1994, 1998; Kim and Choi 1996; Kitaoka et al. 1996), interleukin-2 (Kolosov et al 1992) and interleukin-6 (Volyanik et al. 1993) have all successfully been synthesized in CFCF experiments during the decade after the discovery. The yields ranged from about one hundred micrograms to over a milligram of synthesized protein per ml of the reaction mixture.

The unique features of the CFCF system have been verified in many experiments. The shortage in energy supply in batch and the maintenance of the system activity by substrate supply and product removal has been documented (Matveev et al. 1996, Kitaoka et al. 1996, Kim and Choi 1996). The retention of the translation machinery components in the flow reactor, even with the use of a large pore membrane, that was observed in the original CFCF experiments (Spirin et al. 1988; Spirin 1991), has been confirmed in both eukaryotic and bacterial systems (Kudlicki et al. 1992, Endo et al. 1993, Kitaoka et al. 1996). The compartmentalization of the translation complex components on ribosomes made it possible to use condensed cell-free extracts prepared either by sedimentation (Kudlicki et al. 1992), or by ultrafiltration with a XM-300 membrane (Yamane et al. 1995) or Centricon C10 (Kim et al. 1996). Moreover, the condensation of a cell-free extract in these ways decreased the level of unfavorable activities, such as RNAases and phosphatases, as a result of remaining in supernatant or draining through the pores respectively (Kudlicki et al. 1992, Yamane et al. 1995). A similar draining process seems to take place during a CFCF run: the translation mixture is automatically cleaned up by the flow from various foreign components (i.e. those not associated with the translation machinery complexes), including degrading enzymes. This washing-out effect, together with the continuous supply of substrates, provides for the long lifetime of the CFCF system. The washing-out of RNAases may explain the unusually high stability of mRNA observed in the CFCF systems (Spirin et al. 1988, Endo et al. 1993, Alexandrov et al. 1996, Kitaoka et al. 1996). It should be noted that the leakage of essential translation components reported in some cases could be the consequence of the system inactivation, rather than the cause of the system halt; inactive components can leave the dynamic translation complexes.

Continuous removal of synthesized proteins from the reaction mixture by flow can be an important advantage of the CFCF system: this allows outflow fractions to contain a relatively pure and soluble product. Thus, a protein product of 80–85% purity was obtained in the outflow of a bacterial CFCF system synthesizing DHFR (Kudlicki et al. 1992) and in a wheat germ CFCF system synthesizing interleukin-6 (Volyanik et al. 1993). The synthesis of human interleukin-2 was performed in the wheat germ CFCF system with high yield and without product aggregation (Kolosov et al. 1992), whereas synthesis in batch resulted in the formation of complex protein aggregates and the inhibition of translation as the product concentration increased. (The aggregation because of the protein product accumulation in the reaction mixture can also happen in CECF systems, especially when small-pore or simple dialysis membranes are used.)

At the same time, aggregation and membrane clogging have often been claimed to be the major drawbacks of the CFCF system. These phenomena were seen to take place in certain cases, mainly as a result of the lowering of the pH over the course of the reaction, or the instability (denaturation) of the product or cell-free extract proteins, etc. In many cases, the remedy can be found in increasing the buffer capacity, adjusting the reaction conditions to prevent product misfolding (reduction of temperature, addition non-denaturing detergents, presence of chaperones, etc.), thorough clarification of the extract or the reaction mixture before starting the reaction, and so on.

Continuous-Exchange Cell-Free Systems

Dialysis bag is the simplest device in which the continuous supply with substrates and the removal of low-molecular-weight products by exchange with a feeding solution can be accomplished (Alakhov et al. 1995). Using a standard dialysis bag, the Promega Corporation group performed the synthesis of firefly luciferase (62 kDa) in the bacterial (*E. coli*) transcription-translation cell-free system over 20 hours (Davis et al. 1996). The product yield assayed by SDS-PAGE and Western blot analysis was 120–240 µg of protein per ml of reaction mixture: 10–20 times the yield of a batch reaction. However, the yield of luciferase activity was much lower, the larger part of the product being shown to be a truncated protein tending to aggregate. Kim and Choi (1996) faced problems in performing the CFCF system, such as leakage of translational components (see previous section), and developed the "semi-continuous" (dialysis) transcription-translation cell-free system with *E. coli* S30 extract. They reported the synthesis of 1.2 mg of chloramphenicol acetyltransferase (CAT, 26 kDa) in a 1 ml dialysis reactor. The system displayed a steady synthesis rate for at least 14 hours. The product yield (quantified by ELISA) exceeded 10–12 times the yield of the analogous batch system. More recently, the highly productive bacterial CECF system has been developed by the Yokoyama group (Kigawa et al. 1999). Three favorable parameters have been combined in this system: first, the S30 extract condensed by dialysis against PEG 8000 was used; second, the energy substrate was changed for creatine phosphate; third, continuous mode was performed by placing the reaction mixture in a dialyser unit. Concentrations of all the components were optimized. As a result of these modifications, the productivity of the system increased and the syntheses of CAT and Ras proteins amounted to 6 mg per ml over 21 hours of reactions. It is worth noting that the efficiency of batch reactions with the above modifications was also high. The CECF mode extended the reaction time to 12 hours and gave a further 10–15-fold increase in productivity. The specific goal of this work was the preparative synthesis of a stable-isotope labeled protein for NMR spectroscopy. Thus, this demonstrated a practical application of the CECF system for protein synthesis: the synthesis of Ras protein was performed with ^{13}C/ ^{15}N amino acids and NMR spectra of the purified protein were recorded.

   The syntheses of particular products in the CECF system have been recently demonstrated in two communications from our laboratory. One case is the synthesis of HIV Nef, the protein necessary for efficient replication of the virus and for AIDS induction, and thus very promising as an antigen for anti-AIDS immunization. It was performed in the bacterial CECF transcription-translation system (Chekulayeva et al. 2001). The protein was synthesised in the form fused with GFP through a cleavable spacer. The synthesis of the fusion protein was visualized by GFP moiety fluorescence, while the Nef moiety was tested by Western blot analysis with anti-Nef antibodies. The synthesis of this fusion protein in the CECF system may provide some advantages for practical production of the immunologically active Nef.

A similar series of experiments was performed with the synthesis of an antibacterial polypeptide, Cecropin P1 (31 amino acid residues), fused with GFP (Martemyanov et al. 2001). The point is that many difficulties are encountered on the way of the *in vivo* expression of genes encoding for alien polypeptides of such size, especially as a result of their cytotoxicity and the sensitivity to proteolytic degradation. These difficulties can be eliminated or reduced by the use of the cell-free format of gene expression and fused forms of vulnerable polypeptides synthesized. The synthesis was performed in the bacterial CECF transcription-translation system and monitored by GFP fluorescence. After the polypeptide was split off from GFP, it exhibited the antibacterial activity. Thus, successful synthesis of a functionally active antibacterial polypeptide in a cell-free system was demonstrated.

A highly productive eukaryotic CECF system has been developed by the Endo group (Madin et al. 2000; see also Madin et al. this volume). The principal modification in this system consists of the preparation of wheat germ extract. Wheat embryos were hand-selected and washed with detergent in ultrasonic cleaner before disintegration. In this way, embryos were cleaned of contaminating endosperm material, which contains activities inhibiting the cell-free system. The removal of tritin, a ribosome inactivation protein (RIP), was shown. The CECF translation system with the extract prepared from the cleansed wheat embryos reached yields of 0.7 to 4 mg/ml for several functionally active proteins, such as DHFR, GFP, luciferase, and RNA replicase of tobacco mosaic virus. The system was active for several days, provided that the feeding solution was changed and mRNA was added repeatedly. The valuable feature of the system is that the syntheses of the long polypeptide chains of firefly luciferase (62 kDa) and TMV replicase (126 kDa) were quite productive, with little truncated by-products. Synthesis of GFP in the transcription-translation CECF system performed in our laboratory with wheat germ extract prepared from washed embryos is demonstrated in Fig. 3.

One parameter is nearly the same for all of the above-listed systems: the productivity of the CECF reactions is 10 to 20 times higher than that of the corresponding batch reactions, independent of the final productivity. Therefore, the CECF systems that display high yields are based on the improved components and reaction conditions that are also valid for batch-type cell-free systems. A number of modifications proposed earlier for batch systems have been found to be very useful in continuous systems. For instance, the use of condensed extracts that demonstrated some advantage in batch (Nakano et al. 1994) proved to be advantageous in the CECF system (Kim et al. 1996, Kigawa et al. 1999), significantly increasing the synthesis rate per unit of reaction volume. The removal of tritin from wheat germ extract (Madin et al. 2000) demonstrated that elimination of the enzymatic activity that causes the damage to the components of a cell-free system is especially rewarding in long running continuous systems.

Some other improvements made for batch cell-free systems deserve attention, with the object of possible application to continuous systems. The depletion of

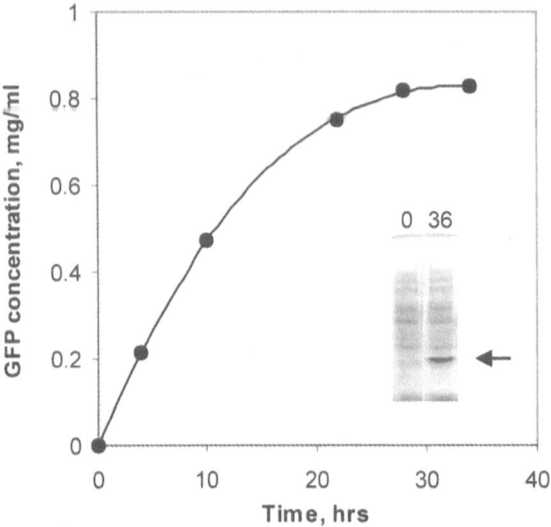

Fig. 3. Synthesis of GFP in continuous-exchange (dialysis) cell-free transcription-translation system with wheat germ extract. Synthesis was performed in a reactor with a flat dialysis membrane (cut-off 12000–14000) and the fixed reaction volume of 50 µl. Reaction mixture contained 36 OU_{260}/ml of wheat germ extract (30% v/v), 7000 U/ml T7 RNA polymerase, 100 µg/ml linearized plasmid with the red-shift GFP gene, 500 U/ml RNase inhibitor (HPRI), 100 µg/ml creatine phosphokinase (CPK), 50 µg/ml yeast total tRNA, 0.2 mM each amino acid, 1 mM ATP, 0.6 mM GTP, 0.4 mM CTP and UTP each, 16 mM creatine phosphate in 40 mM HEPES-KOH buffer pH 7.6 with 4 mM $Mg(OAc)_2$, 100 mM KOAc, 3 mM NaN_3, 2.5 mM DTT, 0.25 mM spermidine, 2% glycerol. Feeding solution (0.5 ml) contained the same components except wheat germ extract, RNA polymerase, plasmid, CPK and HPRI; $Mg(OAc)_2$ concentration was 2.5 mM. The reactor was kept at 25° C whilst the feeding solution was stirred. The feeding solution was changed for a fresh one at the 10 and 22 hour time points. GFP concentration in the reaction mixture was measured by fluorescence (480 nm/510 nm) relative to the reference GFP solutions. Inset: SDS-PAGE analysis of the ribosome-depleted reaction mixture (1 µl) before the reaction (0) and after the 36 hours run. Ribosomes were removed by centrifugation over the sucrose cushion in a TLS 55 rotor. The gel was stained by Coomassie G-250. Arrow indicates the position of GFP band. Wheat germ extract was prepared according to Madin et al. (2000) with minor modifications. The plasmid with GFP gene in a eukaryotic context was kindly provided by Dr. V. N. Ksenzenko

uncoupled phosphatase activity (Kawarasaki et al. 1998, Kang at al. 2000) saves energy resources and thus may reduce the required substrate supply rate and the final cost of the continuous systems. The new energy supply system proposed by Kim and Swartz (1999; see also Kim and Swartz, this volume) eliminates the inhibitory accumulation of inorganic phosphate in the reaction mixture. The uncoupled consumption of certain amino acids was observed in several cases, and the selective addition of arginine, tryptophan and cysteine (Kim and Swartz 2000), or just cysteine (Kim and Choi 2001), or methionine (Patnaik and Swartz 1998) in a bacterial translation system was shown to significantly improve the batch reaction. The proper selection of mRNA constructs for efficient translation, especially in the case of eukaryotic systems, is another wide field for improvement in cell-free synthesis in a continuous format.

The first commercial cell-free gene expression system in the CECF format, the so-called Rapid Translation System (RTS 500) from Roche Molecular Biochemicals, has recently been launched on to the market. It is a real sign that CECF protein synthesis technology has came out of the laboratory and is ready for the use in practical applications. The RTS 500 system comprises a reagent kit and a CECF reactor; it provides protein synthesis in the milligram range.

CFCF versus CECF

There are thus two versions of the continuous system: the flow version (CFCF system) and the dialysis version (CECF system). In the flow (CFCF) version, all the products, including the protein synthesized, if it is small enough, are removed from the reaction compartment. In the dialysis (CECF) version, the protein synthesized is accumulating in the reaction compartment, whereas low-molecular-weight products, such as phosphates, AMP, GDP and dephosphorylated energy substrates, are leaving through a dialysis membrane. As compared with the standard (batch) cell-free translation and transcription-translation systems, the CFCF and CECF systems have at least two principal advantages: (1) Long lifetimes: the systems are active, i.e. capable of synthesizing proteins, for many hours, sometimes days. (2) Large preparative yield: from 0.1 to 10 mg of an individual protein per ml can be produced in a CFCF or CECF reactor. The productivity of the systems can approach the productivity of the *in vivo* expression methods (see for example, Kigawa et al. 1999). At the same time, the cell-free systems provide better purity of the product in the reaction mixture even in the case of the CECF system, and even higher purity in the outflow from the CFCF reactor.

If the two versions of the continuous systems are to be compared with each other, one advantage of the CECF system is obvious: it is much simpler. The consequence of this is its lower cost and simple performance. On the other hand, the CFCF system has several undoubted advantages, as follows:
(1) Purity of protein product: the protein synthesized is selectively flowing out from the reactor, exhibiting up to 85% purity in the effluent (see for example, Spirin 1991).
(2) Longer lifetime: the active flow provides better and more adjustable conditions, as compared with passive exchange, for the substrate replenishment and product removal; in addition, some of the degradative enzymes (proteases and nucleases) are washed off from the incubation mixture during the initial period of the protein synthesis run.
(3) Scaling-up: whereas the effective exchange in the dialysis system strongly depends on the ratio of reaction volume to membrane surface area and so scaling the CECF reactor up is difficult, the CFCF bioreactor, where the exchange is forced by flow, can be scaled-up without problems of this sort.
(4) Control of the reaction mixture: both the inflow and the outflow of the reactor can be programmed, the composition of the reaction mixture can be modified during the process, and the system can be easily automated.

(5) Production monitoring: both the quantity and quality of the protein synthesized, as well as the protein folding, can be checked in the effluent fractions over the course of the synthesis.

One would expect that all these merits will be important when large-scale technological applications of the cell-free systems come on to the agenda in the future.

Problems

Preparation of a cell-free extract eliminates the major part of cell control mechanisms. This allows arbitrary changing of the expression system conditions, thus piloting the process of protein synthesis. However, deregulation of metabolic pathways, damage of some cellular structures and the resultant liberation or induction of a number of destructive activities during cell disintegration cause a series of other problems to appear. Among them, uncoupled energy consumption and mRNA degradation seem to be the most serious in the case of the long-term CFCF and CECF systems.

Energy Consumption

Nucleoside triphosphates (NTPs) and NTP-regenerating substrates (high-energy phosphate donors, such as PEP, CP and AcP) have been shown to be the primary components that are exhausted in batch cell-free protein-synthesizing systems of different origin (see for example Yamane et al. 1995; Matveev et al. 1996; Kitaoka et al. 1996). As a result of NTPase and phosphatase activity in cell extracts, rapid uncoupled hydrolysis of NTPs and NTP-regenerating substrates occurs in the incubation mixture, in addition to their productive consumption during protein synthesis. The energy potential of a batch system drops to zero in 30 to 60 minutes and thus the protein synthesis ceases (Yao et al. 1997; our unpublished data). Moreover, inorganic phosphate (P_i) accumulating in the reaction mixture can inhibit the translation system by itself. In the continuous systems the problem is automatically solved to a significant degree: the energy substrates are constantly supplied and the products, including P_i, are removed by flow (CFCF) or by diffusion (CECF), and therefore the steady-state concentrations of the substrates and products are maintained during the run. Nevertheless, the problem still exists if the rate of flow or exchange is not sufficient to compensate both the coupled and uncoupled consumption of the energy compounds in the reaction mixture. That is why parameters such as the flow rate, the membrane area, pore size and permeability, and the initial concentrations of the energy substrates in the feeding solution must be properly adjusted.

In the CECF (dialysis) systems, the feeding solution can be changed at certain time intervals to maintain the initial energy compound concentration outside the reaction compartment. Higher initial concentrations of the energy substrates may be used in cell-free systems of different types, if these substrate concentrations are not inhibitory for the system, as they were not in cases utilizing acetyl phosphate (Ryabova et al. 1995) or creatine phosphate (Kigawa et al. 1999) in bacterial systems. The presence of sodium azide results in partial inhibition of NTPase activity, at least in crude bacterial extracts. In the CFCF system, the level of phosphatase and NTPase activity in the reaction compartment decreases as they are washed-out during the run. New approaches with the depletion of phosphatase activity from cell extracts (Kawarasaki et al. 1998; Kang at al. 2000) may be especially promising for CFCF and CECF systems: the absence of phosphatase will reduce the consumption of energy substrates and therefore the required substrate supply/exchange rate. With regard to this, the reconstitution of a translation or transcription-translation mixture from all purified components (see Ueda et al., this volume) would be the best solution to the problem, provided that a sufficiently high protein-synthesizing capacity could be attained in such a system.

A novel NTP regeneration system that avoids accumulation of inorganic phosphate has recently been proposed (Kim and Swartz 1999; see also Kim and Swartz, this volume). In this system, acetyl phosphate (AcP) is generated by pyruvate oxidase from pyruvate and inorganic phosphate directly in the reaction mixture:

$$\text{Pyruvate} + P_i + O_2 + H_2O \rightarrow \text{Acetyl phosphate} + CO_2 + H_2O_2.$$

Acetyl phosphate is then used by acetate kinase to restore NTPs from NDPs, and hydrogen peroxide is decomposed by catalase. Phosphate is thus recycled, whereas acetate, the only accumulating product, is not inhibitory for the system. The pyruvate-pyruvate oxidase regeneration system has been shown to prolong the *E. coli* cell-free transcription-translation system in the batch format. Undoubtedly this regeneration system might be useful for use in continuous cell-free systems as well.

mRNA Degradation and Sustenance

Maintenance of the appropriate mRNA concentration during the long runtime is another key point relating to the continuous cell-free expression systems. Various RNAase activities present in cell extracts usually restrict the lifetime of mRNA. The problem can be solved in several ways. One way is the periodical re-addition of mRNA into the reaction translation mixture, as has been performed over a 60 hour run of CECF synthesis of dihydrofolate reductase (DHFR) in wheat germ extract (Madin et al. 2000).

Coupled and combined transcription-translation systems where mRNA is continuously renewed by ongoing transcription of DNA template are more preferable than translation systems with pre-synthesized mRNA, especially for use in the long-term CECF and CFCF formats. The transcription-translation mode has practically no alternative for the continuous systems based on crude bacterial extracts (S-30). The transcription in the reaction mixture may be executed from an appropriate promoter either by the endogenous *E. coli* RNA polymerase, or by an added phage RNA polymerase. Transcription by phage RNA polymerases can be also combined with the eukaryotic translation systems based on either wheat germ extract or rabbit reticulocyte lysate (Spirin 1991; Craig et al. 1992). Circular plasmid DNA is commonly used as template for transcription in the prokaryotic system. In the wheat germ system with lower nuclease activities both plasmid DNAs and linear PCR fragments serve well. As the excess of mRNA may inhibit translation in eukaryotic systems, the rate of transcription must be adjusted to the level that prevents overproduction of mRNA. In the continuous systems, the transcription rate can easily be regulated during a run by changing the Mg^{2+} and NTP concentrations via dialysis or flow (Biryukov et al. 2000). It is advantageous to start the combined transcription-translation run with a higher Mg^{2+} concentration when transcription is at a high rate and mRNA accumulates, and then continue with a lower Mg^{2+} concentration that provides a modest level of mRNA synthesis, compensating the degradation. An example of GFP synthesis in the CECF system using this approach is given in Fig. 3.

Translational efficiency and stability of mRNA depends on its structure. The presence of the classic Shine-Dalgarno sequence and the *epsilon* enhancer of T7 phage gene 10 upstream of the initiation codon are favorable for efficient initiation of translation in prokaryotic cell-free systems. A stable hairpin at the 3'-end of mRNA contributes to the protection of mRNA against 3'-exonucleolytic degradation. Significant stability of mRNA can be achieved by insertion of the coding sequence between mutually complementary 5'- and 3'-terminal untranslated sequences forming a stable stem helix (Ugarov et al. 1994).

In the eukaryotic systems, a useful element for the efficient initiation of translation is the so-called Kozak sequence context around the initiation AUG codon, such as GCCACCAUGG in mammalian systems (Kozak 1989). The 5'-cap structure and the 3'-poly(A) tail increase both the translation initiation efficiency and the stability of eukaryotic mRNAs. However, these elements increase the cost and can be used only in translation systems with isolated or pre-synthesized mRNA. Alternative enhancing and stabilizing elements have been found in some viral RNAs. Thus, 5'- and 3'-untranslated regions (UTRs) of satellite tobacco necrosis virus (STNV) RNA and barley yellow dwarf virus (BYDV-PAV) RNA can serve as powerful translation enhancers in the absence of the cap-structure and poly(A) tail (Danthinne et al. 1993; Timmer et al. 1993; Wang and Miller 1995), whereas their long 3'-UTR sequences may contribute to the stability of mRNA. The use of viral UTRs may significantly improve the performance of the eukaryotic continuous systems.

Reactors

Continuous-Flow Reactors

Following the use of the standard micro-ultrafiltration cell (model 8MC, Amicon), several specialized types of flow reactors were invented for use in the CFCF systems (see for example Spirin 1991; Kigawa and Yokoyama 1991; Baranov and Spirin 1993). The reactors with a flat ultrafiltration membrane (30 to 300 kDa cutoff) (Fig. 4A) are simple in design and ensure the maintenance of a stable reaction volume during the run. Upright flow of feeding solution and permanent stirring of the reaction mixture are applied to minimize membrane clogging. Reactors of

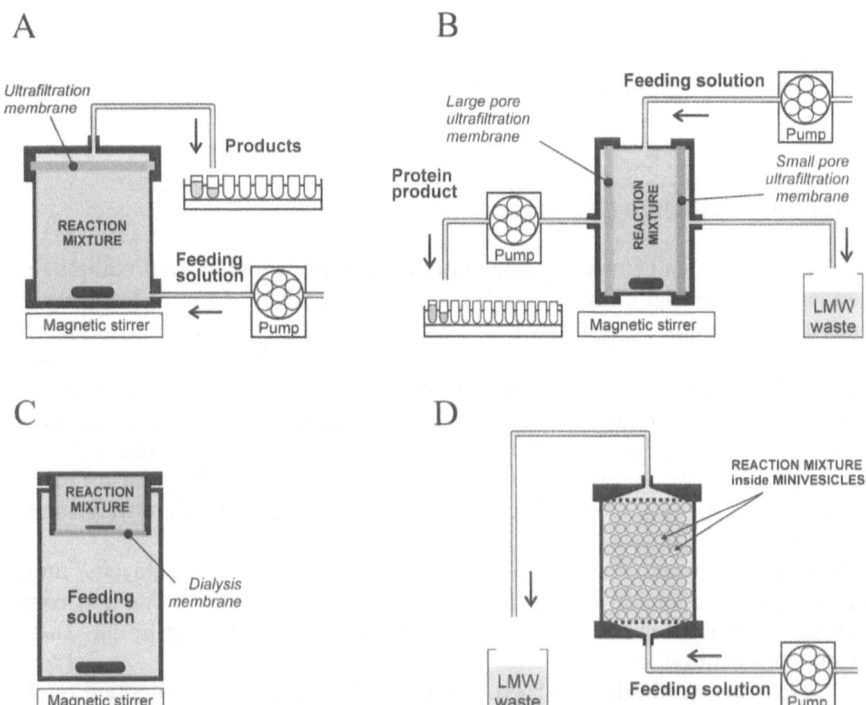

Fig. 4. Schematic drawings of reactor for continuous-flow (A, B) and continuous-exchange (C, D) cell-free expression systems. A, direct-flow CFCF reactor with substrates supplied and products removed by the flow of feeding solution; B, Y-flow CFCF reactor with two ultrafiltration membranes of different pore size providing separate protein product and low molecular weight (LMW) waste outflows (Biryukov et al. 1999); C, CECF reactor with a flat dialysis membrane separating the reacton and feeding solution compartments and providing the diffusion exchange of LMW substrates and products; D, flow-exchange column reactor packed with semi-permeable mini-vesicles containing the reaction mixture; feeding solution is changed by flow through the column, and substrates are supplied by diffusion into the mini-vesicles (Spirin 1991)

this type have usually been utilized in experiments with CFCF systems. The hollow fiber reactors were proposed to have a better surface-to-volume ratio for the high flow rate of feeding solution and the effective exchange of substrates/products (Spirin 1991; Alakhov et al. 1995; Yamamoto et al. 1996). The translation mixture can be placed either in the lumen of the hollow fiber or outside the lumen, while the feeding solution is passed outside or inside the fiber respectively. The hollow fiber reactors are somewhat intricate for laboratory practice, but seem to be of interest for future large-scale industrial applications.

The synthesized protein comes out from the CFCF reactor in a large volume of the effluent, i.e. in the form of a diluted solution. To reduce the dilution effect, the so-called Y-flow reactor with a split outflow has been proposed (Biryukov et al. 1999). The Y-flow reactor (Fig. 4B) has two membranes with different pore sizes. The low-molecular-weight products ("waste") are removed through a small-pore membrane at a high rate, whereas the protein synthesized is collected through a large-pore membrane at a low rate (that is, as a concentrated solution). The protein product flow is controlled by a separate pump. The syntheses of GFP and CAT in the Y-flow reactor are demonstrated in Fig. 2.

Continuous-Exchange (Dialysis) Reactors

In principle, the CECF system does not require any sophisticated equipment. Standard commercial dialysers, such as the DispoDialyzer from Spectrum or a common homemade dialysis bag have been successfully exploited in a number of experiments. A simple device comprising a reservoir with a flat dialysis membrane (Fig 4C) can be also used (see for example Kim and Choi 1996). The pore size of the reactor membranes are usually in the range of 10 to 50 kDa. A better performance of the reactor with a large pore membrane has been reported (Kigawa et al. 1999), although in most cases the standard 10–12 kDa cut-off membrane was exploited and no advantages of a large pore membrane were found (see, e.g., Davis et al. 1996).

In order to meet some demands of scientists and biotechnologists, more sophisticated versions of the CECF reactor are being developed. The first commercial CECF reactor has been recently launched to the market by the Roche Diagnostics GmbH.

The reactors combining both exchange and flow deserve special attention. In one version, the translation reaction mixture is encapsulated into polysaccharide mini-vesicles that can be packed into a column, and the flow of feeding solution is passed through the column (Fig. 4D); the product-substrate exchange across the capsule walls takes place during the flow (Spirin 1991). An opposite approach is implemented in the so-called matrix reactor (see Buchberger et al. this volume), where the molecular sieve beads serve as the feeding solution compartment and the reaction mixture passes through the column. Reactors of this type are expected to provide highly effective substrate/product exchange and can be easily scaled up.

Conclusion

Biotechnology is reaching a landmark: over the past decade, protein biosynthesis, after being the first genetic process to be reproduced without the cell, has been developed into *in vitro* production of individual proteins on preparative scale. Continuous cell-free methodology laid the basis for long-lived and highly productive systems for protein synthesis. Recent improvements in cell-free translation protocols give an additional boost to the productivity of the continuous systems. It happened right on time: an enormous amount of genetic information waits to be read and interpreted in terms of the identity of corresponding proteins. Structural analyses of proteins by the NMR technique demands stable isotope-labeled proteins. Cell-free expression systems provide unrestricted possibilities for these and other applications. The CECF systems with simple dialysis reactors are already ahead and can begin to compete with *in vivo* expression systems in terms of productivity and product purity. The CFCF reactors with more sophisticated designs are being prepared to meet the demands of further complex applications.

Thus, now all the pathway of the genetic information from gene to protein can be practically realized in a cell-free environment. Many limitations of the *in vivo* expression systems can be avoided. Moreover, the cell-free biotechnology provides much better control of genetic processes and results as compared with *in vivo* manipulations. On the whole, the cell-free biotechnology seems to be ecologically safe and devoid of any risks of uncontrolled dissemination of genetic information and serious environmental contamination.

References

Alakhov YB, Baranov VI, Ovodov SJ, Ryabova LA, Spirin AS, Morozov IY (1995) Method of preparing polypeptides in cell-free translation system. United States Patent # 5,478,730

Alexandrov A, Kolosova I, Kolosov M (1996) mRNA stabilization in continuous flow translation system. Biochem Mol Biol Intern 38:1111–1116

Baranov VI, Spirin A.S (1993) Gene expression in cell-free systems on preparative scale. Methods Enzymol 217:123–142

Baranov VI, Morozov IY, Ortlepp SA, Spirin AS (1989) Gene expression in a cell-free system on the preparative scale. Gene 84:463–466

Biryukov SV, Simonenko PN, Shirokov VA, Majorov SG, Spirin AS (1999) Method of preparing polypeptides in cell-free system and device for its realization. Patent pending, PCT application:WO 99/50436

Biryukov SV, Simonenko PN, Shirokov VA, Spirin AS (2000) Method for synthesis of polypeptides in cell-free systems. Patent pending, PCT application:WO 00/58493

Chekulayeva MN, Kurnasov OV, Shirokov VA., Spirin AS (2001) Continuous-exchange cell-free protein-synthesizing system: Synthesis of HIV-1 antigen Nef. Biochem Biophys Res Commun 280:914–917

Chetverin AB, Spirin AS (1995) RQ RNA vectors: Prospects for cell-free gene amplification, expression, and cloning. Prog Nucleic Acid Res Mol Biol 51:225–270

Craig D, Howell MT, Gibbs CL, Hunt T, Jackson RJ (1992) Plasmid cDNA-directed synthesis in a coupled eukaryotic *in vitro* transcription-translation system. Nucleic Acids Res 20: 4987–4995

Danthinne X, Seurinck J, Meulewaeter F, van Montagu M, Cornelissen M (1993) The 3' untranslated region of satellite tobacco necrosis virus RNA stimulates translation *in vitro*. Mol Cell Biol 13:3340–3349

Davis J, Thompson D, Beckler GS (1996) Large scale dialysis cell-free system. Promega Notes Magazine, No 56:14–21

Endo Y, Otsuzuki S, Ito K, Miura K (1992) Production of an enzymatic active protein using a continuous flow cell-free system. J Biotech 25:221–230.

Endo Y, Oka T, Ogata K, Natori Y (1993) Production of dihydrofolate reductase by an improved continuous flow cell-free translation system using wheat germ extract. Tokishima J Exp Med 40:13–17

Erickson AH, Blobel G (1983) Cell-free translation of messenger RNA in a wheat germ system. Methods Enzymol 96:38–50

Kang SH, Oh TJ, Kim RG, Kang TJ, Hwang SH, Lee EY, Choi CY (2000) An efficient cell-free protein synthesis system using periplasmic phosphatase-removed S30 extract. J Microbiol Methods 43:91–96

Kawarasaki Y, Nakano H, Yamane Y (1998) Phosphatase-immunodepleted cell-free protein synthesis system. J Biotechnol 61:199–208

Kigawa T, Yokoyama S (1991) A continuous cell-free protein synthesis system for coupled tran-scription-translation. J Biochem (Japan) 110:166–168

Kigawa T, Yabuki T, Yoshida Y, Tsutsui M, Ito Y, Shibata T, Yokoyama S (1999). Cell-free produc-tion and stable-isotope labeling of milligram quantities of proteins. FEBS Lett 442:15–19

Kim DM, Choi CY (1996) A semi-continuous prokaryotic coupled transcription-translation sys-tem using a dialysis membrane. Biotechnol Prog 12:645–649

Kim DM, Swartz JR (1999) Prolonging cell-free protein synthesis with a novel ATP·regeneration system. Biotech Bioengineering 66:180–188

Kim DM, Swartz JR (2000) Prolonging cell-free protein synthesis by selective reagent additions. Biotech Prog 16:385–390

Kim RG, Choi CY (2001) Expression-independent consumption of substrates in cell-free expres-sion system from Escherichia coli. J Biotechnol 84:27–32

Kim DM, Kigawa T, Choi CY, Yokoyama S (1996) A highly efficient cell-free protein synthesis sys-tem from *Escherichia coli*. Eur J Biochem 239:881–886

Kitaoka Y, Nishimura N, Niwano M (1996) Cooperativity of stabilized mRNA and enhanced translation activity in the cell-free system. J Biotech 48:1–8

Kolosov MI, Kolosova IM, Alakhov VY, Ovodov SY, Alakhov YB (1992) Preparative in vitro syn-thesis of bioactive human interleukin-2 in a continuous flow translation system. Biotech Appl Biochem 16:125–133

Kozak M (1989) Context effects and inefficient initiation at the non-AUG codons in eucaryotic cell-free translation systems. Mol Cell Biol 9:5073–5080

Kudlicki W, Kramer G, Hardesty B (1992) High efficiency cell-free synthesis of proteins: refine-ment of the coupled transcription/translation system. Anal Biochem 206:389–393

Madin K, Sawasaki T, Ogasawara T, Endo Y (2000) A highly efficient and robust cell-free protein synthesis system prepared from wheat embryos: Plants apparently contain a suicide system directed at ribosomes. Proc Natl Acad Sci USA 97:559–564

Martemyanov K.A, Shirokov VA, Kurnasov OV, Gudkov AT, Spirin AS (2001) Cell-free production of biologically active polypeptides: application to the synthesis of antibacterial polypeptide cecropin. Protein Expr Purif 21:456–461

Morozov IY, Ugarov VI, Chetverin AB, Spirin AS (1993) Synergism in replication and translation of messenger RNA in a cell-free system. Proc Natl Acad Sci USA 90:9325–9329

Matveev SV, Vinokurov LM, Shaloiko LA, Matveeva EA, Alakhov YB (1996) Effect of ATP the lev-el on the overall protein biosynthesis rate in a wheat germ cell-free system. Biochim Biophys Acta 1293:207–212

Nakano H, Tanaka T, Kawarasaki Y, Yamane T (1994) An increased rate of cell-free protein syn-thesis by condensing wheat germ extract with ultrafiltration membranes. Biosci Biotechnol Biochem 58:631–634

Nishimura N, Kitaoka Y, Mimura A, Takahara Y (1993) Continuous protein synthesis system with Escherichia coli S30 extract containing endogenous T7 RNA polymerase. Biotech Lett 15:785–790

Nishimura N, Kitaoka Y, Niwano M (1995) Cell-free system derived from heat-shocked Escherichia coli: Synthesis of enzyme protein possessing higher specific activity. J Ferment Bioeng 79:131–135

Patnaik R, Swartz JR (1998) E.coli-based in vitro transcription/translation: in vivo-specific syn-thesis rates and high yields in a batch system. BioTechniques 24:862–868

Pratt JM (1984) Coupled transcription-translation in prokaryotic cell-free systems. In: Hames BD, Higgins SJ (eds) Transcription and translation: a practical approach. IRL Press, New-York, pp 179–209

Roberts BE, Paterson BM (1973) Efficient translation of tobacco mosaic virus RNA and rabbit globin 9S RNA in a cell-free system from commercial wheat germ. Proc Natl Acad Sci USA 70:2330–2334

Ryabova LA, Ortlepp SA, Baranov VI (1989) Preparative synthesis of globin in a continuous cell-free translation system from rabbit reticulocytes. Nucleic Acids Res 17:4412

Ryabova L, Volianik E, Kurnasov O, Spirin AS, Wu Y, Kramer FR (1994) Coupled replication-translation of amplifiable messenger RNA: A cell-free protein synthesis system that mimics viral infection. J Biol Chem 269:1501–1505

Ryabova LA, Vinokurov LM, Shekhovtsova EA, Alakhov YB, Spirin AS (1995) Acetyl phosphate as an energy source for bacterial cell-free translation system. Anal Biochem 226:184–186

Ryabova LA, Morozov IY, Spirin AS (1998) Continuous-flow cell-free translation, transcription-translation, and replication-translation systems. In: Martin R (ed) Methods in Molecular Biology vol 77: Protein Synthesis: Methods and Protocols. Humana Press, Totowa, NJ, pp 179–193

Spirin AS (1991) Cell-free protein synthesis bioreactor. In: Todd P, Sikdar SK, Beer M (eds) Frontiers in Bioprocessing II. American Chemical Society, Washington, DC, pp 31–43

Spirin AS, Baranov VI, Ryabova LA, Ovodov SY, Alakhov YB (1988) A continuous cell-free translation system capable of producing polypeptides in high yield. Science 242:1162–1164

Timmer R, Benkowski LA, Schodin D, Lax SR, Metz AM, Ravel JM, Browning KS (1993) The 5' and 3' untranslated region of satellite tobacco necrosis virus RNA affect translational efficiency and dependence on a 5' cap structure. J Biol Chem 268:9504–9510

Ugarov VI, Morozov IYu, Jung GV, Chetverin AB, Spirin AS (1994) Expression and stability of recombinant RQ-mRNA in cell-free translation systems. FEBS Lett 341:131–134

Volyanik EV, Dalley A, McKay IA, Keigh I, Williams NS, Bustin SA (1993) Synthesis of preparative amounts of biologically active interleukin-6 using a continuous-flow cell-free translation system. Anal Biochem 214:289–294

Yamamoto YI, Nagahori H, Yao S, Zhang ST, Suzuki E (1996) Hollow fiber reactor for continuous flow cell-free protein production. J Chem Eng Japan 6:1047–1050

Yamane T, Kawarasaki Y, Nakano H (1995) In vitro protein biosynthesis using ribosome and foreign mRNA. An approach to construct a protein biosynthesizer. Annals NY Acad Sci 750: 146–157

Yao SL, Shen XC, Suzuki E (1997) Biochemical energy consumption by wheat germ extract during protein synthesis. J Ferment Bioeng 84:7–13

Zubay G (1973) In vitro synthesis of protein in microbial systems. Annu Rev Genet 7:267–287

Chapter 9

Highly Productive Plant Continuous Cell-Free System

9

Kairat Madin, Tatsuya Sawasaki, Yaeta Endo*

Introduction

The development of a system capable of synthesizing any desired protein on a preparative scale is one of the most important endeavors in biotechnology today. Three strategies are currently being used: chemical synthesis, *in vivo* expression, and cell-free protein synthesis. The first two methods have severe limitations: chemical synthesis is not feasible for the synthesis of long peptides because of low yield, and *in vivo* expression can produce only those proteins that do not affect the physiology of the host cell [1–3]. Cell-free translation systems, in contrast, can synthesize proteins with high speed and accuracy, approaching *in vivo* rates [4–5], and they can express proteins that would interfere with cell physiology. However, they are relatively inefficient because of their instability[6].

As cell free systems nonetheless have great potential for large scale protein synthesis, many efforts have been made to increase their efficiency. Spirin *et al.* [7] proposed a continuous flow cell-free translation system, in which a solution containing amino acids and energy sources is supplied to the reaction chamber through a filtration membrane. This design is significantly more efficient than conventional batch systems: The reaction works for tens of hours and produces hundreds of micrograms per ml reaction volume [7–9]. Recently, several modified versions of the Spirin system have been reported [10–13]. Kigawa *et al.* showed that by using a dialysis membrane to facilitate the continuous supply of substrates and removal of byproducts, an *E. coli* coupled transcription-translation system yields as much as 6 mg protein per ml reaction volume [12]. This high productivity can, however, only be expected with fairly small proteins such as Ras protein (21kDa) or chloramphenicol acetyltransferase (26 kDa). The problem with larger proteins is that with the increasing molecular weight of the mRNAs their degradation by endogenous *E. coli* ribonuclease(s) also increases. Kawarasaki *et al.* showed that in a wheat germ cell-free system translational efficiency increases after neutralization of endogenous ribonucleases and phosphatases with copper ions and anti-phosphatase antibodies [13]. For their improvements these groups focused on modifying the reaction chamber and/or optimizing the reaction conditions while using conventional extracts. We used a different approach, instead focusing on clarifying the nature of the instability of the extracts.

* Yaeta Endo, e-mail: yendo@en3.ehime-u.ac.jp

We concentrated on wheat germ cell-free systems because they have numerous advantages such as low cost, easy availability in large amounts, low endogenous incorporation, and the capacity to synthesize high-molecular-weight proteins. Moreover they are eukaryotic systems and hence more suitable for the expression of eukaryotic proteins. After we discovered that the mechanism of action of the ricin toxin is ribosome inactivation [14–16], many other ribosome-inactivating proteins (RIPs) with identical mechanism of action have been found in higher plants [17]. Most commonly these toxins are single-chain proteins, and they inhibit protein synthesis by removing a single adenine residue in a universally conserved stem-loop structure of 28S ribosomal RNA [14–17]. Although the biological function of the RIPs is not known, it is generally believed that they are important for cell defense [17]. The most widely studied example is an antiviral effect during infection by several plant viruses [18]. As originally proposed by Ready et al. [19], the explanation for the antiviral activity of RIPs is that when a cell wall is damaged, the RIP is released into the cytosol where it inactivates ribosomes, thereby preventing virus replication. Tritin, found in wheat seeds and thought to be localized mainly in the endosperm, is such a single-chain RIP [20]. Recently we have found a new class of enzyme in wheat embryos that acts on the depurinated embryonic ribosomes brought by tritin action: ribosomal RNA apurinic site specific lyase (RALyase), and have proposed a possible biological role of the RIP and RALyase, as a complex self-defense mechanism by their total inactivation of depurinated ribosomes [21]. Similar mechanisms of ribosome-suicide systems have been observed in several other plants [21].

In order to improve protein synthesis in wheat germ cell-free systems, we started with the hypothesis that the embryonic ribosomes are in fact susceptible to tritin. In this case, contamination of wheat germ preparations with tritin-containing endosperm fragments would be fatal. Accordingly, we prepared our cell-free system from extensively washed embryos, and found that the system did indeed became more active (22). In a batch system we observed continuous translation for 4 hours, and sucrose density gradient analysis showed formation of large polysomes, indicating high protein synthesis activity. When the reaction was performed in a dialysis bag, enabling the continuous supply of substrates together with the continuous removal of small byproducts, translation proceeded for more than 60 hours, yielding 1 to 4 mg of enzymatically active proteins, and 0.6 mg of a 126 kDa TMV protein, per ml reaction. Our results demonstrate that plants contain endogenous inhibitors of translation, and that after their elimination the translational apparatus is very stable. Our method is useful for the preparation of large amounts of active protein as well as for the study of protein synthesis itself.

In addition to the benefit of a better protein synthesis system, these results shed new light on the translational apparatus itself: While it is usually seen as a rather fragile apparatus, it appears instead to be very stable; so stable, in fact, that plants seem to have developed a suicide mechanism (the RIPs) directed against the translational apparatus, further emphasizing its crucial role in cell physiology.

We believe that the strategy we followed to improve the wheat cell-free system – elimination of endogenous translational inhibitors – is equally applicable for other systems.

Removal of contaminants leads to a more active
cell-free protein synthesis system

Since the first report of solvent flotation for the enrichment of viable, intact embryos from wheat seeds by Johnston *et al.* [23], this method has commonly been used for the preparation of wheat embryos. In this procedure, wheat seeds were first ground in a mill (Roter Speed Mill model pulverisette 14, Fritsch, Germany), then sieved through a 710–850 mm mesh. Embryos were selected with the solvent flotation method of Erickson and Blobel, using a solvent containing cyclohexane and carbon tetrachloride (240:600, v/v) [24]. We first addressed the possibility of tritin contamination originating from endosperm as the critical reason for the instability of wheat germ cell-free systems. If wheat germs are isolated from dry wheat seeds by conventional procedures [24], microscopic examination reveals that the sample contains embryos as well as some white material and a number of white and brownish granules (Fig. 1A). Analysis of ribosomal RNAs from a protein synthesis reaction prepared from such a sample by the procedure [24] shows that the depurination of ribosomes occurs, which is consistent with earlier reports [21–22] (Fig.1B). After 4h of incubation, 24% of the ribosome population has been depurinated, as judged by the aniline-dependent formation of a specific RNA fragment (arrow). Furthermore, even at the start of the incubation 7% of the population has already been depurinated. The site of depurination was confirmed by direct sequencing of the fragment to be in the universally conserved sarcin/ricin domain of 28S rRNA (data not shown). When RNA was extracted directly from embryos by guanidine isothiocyanate-phenol, little formation of the aniline-induced fragment was observed (lanes 7, 8). Thus, depurination must have occurred during the extract preparation, and then continued during the protein synthesis reaction.

The observed extent of depurination constitutes a considerable damage to protein synthesis, since inactivation of any one ribosome among the actively translating ribosomes on an mRNA results in blockage of the respective polyribosome and cessation of translation [16]. Attempts were made to neutralize the depurinating enzyme with synthetic RNA aptamers that tightly bind to the RIP [25], but these attempts failed. Instead, careful selection and subsequent extensive washing of the embryos yielded better results. In order to follow this, damaged embryos and contaminants were discarded after flotation, and intact embryos were dried overnight in a fume hood. To remove contaminating endosperm, the embryos were washed 3 times with 10 volumes of water with vigorous stirring, and then sonicated for 3 minutes in a 0.5% solution of NP-40 using a Bronson model 2210 sonicator (Yamato, Japan). Finally, the embryos were washed once more in the sonicator with sterile water. These embryos had few contaminants (lower panel, Fig. 1A), and when the depurination assay was performed, no aniline-induced cleavage was detectable (lanes 10–12), indicating minimal if any depurination during preparation as well as incubation.

The protein synthesis was programmed with capped mRNA encoding dihydrofolate reductase (dhfr), synthesized by in vitro transcription of linearized plasmid pSP65 carrying the gene under SP6 RNA polymerase promoter control

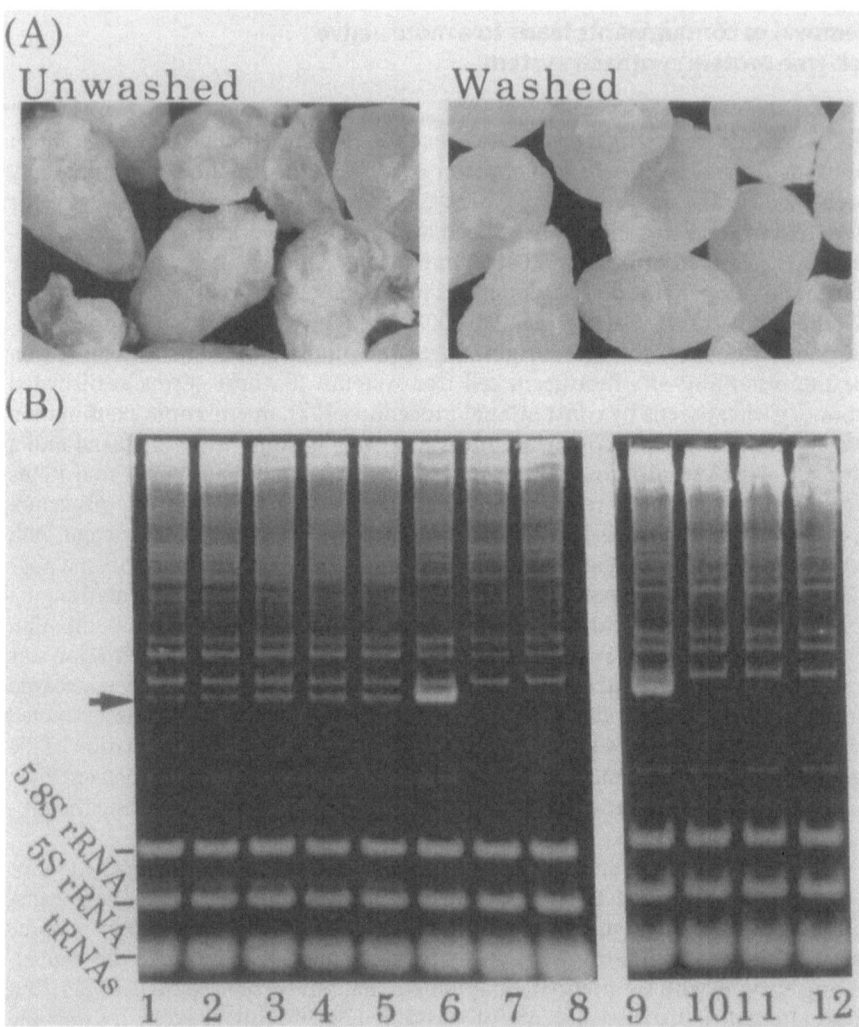

Fig. 1. Removal of tritin from embryos. Extracts were prepared from unwashed or washed embryos (A) and the depurination assay was performed (B). Translation mixtures prepared with the extract from unwashed embryos were incubated for 0, 1, 2, 3, 4 hrs (lanes 1–5 respectively); mixtures with washed embryos were incubated for 0, 2, 4 hrs (lanes 10–12, respectively). Isolated RNA was treated with acid/aniline, then separated on 4.5% polyacrylamide gels. Additionally, RNA was directly extracted from embryos with guanidine isothiocyanate-phenol and analyzed as above before (lane 7) and after (lane 8) treatment with acid/aniline. For the fragment marker (lanes 6 and 9), incubation was carried out in the presence of gypsophilin, a highly active RIP from *Gypsophila elegance*; the arrow indicates the aniline-induced fragment

[9]. The transcript is 1079 nucleotides long and consists of the sequence m^{7}Gppp-GAAUACACGGAAUUCGAGCUCGCCCGGGAAAU-CUCA<u>AUG</u> (the underlined sequence is the initiation codon) at its 5' end, a 477 nucleotide long coding sequence, and a 3' non-coding region of 565 nucleotides with a poly(A) tail of 100 adenosines [9]. As shown in Fig. 2, the cell-free system prepared from washed embryos has much higher translational activity than the conventional system (comp. Fig. 2A and B). When programmed with mRNA coding for dihydrofolate reductase (dhfr) almost linear kinetics in dhfr synthesis are seen over 4 hrs in a system containing 24% extract, as opposed to the regular system which ceased to function after 1.5 hrs. When the content of washed extract in the reaction volume was increased to 48%, amino acid incorporation occurred initially at a rate twice that with 24% extract, but then stopped after 1 hr. However, this halting is caused by a shortage of substrates rather than an irreversible inactivation of ribosomes or factors necessary for translation: Addition of amino acids, ATP and GTP after cessation of the reaction (arrow) restarted translation with kinetics similar to the initial rate. In contrast, if conventional extract was added to 48%, protein synthesis actually decreased compared with the 24% extract reaction. Furthermore, the halting of protein synthesis in the reaction with 24% extract can not be reversed

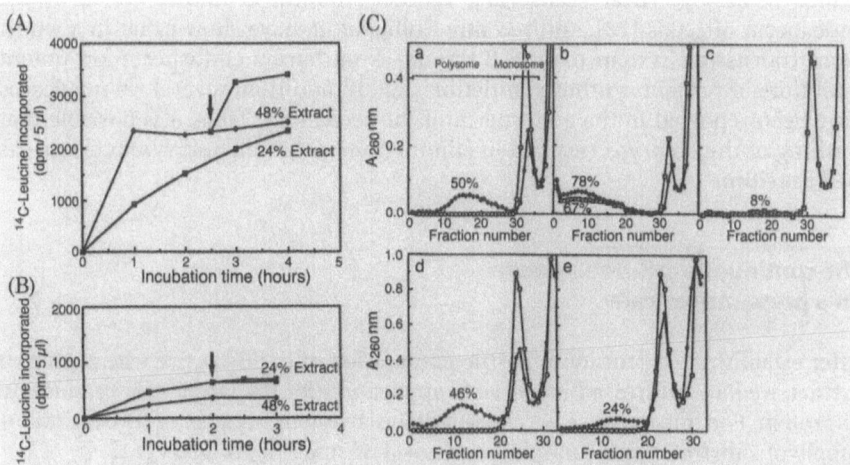

Fig. 2. Protein synthesis with an extract prepared from washed embryos. The batch system contains either 12 μl (24%) or 24 μl (48%) of extracts from washed (A) or unwashed wheat embryos (B). Protein synthesis was measured as hot tricholoroacetic acid insoluble radioactivity. Arrows show addition of substrates. Panel C shows the polysome profiles of 15 μl reaction mixture aliquots loaded onto a linear 10% to 45% sucrose gradient in 25 mM Tris-HCl, pH 7.6, 100 mM KCl, and 5 mM MgCl$_2$. After centrifugation, fractions were collected from the bottom of the tubes and measured at 260 nm. Incubation times were 0 (open circles in a), 1 hr (closed circles in a), 2 hrs (b) in the absence (open circles) or presence (closed circles) of 0.4 μM cycloheximide. In (c), the translation system prepared from unwashed embryos was incubated for 2 hrs. In (d) and (e), aliquots from the dialysis system were withdrawn after 48 hrs and 60 hrs and were incubated in the presence of 0.4 μM cycloheximide for another 60 min at 26°C (closed circles). Similar analyses of the samples were carried out in the absence of mRNA (open circles) as negative controls

by addition of more substrate, indicating an irreversible damage by contaminants from endosperm (Fig. 2B).

High protein synthesis activity of the system with washed embryos can also be demonstrated by sucrose density gradient analysis (Fig. 2C). Significant formation of polysomes was observed after 1 hr of incubation, and at 2 hrs a shift to heavier polysomes with a concomitant decrease of 80S monosomes was seen (a, b). In the presence of low concentrations of cycloheximide polysome formation is a measure of translational initiation [26]. A concentration of cycloheximide of 0.4 μM reduced the incorporation of [^{14}C]leucine to 21% of the control (data not shown), and resulted in an accumulation of large polysomes, with 78% of ribosomes in polysomes (open circle in b). A similar analysis of cell-free reactions prepared with regular extracts [24], but done in the absence of cycloheximide, did not show significant polysome formation (c). The high efficiency of our system therefore can be attributed to at least two factors: First, high initiation, elongation and termination rates (efficient usage and recycling of ribosomes); and second, low endogenous ribonuclease activity (retention of heavy polysomes for a prolonged time).

There is an additional explanation for the dramatic improvement of protein synthesis after washing of the embryos. Thionins are a group of small basic and cysteine rich proteins, originally purified as antifungal proteins from a variety of plants, including wheat seeds [27]. Wheat γ-thionin is known to be in the endosperm of seeds [28], and recently Collila *et al.* have shown that in a wheat germ translation system α- and β-thionin from barley endosperm are potent inhibitors of protein synthesis initiation [29]. In addition, several ribonucleases have been reported in the endosperm of the seeds [30]. Thus, it is possible that washing of the embryos resulted in elimination of thionin and ribonucleases as well as tritin.

The continuous cell-free system on a preparative scale

After establishing a procedure for the preparation of highly active wheat embryo extract, we have addressed its possible application for the large scale production of protein. For this purpose we chose a dialysis system because of its continuous supply of substrates and continuous removal of small byproducts [12].

With dhfr mRNA as template, protein synthesis works efficiently, as demonstrated by a Coomassie blue stained gel (Fig. 3A, arrow). Densitometric quantitation, as well as a direct determination of purified dhfr, revealed that the reaction proceeded for up to 60 hrs, yielding 4 mg of enzyme in 1 ml reaction (Fig. 3C). This yield was achieved when the system was supplemented with fresh mRNA every 24 hrs; without the addition of fresh mRNA the reaction ceased after 24 hrs and yielded 1 mg of dhfr (Fig. 3B and open circles in C). When aliquots of the reaction mixtures were withdrawn after 48 and 60 hrs and then incubated in the presence of a low dose of cycloheximide for an additional 1 hr, sucrose gradient centrifugation revealed polysome formation (Fig. 2C, d, e). This is a direct indication of a robust system with high translational activity. The product has a similar specific activity as the authentic enzyme, 15.3 vs. 19.1 units/mg [9].

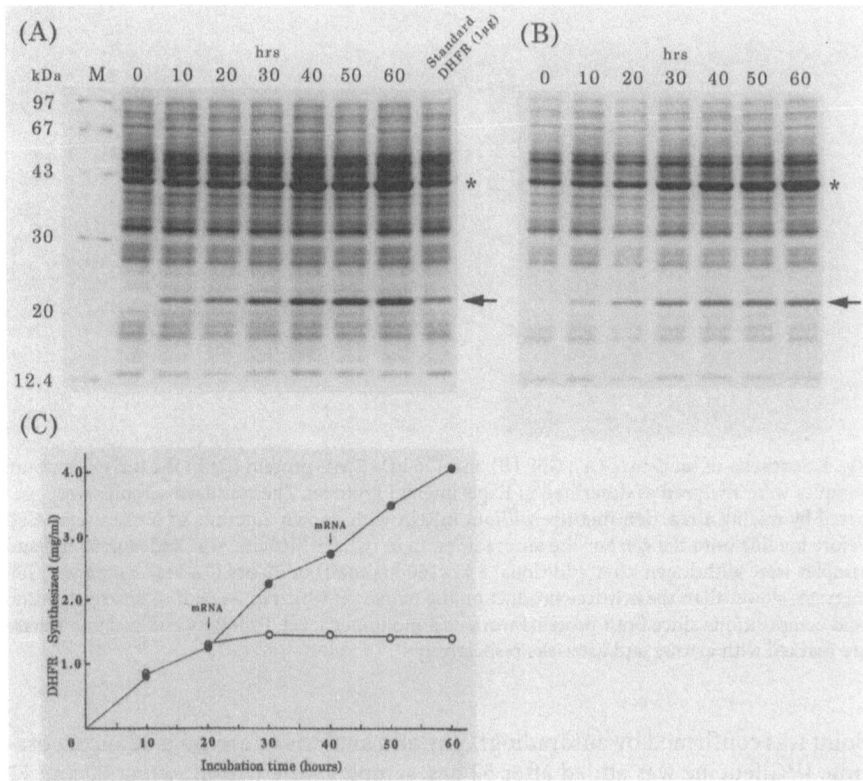

Fig. 3. Protein synthesis in the dialysis system. Upper panels: Coomassie blue stained SDS poly-acrylamide gels showing dhfr synthesis with (A) or without (B) addition of new mRNA. Arrows and asterisks mark dhfr and creatine kinase, respectively. The standard sample was prepared by mixing a reaction mixture without mRNA with known amounts of dhfr before loading onto the gel. Panel C: Amounts of dhfr synthesized as determined from densitometric scans of the gels in A (closed circles) and B (open circles)

To see the efficiency of the system, we next performed syntheses of larger proteins. Coding sequences for GFP (717 nts) [31] and luciferase (1650 nts) were cloned into the above plasmid in such a way that the 5'- and 3'-UTR of dhfr were preserved. Capped TMV RNA (6388 nucleotides) was transcribed from lin-earized plasmid pTLW3 carrying the genome under T7 RNA polymerase pro-moter control [32]. As shown in Figure 4, the system also synthesized proteins of higher molecular weight in a preparative scale: 1.1 mg of luciferase (65 kDa), 1.2 mg of GFP (27 kDa). These proteins have the same or even higher specific activity compared with commercially available recombinant forms (Fig. 4A, B; see Experimental Protocols). Furthermore, the 126 kDa replicase of TMV, a major genome product [33] during infection, was produced with a yield of as much as 0.6 mg (Fig. 4). The synthesis proceeded for up to 72 hrs as shown by the increase in intensity of the Coomassie brilliant blue stained bands. This

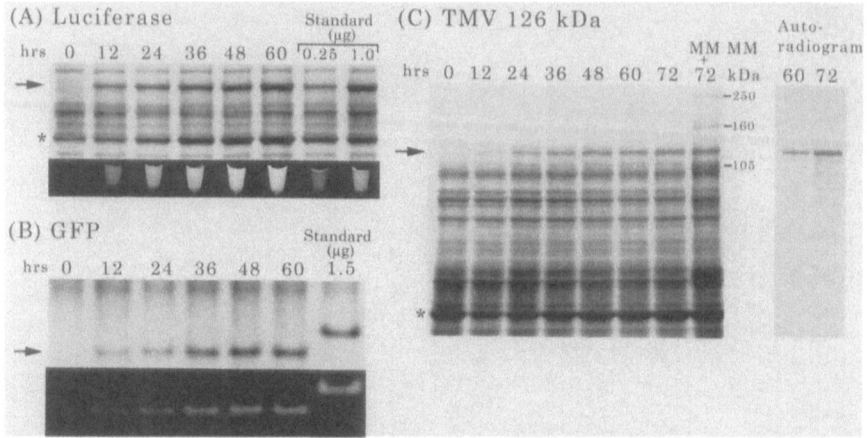

Fig. 4. Synthesis of luciferase (A), GFP (B) and 126 kDa TMV protein (C) in the dialysis system. Samples were analyzed as described in Experimental Protocol. The standard samples were prepared by mixing a reaction mixture without mRNA with known amounts of luciferase or GFP before loading onto the gel. For the autoradiogram in (C), [^{14}C]leucine was added at 52 hrs and samples were withdrawn after additional 8 hrs (60 hrs total) or 20 hrs (72 hrs). Authentic GFP migrates slower than the cell-free product on the native gel which as a result of different amino acid compositions since both proteins work as a monomer form. Products and creatine kinase are marked with arrows and asterisks respectively

point was confirmed by autoradiography and analysis of amino acid incorporation: [^{14}C]leucine was added after 52 hrs, samples were withdrawn at 60 and 72 hrs, and the samples analyzed by SDS gel electrophoresis and autoradiography (Fig. 4). Densitometric quantitation of the bands showed linear synthesis: the photostimulated luminescene (PSL) of the sample after 8 hrs of synthesis (at the 60 hr time point) was 186, and after 20 hrs (at the 72 hr point) was 465, even though the rate of protein synthesis, as measured by leucine incorporation, was 21% of the rate at the beginning of incubation. This is more direct evidence of the robustness of the system and its efficiency in synthesizing even a 126 kDa protein for 3 days.

Discussion

We show here that removal of contaminated endosperm, which contains protein synthesis inhibitor(s), from the embryo fraction improves protein synthesis in a wheat germ cell-free system. The improvement is likely to be through the increased translational activity resulting from elimination of inhibitors of the translation (e.g. the thionins), and ribonucleases, as well as the RIP tritin. It is generally believed that cell-free translation systems are inherently unstable, but our results demonstrate the opposite: the translational apparatus appears to be very stable, *in vitro* and presumably also *in vivo*. We believe that our results shed light on the biological function of the nearly ubiquitous plant RIPs. We propose that during evolution plants acquired a suicide system useful to prevent even more damage, and that because of its stability the translational machinery is the most important target of a suicide system. Viral attack would be one instance in which this suicide mechanism is employed. Ribosomes are also a popular target for antibiotics, emphasizing their central role in cell metabolism. The observed high stability of the translational apparatus might be an essential requirement for the evolution of life: certain basic physiological processes such as protein synthesis might be required to function even in adverse conditions.

The structures of 5'- and 3'-untranslated regions are important for the efficiency of initiation and termination and also for the stability of mRNA [34]. The mRNA constructs used here were not optimized in this respect, and we believe that the yields in our experiments do not therefore necessarily reflect maximum capacity. Efficient mRNA translation and its regulation requires a series of protein-mRNA and protein-protein interactions [34], and Wells *et al.* have recently shown the circularization of mRNA *in vitro* [35]. Our method thus provides, in addition to its protein synthesis capacity, an opportunity to study translation itself, including the phenomenon of circular mRNA or the characterization of untranslated regions of mRNA in terms of efficient initiation or stability.

The strategy which we followed to improve the wheat cell-free system, i. e. the elimination of the translational suicide system, is applicable to other systems as well. For instance, the widely used cell-free system from *E. coli* contains high ribonuclease activity, and is hampered by a low efficiency in the translation of large mRNAs. *E. coli* systems are limited when selecting large polypeptides for polysome display because of significant levels of template degradation.

In conclusion, our protein synthesis system has several advantages compared with existing systems, in addition to its high efficiency: Whereas in a prokaryote such as *E. coli* additional steps are crucial for the folding after the synthesis, in an eukaryote cotranslational folding has been proposed by Netzer and Hartl [36]. Hence, our system as an eukaryotic system is more amenable to the production of eukaryotic proteins, which are generally composed of more domains in structurally folded and functionally active forms; the system can produce high molecular weight proteins; because of little template degradation it is useful for polysome display [37]; and proteins that would normally interfere with cell physiology can be synthesized. Additionally, it should be a useful tool in the study of translation itself.

References

1. Golf SA, Goldberg A L (1987) An increased content of protease La, the lon gene product, increases protein degradation and blocks growth in *E. coli*. J Biol Chem 262: 4508–4515
2. Chrunyk BA, Evans J, Lillquist J, Young P, Wetzel R (1993) Inclusion body formation and protein stability in sequence variants of interleukin-1β. J Biol Chem 268: 18053–18061
3. Henrich B, Lubitz W, Plapp R (1982) Lysis of *E. coli* by induction of cloned φX174 gene. Mol Gen Genet 185: 493–497
4. Kurland C G (1982) Translational accuracy *in vitro*. Cell 28: 201–202
5. Pavlov MY, Ehrenberg M (1996) Rate of translation of natural mRNA in an optimized *in vitro* system. Arch Biochem Biophys 328: 9–16

6. Roberts BE, Paterson BM (1973) Efficient translation of tobacco mosaic virus RNA and rabbit globin 9S RNA in a cell-free system from commercial wheat germ. Proc Natl Acad Sci USA 70: 2330–2334
7. Spirin AS, Baranov VI, Ryabova LA, Ovodov Syu, Alakhov YuB (1988) A continuous cell-free tranlation system capable of producing polypeptides in high yield. Science 242: 1162–1164
8. Baranov VI, Morozov IYu, Ortlepp SA, Spirin AS (1989) Gene expression in a cell-free system on the preparative scale. Gene 84: 463–466
9. Endo Y, Otsuzuki S, Ito K, Miura K (1992) Production of an enzymatic active protein using a continuous flow cell-free translation system. J Biotechnol 25: 221–230
10. Kigawa T, Yokoyama S (1991) A continuous cell-free protein synthesis system for coupled transcription-translation. J Biochem 110: 166–168
11. Kim DM., Kigawa T, Choi CY, Yokoyama S (1996) A highly efficient cell-free protein synthesis system from *E. coli*. Eur J Biochem 239: 881–886
12. Kigawa T, Yabuki T, Yoshida Y, Tsutsui M, Ito Y, Shibata T, Yokoyama S (1999) Cell-free production and stable-isotope labeling of milligram quantities of protein. FEBS Lett 442: 15–19
13. Kawarasaki Y, Kawai T, Nakano H, Yamane T (1995) A long-lived batch reaction system of cell-free protein synthesis. Anal Biochem 226: 320–324
14. Endo Y, Mitsui K, Motizuki M, Tsurugi K (1987) The mechanism of action of ricin and related toxic lectins on eukaryotic ribosomes. J Biol Chem 262: 5908–5912
15. Endo Y, Tsurugi K (1987) RNA N-glycosidase activity of ricin A-chain. J Biol Chem 262: 8128–8130
16. Wool IG, Glück A, Endo Y (1992) Ribotoxin recognition of ribosomal RNA and a proposal for the mechanism of translation. Trends Biochem Sci 17: 266–269
17. Barbieri L, Battelli MG, Stirpe F (1993) Ribosome-inactivating proteins from plants. Biochim Biophys Acta 1154: 237–282
18. Taylor S, Massiah A, Lomonossoff G, Robert LM, Lord JM, Hartely MR (1994) Correlation between the activities of five ribosome-inactivating proteins in depurination of tobacco ribosomes and inhibition of tobacco mosaic virus infection. Plant J 5: 827–853
19. Ready MP, Brown DT, Robertus JD (1986) Extracellular localization of pokeweed antiviral protein. Proc Natl Acad Sci USA 83: 5053–5056
20. Massiah AJ, Hartely MR (1995) Wheat ribosome-inactivating proteins: Seed and leaf forms with different specificities and cofactor requirements. Planta 197: 633–640
21. Ogasawara T, Sawasaki T, Morishita R, Ozawa A, Madin K, Endo Y (1999) A new class of enzyme acting on damaged ribosomes: ribosomal RNA apurinic site specific lyase found in wheat germ. The EMBO J 18: 6522–6531
22. Madin K, Sawasaki T, Ogasawara T, Endo Y (2000) A highly efficient and robust cell-free protein synthesis system prepared from wheat embryos: Plants apparently contain a suicide system directed at ribosomes. Proc Natl Acad Sci USA 97: 559–564
23. Johnston FB, Stern H (1957) Mass-isolation of viable wheat embryos. Nature 179: 160–161
24. Erickson AH, Blobel G (1983) Cell-free translation of messenger RNA in a wheat germ system. Meths Enzymol 96: 38–50
25. Hirao I, Yoshinari S, Yokoyama S, Endo Y, Ellington AD (1997) RNA aptamers that bind to and inhibit the ribosome-inactivating protein, pepocin. Nucleic Acids Symp Ser 37: 283–284
26. Lodish HF, Housman D, Jacobsen M (1971) Initiation of haemoglobin synthesis. Specific inhibition by antibiotics and bacteriophage ribonucleic acid. Biochemistry 10: 2348–2356
27. Bohlmann H (1994) The role of thionins in plant protection. Critical Rev in Plant Sciences. 13: 1–16
28. Colilla FJ, Rocher A, Mendez E (1990) Gamma-purothionins: amino acid sequence of two polypeptides of a new family of thionins from wheat endosperm. FEBS Lett 270: 191–194
29. Brummer J, Thole H, Kloppstech K (1994) Hordothionins inhibit protein synthesis at the level of initiation in the wheat germ system. Eur J Biochem 219: 425–433
30. Matsushita S (1959) On the protein formation and changes of the amounts of the ribonucleic acid and ribonuclease activity in the grains during the ripening process of wheat. Memoirs of the Res. Inst. for Food Science, Kyoto Univ. No. 19: 1–4
31. Chiu WL, Niwa Y, Zeng W, Hirano T, Kobayashi H, Sheet J (1996) Engineered GFP as a vital reporter in plants. Current Biology 6: 325–330

32. Hamamoto H, Sugiyama K, Nakagawa N, Hashida E, Matsunaga Y, Takemoto S, Watanabe Y, Okada Y (1999) A new tobacco mosaic virus vector and its use for the systemic production of angiotensin-1-covering enzyme inhibitor in transgenic tobacco and tomato. Biotechnology 11: 930–932
33. Dawson WO (1992) Tobamovirus-plant interaction. Virology 186: 359–367
34. Sachs AB, Sarnow P, Hentze M W (1997) Starting at the beginning, middle, and end: translation initiation in eukaryotes. Cell 89: 831–838
35. Wells SE, Hillner PE, Vale RD, Sachs AB (1998) Circularization of mRNA by eukaryotic translation initiation factors. Molecular Cell 2: 135–140
36. Netzer WJ, Hartl FU (1997) Recombination of protein domains facilitated by co-translational folding in eukaryotes. Nature 388: 343–349
37. Mattheakis LC, Bhatt RR, Dower WJ (1994) An *in vitro* polysome display system for identifying ligands from very large peptide libraries. Proc Natl Acad Sci USA 91: 9022–9026

Matrix Reactor: A New Scalable Reactor Principle for Cell-Free Protein Expression

10

BERND BUCHBERGER, WOLFGANG MUTTER, ALBERT RÖDER*

Introduction

For a long time, cell-free protein synthesis was more or less an analytical method due to low synthesis rate and productivity. However, several key improvements over the last decades have lead to protocols providing real preparative amounts of protein:

1. Coupling of transcription/translation using DNA templates and exogenous phage RNA-polymerases (Zubay 1973, Spirin 1992, Craig et al. 1992)
2. Reactor technologies using continuous supply of substrates/continuous removal of products: Continuous exchange cell-free (CECF) or continuous flow cell-free (CFCF) systems by Spirin et al. (1988)
3. Optimization of biochemistry of lysate systems (Baranov and Spirin 1993, Kim et al. 1996, Kim and Swartz 1999, Madin et al. 2000).

Consequently, modern systems based on *Escherichia coli* lysate or wheat germ lysate provide yields up to 5 mg/ml within 1 or 3 days, respectively (Kim et al. 1996, Madin et al. 2000).

While productivity of the in vitro systems reach levels close to cellular systems, scalability still remains a problem in two ways. For small volumes (10–100 µl), CECF or CFCF technology could hardly be applied because suitable reaction devices are not available. On the other hand, large-scale reactions (10–1000 ml) depend on specially designed reactors composed of membrane surfaces and multiple chambers. These kinds of reactors are difficult to construct and are expensive.

The aim of this work was to establish a new reactor principle, which combines the advantages of CECF (or CFCF) and which is easy to scale from microliter to liter dimensions.

* Bernd Buchberger, Wolfgang Mutter, Albert Röder
  Roche Molecular Biochemicals, Nonnenwald 2, 82372 Penzberg, Germany

Matrix Reactor Principle

Basically, CECF/CFCF reactors are composed of two or more compartments separated by one or several porous membrane(s). A CECF reactor consists of a reaction compartment and a feeding compartment divided by a semi-permeable membrane. Critical parameters are the surface area of the porous membrane relative to the reaction volume and diffusion distances within the individual compartments. Both could be limiting for an efficient supply of substrates and the efficient removal of reaction by-products.

The matrix-reactor concept compares to the CECF principle (Fig. 1), although it differs completely in reactor design. The reactor volume is totally or partially filled with a porous, spherical matrix. This matrix, which is characterized by a defined exclusion molecular weight, divides the reactor space into an inner matrix space (IMS), which can only be penetrated by small molecules (MW <5 kDa) and a external matrix space (EMS), where both large and small molecules are distributed. Projected to the functional entities of a CECF reactor, the EMS compares to the reaction compartment, where all the macromolecular constituents of the transcription/translation machinery (e.g., ribosomes, mRNA, tRNAs, RNA polymerase, enzymes, factors, etc.) and low-molecular-weight substances (buffer, salt, substrates, cofactors, etc.) are present. The IMS compares to the feeding compartment, a reservoir for substrates and a diffusion space for potentially inhibitory reaction by-products.

Two advantages of using a three-dimensional matrix space instead of a membrane-type reactor are obvious. First, the exchange area between the reaction compartment (EMS) and feeding compartment (IMS) is very large and can be easily varied with particle size. Second, the diffusion distances within and between the compartments are very small (again dependent on particle size). Both parameters ensure that mass exchange between IMS and EMS are, at any time, close to equilibrium.

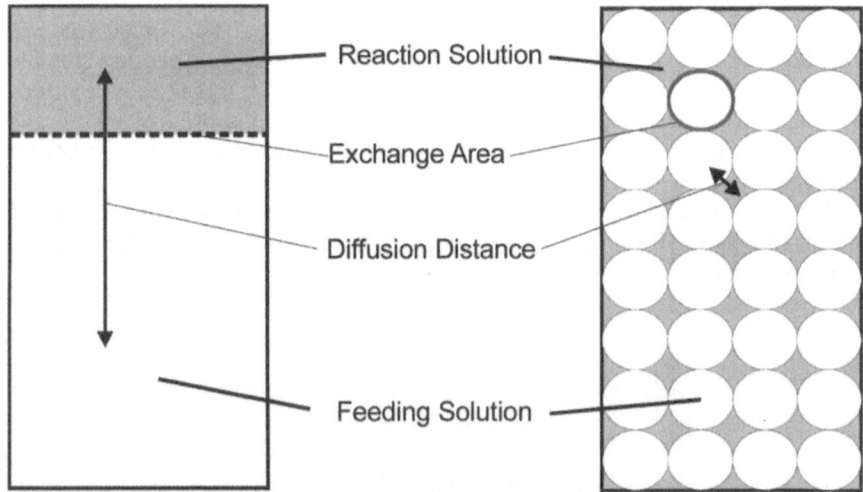

Fig. 1. Comparison of matrix reactor principle with a CECF-type reactor

Methods

Vector pIVEX2.3 GFP

Green fluorescent protein (GFP) from *Aequoria victori* in the form of a mutant GFPcycle3 (27 kDa) (Crameri et al. 1996) was cloned into a pIVEX2.3 cloning vector (Roche Diagnostics, Mannheim, Germany).

E. coli S30 Lysate

The lysate was prepared using an *E. coli* A19 strain by a modified method according to Zubay (1973). Modified lysate buffer: 100 mM HEPES-KOH pH 7.6/30°C, 14 mM magnesium acetate, 60 mM potassium acetate, 0.5 mM dithiothreitol. Differences in the yields in the various experiments are due to the different lysate preparations.

Reaction Solution

The reaction solution was 185 mM potassium acetate, 15 mM magnesium acetate, 4% glycerol, 2.06 mM ATP, 1.02 mM CTP, 1.64 mM GTP, 1.02 mM UTP, 257 µM of each amino acid (all 20 naturally occurring amino acids), 10.8 µg/ml folinic (acid), 1.03 mM EDTA, 100 mM HEPES-KOH pH 7.6/30°C, 1 µg/ml rifampicin, 0.03% sodium azide, 40 mM acetyl phosphate, 480 µg/ml tRNA from *E. coli* MRE600, 2 mM dithiothreitol, 10 mM mercaptoethane sulfonic acid, 70 mM KOH, 0.1 U/µl RNase inhibitor, 15 µg/ml plasmid, 220 µl/ml *E. coli* A19 lysate, 2 U/µl T7-RNA polymerase.

Feeding Solution

The feeding solution was 185 mM potassium acetate, 15 mM magnesium acetate, 4% glycerol, 2.06 mM ATP, 1.02 mM CTP, 1.64 mM GTP, 1.02 mM UTP, 257 µM of each amino acid (all 20 naturally occurring amino acids), 10.8 µg/ml folinic (acid), 1.03 mM EDTA, 100 mM HEPES-KOH pH 7.5/30°C, 1 µg/ml rifampicin, 0.03% sodium azide, 40 mM acetyl phosphate, 2 mM dithiothreitol, 10 mM mercaptoethane sulfonic acid, 70 mM KOH.

Equilibrated Matrix

To equilibrate the matrix, Sephadex G25, was hydrated and washed according to the manufacturer's instructions. The swollen material was re-suspended three times in a threefold volume of feeding buffer (10 min each) and sucked dry under a slight underpressure to remove the buffer.

Buffer A

Buffer A was 100 mM HEPES-KOH pH 7.6/30°C, 14 mM magnesium acetate, 60 mM potassium acetate, 0.5 mM dithiothreitol.

GFP determination

GFP requires the presence of adequate amounts of oxygen to form the fluorophore. Hence, all samples were "matured" for 12–32 h at 4°C before measurement. The samples were measured using a spectral fluorimeter from the Kontron Instruments (Lohof, Germany), type Tegimenta SFM-25. The excitation was at a wavelength of 395 nm; the emission was determined at a wavelength of 510 nm. Recombinant GFP from Roche Molecular Diagnostics (Mannheim, Germany) was used as the standard. For the measurement, samples were diluted 1:50 to 1:400, depending on the concentration with 100 mM HEPES-KOH pH 7.6/30°C, 14 mM magnesium acetate, 60 mM potassium acetate, 0.5 mM dithiothreitol.

Results

Matrix-Fed Small-Scale Reaction

Standard batch reactions are known to be limited by the availability of substrates and the accumulation of inhibitory reaction by-products. Reaction times are limited to 1–2 h, and the yields could only reach levels of several hundred µg/ml. We used an optimized *E. coli* lysate system, according to Spirin, to evaluate the effect of Sephadex G25 beads, equilibrated with feeding buffer, on the kinetics and yield of small-scale reactions. To optimize the amount of matrix, we added between 0 mg and 150 mg of the equilibrated Sephadex G25 to various volumes of standard reaction (Fig. 2). The reaction was performed at 30°C for 5 h using pIVEX2.3 GFP as a template vector. For analysis, all samples were diluted to a fixed volume (500 µl) with Buffer A. The synthesized GFP was quantified via fluorescence. It was observed that with an increasing ratio of matrix, the yield improves, reaching a plateau at 100 mg matrix per 50 µl. Compared to the control (standard batch reaction), the improvement in this experiment was about 2.5-fold (64 µg/ml to 160 µg/ml). Analyzing synthesis rate, we find the batch reaction shows almost linear reaction for about 2 h, whereas the matrix-fed reaction shows an extended synthesis time of about 6 h (Fig. 3). The initial rate of the matrix-fed reaction is comparable to the batch system.

Matrix Column Reactor

The above described matrix-fed reaction is still a static system (like batch or CECF) because the ratio of reaction to feeding is fixed, and the conditions are con-

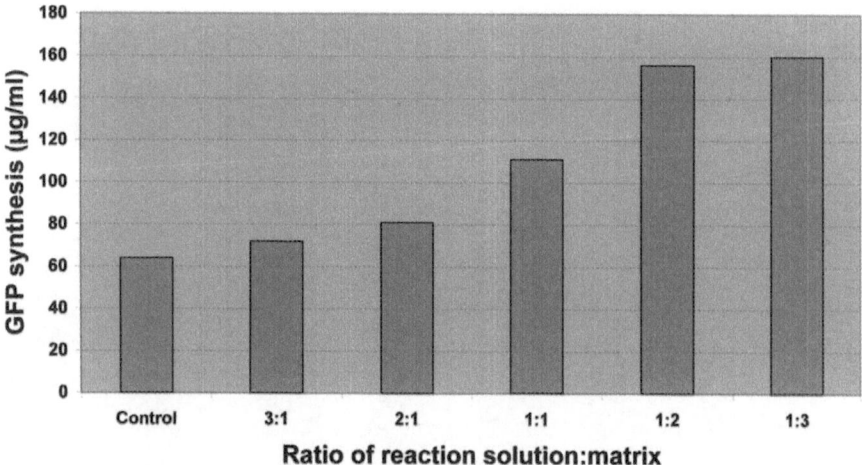

Fig. 2. Application of different amounts of equilibrated matrix compared to batch reaction

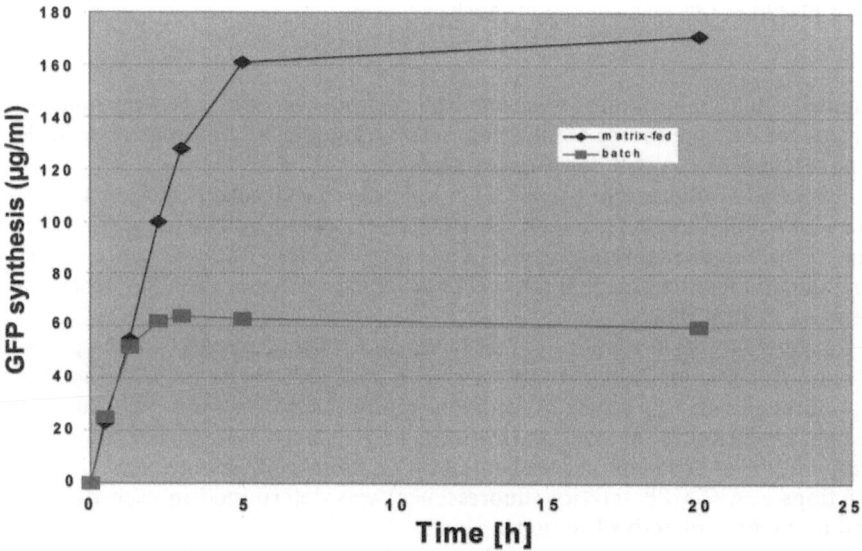

Fig. 3. Kinetics of a matrix-fed reaction compared with a batch reaction

tinuously changing until the point where the reaction stops. Extending the concept of matrix reactor to more steady-state-like conditions, a column approach was used. Commercially available PD10 columns were equilibrated with three volumes of feeding solution at 30°C. A 1-ml reaction solution, activated with pIVEX2.3 GFP template vector, was applied to each of six columns. In a stepwise process, the reaction volume was washed through the columns; every 60 min 0.65 ml Buffer A was

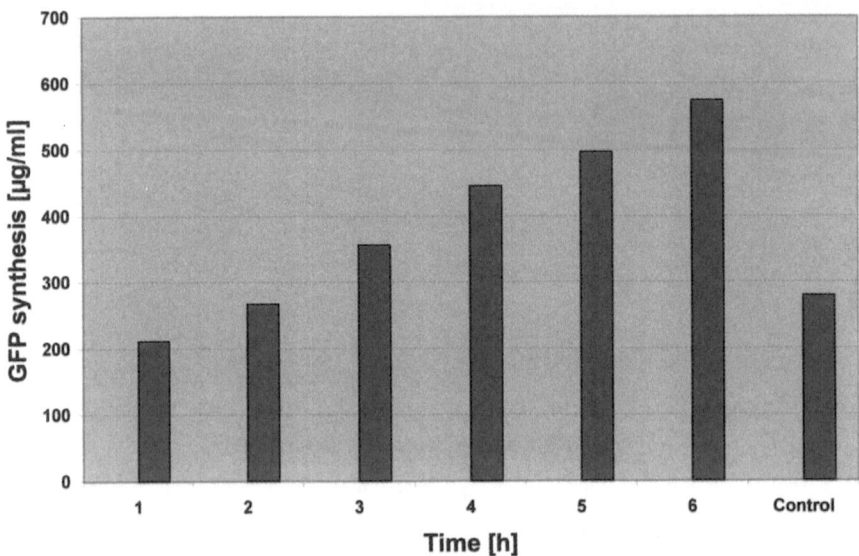

Fig. 4. Kinetics of GFP synthesis on a PD10 column reactor

applied. One of the columns was completely eluted with Buffer A, every hour for 6 h, to create a time course of the synthesis reaction (Fig. 4). As a control, a CECF-type reactor was run with the same reagents.

Synthesis of fluorescent active GFP within the PD10 column showed a linear increase within the 6-h period. In this experiment, the overall yield of GFP in the matrix reactor was substantially higher compared to the CECF reactor (control). One should mention here that the optimum reaction time for the CECF reactor is between 15 and 20 h.

To show further scalability of this method, a 200-ml column (double-jacked column dimension: 2,6 cm × 40 cm) was filled with Sephadex G25 fine and equilibrated with 1.5 volume of feeding solution (12 ml/h). We applied 20 ml of reaction solution to the column (60 ml/h) and eluted at a rate of 12 ml/h. The temperature of the column was kept at 30°C; the eluate was collected in 4-ml fractions at 4°C. GFP activity (fluorescence) was determined in each fraction and in the pool of active fractions.

The elution profile of this run is shown in Fig. 5. In the active fractions (18–25, approximately 27 ml) a GFP concentration of 563 μg/ml was determined. Recalculated to the starting volume of 20 ml, this means a concentration of 759 μg/ml, which is a total protein synthesis of 15.2 mg of GFP. Compared to the control reaction (CECF), the yield of the column reactor in this 20/200 ml scale is approximately 130%.

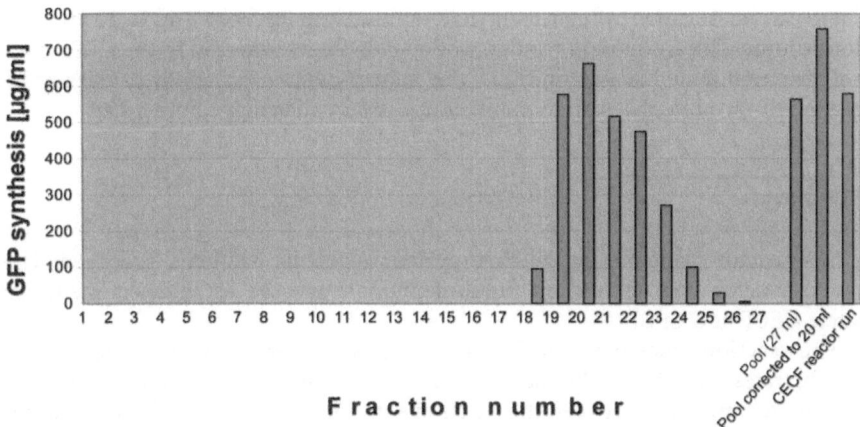

Fig. 5. GFP synthesis on a 200-ml column. Reaction volume 20 ml

Discussion

The key to substantially prolonged cell-free protein synthesis reactions is a continuous supply of substrates and a continuous removal of by-products without dilution of the reaction solution. This is normally archieved in heterogeneous systems using membranous barriers, where the reaction solution is locked into a fixed reactor space, corresponding to a stationary phase, which is then perfused with or dialyzed against the feeding solution. In contradiction, the matrix reactor uses the inverse principle. The three-dimensional porous matrix excludes the catalytic machinery from major parts of the entire reactor space. Therefore, in the matrix reactor, the reaction solution is equivalent to the mobile phase, free to move around the stationary feeding particles.

The matrix-fed reaction is still a static system, comparable to the established CECF principle. Using Sephadex materials, the ratio of EMS:IMS is about 0.4, which means that the system has an extra substrate capacity of 2.5-fold compared to standard batch reaction, if whole EMS is occupied by reaction solution. This is exactly what the experiment showed, where the maximum increase in yield was about 2.5-fold. In addition, this value was achieved close to the optimal ratio of 1:2.5 (reaction solution:matrix). Depending on the nature of matrix, the EMS:EMS-ratio can be different (e.g., polyacrylamide beads, data not shown). On the other hand, this parameter can also be influenced by mixing identical materials with different particle sizes. From a practical point of view, the matrix-fed reaction is a very simple approach to increase the productivity of a given expression system without the need of special reaction devices or instrumentation.

The matrix column reactor is a non-equilibrium system, where the high-molecular-weight reaction components are constantly moved through the column. Comparable to CFCF reactors, fresh substrates are continuously supplied, and reaction by-products are removed, as long as they are of low molecular weight. With variation of pumping rate, the exchange factor can be optimized.

There was no negative influence on yield with scaling up from 1 ml to 20 ml reaction volume. The amount of product was even higher compared to the CECF control reaction. One can imagine that the matrix-reactor principle can easily be applied to variable scales up to many liters, just by adapting column size.

Summary

A new reactor principle for cell-free protein synthesis has been proposed. The matrix-reactor concept differs fundamentally from the established methods: batch, CECF and CFCF.
1. The reaction takes place in the mobile phase (external matrix space, EMS), substrates are supplied by the stationary phase (inner matrix space, IMS).
2. The exchange area between the reaction solution and feeding solution is largely extended compared to membrane reactors, and the problem of membrane clogging is eliminated.
3. Short diffusion distance avoids the need for any sort of agitation.
4. It is a scalable process due to simple reactor conception (standard chromatography column and gel-filtration matrix).

References

Baranov VI, Spirin AS (1993) Gene Expression in Cell-free System on Preparative Scale. Methods Enzym. 217: 123–142
Craig D, Howell MT, Gibbs CL, Hunt T, Jackson RJ (1992) Plasmid cDNA-directed Protein Synthesis in a Coupled Eukaryotic in vitro Transcription-translation System. Nucleic Asids Res. 20: 4987–4995
Kim DM, Kigawa T, Choi CY, Yokoyama S (1996) A Highly Efficient Cell-free Protein Synthesis System from E. coli. Eur. J. Biochem. 239: 881–886
Kim DM, Swartz JR (1999) Prolonging Cell-free Protein Synthesis with a Novel ATP Regeneration System. Biotechnol. Bioeng. 66: 180–188
Madin K, Sawasaki T, Ogasawara T, Endo Y (2000) A Highly Efficient and Robust Cell-free Protein Synthesis System Prepared from Wheat Embryos: Plants Apparently Contain a Suicide System Directed at Ribosomes. Proc Natl Acad Sci USA 97: 559–564
Spirin AS (1991) Cell-free Protein Synthesis Reactor. In: Todd P, Sikdar SK, Beer M (eds) Frontiers in Bioprocessing II. American Chemical Society, Washington, DC, pp 31–43
Spirin AS, Baranov VI, Ryabova LA, Ovodov SY, Alakhov YB (1988) A Continuous Cell-free Translation System Capable of Producing Polypeptides in High Yield. Science 242: 1162–1164
Zubay G (1973) In vitro Synthesis of Protein in Microbial Systems. Annu Rev Genet 7: 267–287
Crameri A, Whitehorn EA, Tata E, Stemmer WPC (1996) Improved Green Fluorescent Protein by Molecular Evolution Using DNA Shuffling. Nature Biotechnology 14: 315–319

PART IV

Protein Folding in Cell-Free Translation System

Co-Translational Protein Folding in Prokaryotic and Eukaryotic Cell-Free Translation Systems

11

VYACHESLAV A. KOLB, AIGAR KOMMER, ALEXANDER S. SPIRIN*

Introduction

Native structure formation of newly synthesized proteins is a key problem relating to the cell-free translation technology that is aimed at the synthesis of biologically active products. At the same time, a lot of experimental data indicate that the folding of newly synthesized proteins differs significantly from the *in vitro* process observed by Anfinsen in his classical experiments on the refolding of ribonuclease (Anfinsen 1973) and studied in detail by many other workers. The difference results mainly from (i) the contribution of multiple cellular components, such as protein disulphide isomerase, peptidylprolyl isomerase, and molecular chaperones that catalyze or assist the folding of synthesized proteins (Gething and Sambrook 1992; Georgopoulos and Welch 1993; Ellis 1994; Hartl 1996; Fink 1999), and (ii) the co-translational mode of the folding, implying that the N-terminal part of a nascent peptide starts its folding as soon as it is synthesized and emerges from the ribosome prior to the formation of the entire polypeptide chain (see refs below). The existence of stepwise protein folding starting from the N-terminal section of a growing polypeptide and progressively proceeding towards the C-terminus was first proposed by Phillips et al. (1967) from theoretical analysis of the folding pattern of hen egg lysozyme. Later, several experimental approaches demonstrated co-translational formation of native or native-like structure of nascent polypeptides, both *in vitro* and *in vivo*. It was shown that growing polypeptides are able to interact with free subunits of the same protein, thus being involved in the assembly of quaternary structures on ribosomes (Kiho and Rich 1964; Gilmore et al. 1996). Several proteins were shown to bind their specific ligands or cofactors while bound to the ribosome as growing polypeptides (Mullet et al. 1990; Kim et al. 1991; Chen et al. 1995; Komar et al. 1997; Hanes and Plückthun 1997). Formation of correct S-S bridges was demonstrated to proceed during polypeptide elongation (Bergman and Kuehl 1979; Peters and Davidson 1982; Chen et al. 1995; Ryabova et al. 1997). Compact protease-resistant domains typical of mature proteins were

* Institute of Protein Research, Russian Academy of Sciences
 142290 Pushchino, Moscow Region, Russia
 e-mail: spirin@vega.protres.ru, telephone: 7(095) 924 04 93, fax: 7(095) 924 04 93

formed during elongation for a number of nascent polypeptides (Frydman et al. 1994; Netzer and Hartl 1997; Frydman et al. 1999; Lin et al. 1998). Finally, it was found that proteins can display their enzymatic activity while still attached to the ribosome (Kolb et al. 1995; Kudlicki et al. 1995; Makeyev et al. 1996; Nicola et al. 1999; Kolb et al. 2000) or immediately upon releasing from it (Kolb et al. 1994; Kolb et al. 2000).

In our study, the folding kinetics of a multidomain protein, *Photinus pyralis* luciferase, during its synthesis in bacterial and eukaryotic cell-free translation systems, was explored and compared with the kinetics of denatured luciferase refolding. A sensitive assay for the luciferase enzymatic activity made possible a continuous recording of the luminescence produced by newly synthesized luciferase and thus the detection of the active (i.e. properly folded) molecules as soon as they were formed in a translation system supplied with the luciferase substrates, luciferin and ATP (Kolb et al. 1994). To measure the duration of the period when full-length enzyme released from the ribosome attains its native structure, continuous recording of luciferase luminescence was applied, together with injection of translation inhibitors during the enzyme synthesis. The delay between the arrest of translation and the cessation of enzyme activity accumulation is the duration of the post-translational phase of luciferase folding. It was found that the arrest of luciferase synthesis results in immediate cessation of active enzyme accumulation both in bacterial and eukaryotic cytosols. This suggests that the enzyme acquires its native structure immediately upon release from the ribosome, without any delay for post-translational folding. Thus the result reveals the co-translational mode of luciferase folding, irrespective of whether the protein is synthesized in eukaryotic or prokaryotic translation systems.

Synthesis of active firefly luciferase in bacterial and eukaryotic cell-free translation systems

Two constructs of luciferase mRNA prepared by *in vitro* transcription with T7 and SP6 polymerases were used to synthesize firefly luciferase in cell-free translation systems. Both mRNAs contained the identical coding sequence for luciferase with its natural stop codon and differed only in their untranslated 5'-regions comprising the ribosome binding site (Kolb et al. 1994; Kolb et al. 2000). Continuous monitoring of the enzymatic activity of luciferase in the course of translation was performed by placing translation reactions into a luminometer cell at 25°C (Kolb et al. 1994). Luciferin and ATP, the substrates of luciferase, were present in the mixtures from the beginning of incubation, allowing the newly synthesized enzyme to display its activity. The light emitted as a result of luciferase-catalyzed reaction was recorded to measure the amount of enzymatically active luciferase throughout the experiment. The translation was initiated by addition of luciferase mRNA. Kinetics curves of active luciferase accumulation are shown in Figures 1c and d. Detectable activity appeared after 8 min incubation in the bacterial translation mixture or after 18 min in the wheat germ extract. The activity accumulation curve reached a plateau by the 40th min and

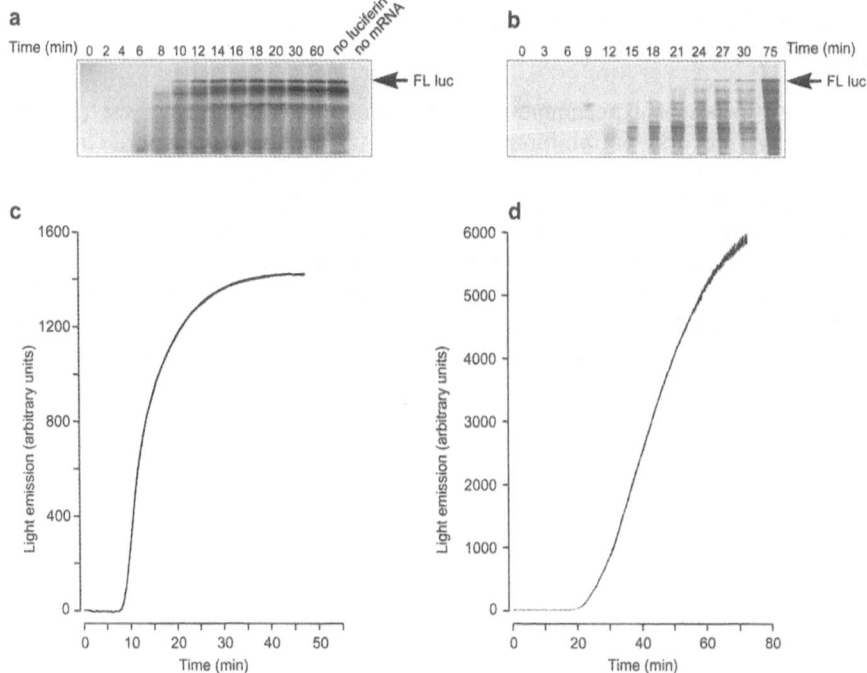

Fig. 1. Time course of luciferase mRNA translation in cell-free systems. Translation was performed at 25°C in the presence of 0.1 mM luciferin, 0.2 mM coenzyme A and [^{35}S]methionine. Composition of *E. coli* S30 translation system (*a* and *c*) was the same as in Kolb et al. 2000, composition of wheat germ translation system (*b* and *d*) is described in Kolb et al. 1994. *a, b,* time course of full-length luciferase accumulation as revealed by 15% SDS-PAGE and autoradiography. 5-μl aliquots were removed after indicated incubation time, treated with 50 mM NaOH for 20 min at 37°C and subjected to electrophoresis. Lanes representing control samples devoid of mRNA and luciferin and incubated for 60 min are marked correspondingly. The position of full-length luciferase (FL luc) is pointed out by arrowheads; *c, d,* kinetics of luciferase activity accumulation in the same translation mixtures as revealed by continuous monitoring of luciferase luminescence

70th min of incubation in the bacterial and eukaryotic translation reactions respectively.

Aliquots from the same incubation mixtures were analyzed by 15 % SDS-PAGE with subsequent autoradiography. It is seen in Figures 1a and b that the full-length luciferase can be detected as an intensive band in the gel by the 10th min of incubation of the bacterial translation reaction or by the 24th min of incubation of the wheat germ translation mixture. Thus, in both cases, the emergence of light-emitting activity correlates with the appearance of the full-length polypeptide. The correlation indicates that there is no significant lag between the formation of the complete luciferase chain and the development of luciferase activity, and thus the protein has acquired its native conformation mainly during translation in both types of translation reactions.

Cessation of active luciferase accumulation
upon arrest of translation

In the next series of experiments, translation of luciferase mRNA was rapidly arrested by injection of inhibitors to the incubation mixture producing the enzyme at maximal rate (the linear region of the active luciferase accumulation curve). The rapid block in translation prevented the further formation of luciferase, so that a further increase in enzymatic activity would reflect the folding of the protein that had already been released from the ribosome (Kolb et al. 1994; Fedorov and Baldwin 1995; Kolb et al. 2000). As is evident from Fig. 2, the addition of such inhibitors as RNase A or thiostrepton to the translation mixtures results in abrupt cessation of the increase in luciferase activity, identically in bac-

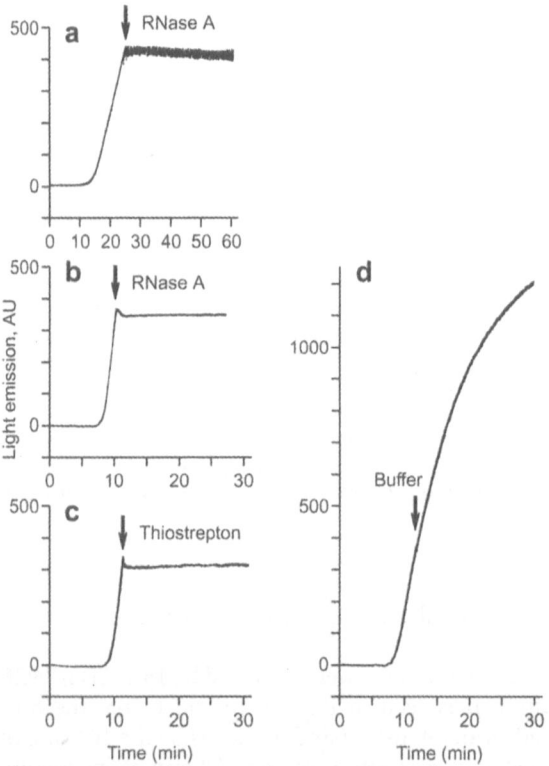

Fig. 2. Time course of luciferase activity accumulation upon arrest of translation. Luciferase synthesis in cell-free translation systems was arrested by injection of either RNase A to final concentration of 0.8 mg/ml, or thiostrepton to concentration of 20 µM. Translation reactions were carried out as described in the legend to Fig.1, except that only unlabeled amino acids were added to the reactions. *a*, injection of RNAase into *E. coli* S30 translation system; *b*, injection of RNase into wheat germ translation system; *c*, injection of thiostrepton into wheat germ translation system; *d*, control with injection of buffer into wheat germ translation system

terial and eukaryotic cell-free systems. The addition of a buffer instead of the inhibitors results only in a slight decrease of the slope of the activity accumulation curve, reflecting the dilution of the mixture (Fig. 2d). The immediate halt of active accumulation upon addition of the translation inhibitors indicates that all luciferase molecules released from the ribosome already had the active conformation.

Luciferase synthesis in the cell-free translation system enriched with chaperones

Similar experiments on translation inhibition by thiostrepton were carried out with bacterial reaction mixtures containing an excess of purified chaperones. The following three combinations of chaperones were used: either 90 nM GroEL 14-mer with 180 nM GroES 7-mer (Fig. 3a), or 1.3 µM DnaK with 200 nM DnaJ and 650 nM GrpE dimer (Fig. 3b), or both the GroEL and the DnaK chaperone systems added together at concentrations listed above (Fig. 3c). The concentration of newly synthesized full-length luciferase in these experiments was estimated to be 1.2 to 2.5 nM. Although chaperones were added in large excess in relation to the newly synthesized polypeptide, the identical patterns of abrupt cessation of active luciferase accumulation were observed in all the three variants of the translation system enriched with chaperones (Fig.3). This result rules out the possibility that the immediate halt in active luciferase accumulation (and thus, in the co-translational mode of the enzyme folding) is because of scarcity of chaperones in the bacterial S30 extracts used.

Refolding of denatured luciferase in prokaryotic and eukaryotic cytosols

In the refolding experiments, a commercial preparation of *Photinus pyralis* luciferase, showing a single protein band with M_r 62K when analysed by SDS-PAGE was used. Refolding of luciferase denatured with buffered 7.8 M urea was performed by 400-fold dilution with translation mixture containing no mRNA. Thus, the refolding occurred at the same conditions and in the presence of the same components, except mRNA, as the folding during translation. The concentration of luciferase in the refolding experiments was also within the same range as in the cell-free translation system. The course of refolding was recorded in a luminometer cell by recovery of luciferase activity as a function of time. Samples of non-denatured active luciferase diluted with the corresponding translation mixtures were taken as controls. The refolding of luciferase in wheat germ cytosol proceeded with the same efficiency, and the same average half-time of 14 min (Fig.4a) as was reported earlier (Kolb et al. 1994). In contrast, the refolding in *E. coli* cytosol was inefficient (Fig.4b), with about 8 % activity recovered after one hour of incubation (a plateau level; data not shown). The duration of the protein incubation under denaturing conditions within time intervals from 10 s to 10 min had no effect on the active enzyme recovery (data not shown). This indicates that

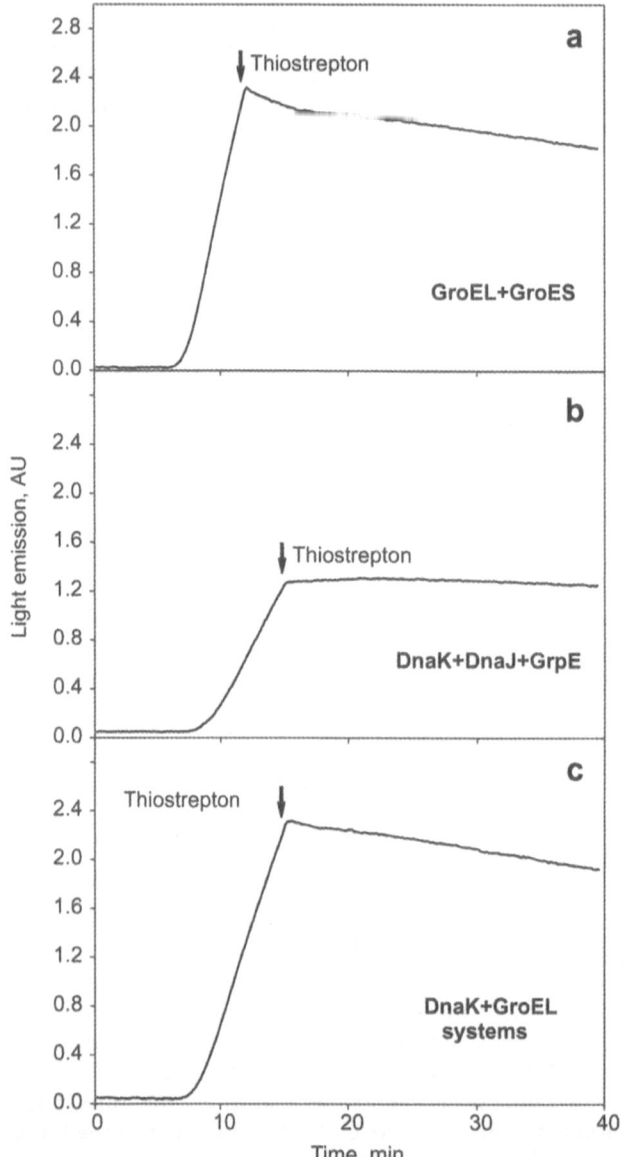

Fig. 3. Time course of luciferase activity accumulation upon arrest of translation in chaperone-enriched translation systems. Synthesis of luciferase was arrested by injection of thiostrepton to final concentration of 20 μM into the *E. coli* S30 extract translation mixtures. Translation reactions were carried out in 25 μl reaction volume containing 40 μg/ml of luciferase mRNA according the description in the legend to Fig.1, except that individual purified chaperones and unlabeled amino acids were added to the reactions. *a*, translation in the presence of 90 nM GroEL 14-mer with 180 nM GroES 7-mer; *b*, translation in the presence of 1300 nM DnaK with 200 nM DnaJ and 650 nM GrpE dimer; *c*, both the GroEL and the DnaK chaperone systems were added together to the translation mixture at the concentrations listed above

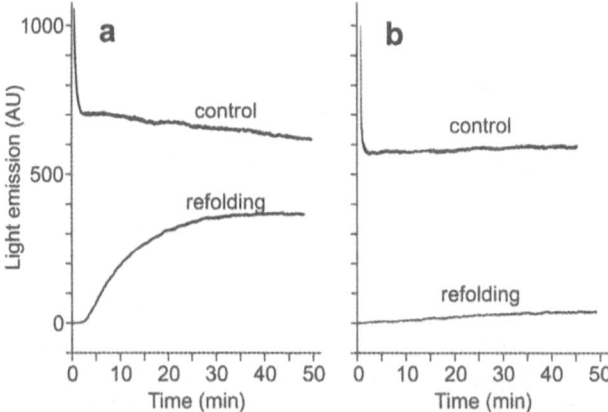

Fig. 4. Refolding of urea-denatured firefly luciferase in cell-free translation systems devoid of mRNA. The translation systems (Kolb et al. 2000) contained 0.1 mM luciferin and 0.2 mM coenzyme A. Refolding was initiated by dilution of a 0.5-μl aliquot containing 20 ng of denatured enzyme with 200 μl of translation system and proceeded at 25°C. Addition of the same amount of native enzyme to the translation system was used as a control. *a*, time course of recovered luciferase activity upon dilution with wheat germ translation system; *b*, time course of recovered luciferase activity upon dilution with *E. coli* S30 translation system

the *cis-trans* isomerisation of peptidyl-prolyl bonds is not the rate-limiting stage of the enzyme refolding. The half-time of refolding in the prokaryotic cytosol is estimated to be about 20 min (Fig.4b). The main conclusion drawn from these experiments is that the refolding of denatured luciferase in prokaryotic cytosol is a slow and inefficient process, and in eukaryotic cytosol this process occurs at a similar rate, although it is more productive.

Taken together, the data on refolding rates, enzymatic activity emergence and blocking of active enzyme accumulation upon arrest of its synthesis, directly show that luciferase starts its folding while it still resides on the translating prokaryotic ribosome, and probably is almost completely folded when leaving the ribosome after termination of translation.

Discussion

Comparing the folding kinetics of firefly luciferase during its synthesis with the kinetics of denatured luciferase refolding, the conclusion can be drawn that the newly synthesized enzyme attains its native structure co-translationally, both in bacterial and eukaryotic cell-free translation systems. Virtually no post-translational folding phase was observed in either extract, as indicated by the absence of the luciferase activity accumulation immediately after the halt of translation.

The conclusion relating to the co-translational mode of luciferase folding in bacterial cell-free extracts is also corroborated by the fact that synthesis of enzymatically active luciferase in prokaryotic cytosol occurs much earlier than the half-time of *in vitro* refolding of denatured enzyme under identical conditions or even at optimal conditions for chaperone-assisted refolding (Buchberger et al. 1996).

The poor refolding capacity in the prokaryotic cell-free translation system, recovering only 8 % of the original activity of luciferase after 50 min incubation at 25°C also suggests a co- rather than posttranslational mechanism for active enzyme formation. On the other hand, the low yield of refolding could indicate a lack of chaperones in the *E. coli* S30 extract, as fast and productive luciferase refolding requires the DnaK chaperone system (Schröder et al. 1993; Buchberger et al. 1996). To rule out the possibility that the co-translational mode of folding was detected as a result of the scarcity of chaperone, the same *E. coli* S30 extract was supplied with an excess of chaperones of both the DnaK and GroEL families. We have found that large excess of chaperones in the prokaryotic translation system did not either extend the duration of posttranslational phase of *de novo* luciferase folding (it remained within a few seconds or less), or improve the active enzyme production (not shown). Finally, in spite of the difference in refolding activity of pro- and eukaryotic cell-free extracts, the equally high level of luciferase specific activity was determined for newly synthesized enzyme, irrespective of whether it was produced in pro- or eukaryotic translation systems (Kolb et al. 2000). Thus, the conclusion can be made that the ribosomal machinery by itself is principally capable of maintaining the correct and fast folding of the polypeptide chain during translation, and no additional molecular chaperones are required for co-translational folding.

The significance of chaperone activity for productive *de novo* folding in cell-free translation systems has been demonstrated for a number of proteins, e.g. the reovirus cell attachment protein (Gilmore et al. 1996), chloramphenicol acetyltransferase (Vysokanov 1995), bovine rhodanese (Kudlicki et al. 1995), heavy meromyosin subfragment (Srikakulam and Winkelmann 1999), actin, tubulins and actin-related proteins (reviewed by Lewis et al. 1996). It should be pointed out, however, that most of these proteins are functionally active when assembled in oligomeric or multimeric complexes. It is likely that the synthesis of properly folded globular proteins, including multidomain ones, dispenses with the need for activity of chaperones in cell extracts used for cell-free translation systems. At the very least, the presence of chaperones appears not to be a critical factor, provided that the proteins fold co-translationally.

The results presented here demonstrate the similarity of protein folding mode in two kingdoms of life. The co-translational protein folding thus seems to be a universal mode for the majority of proteins to attain their unique tertiary structure. The example of firefly luciferase shows that successful synthesis of functionally active eukaryotic proteins is principally compatible with cell-free systems based on prokaryotic extracts.

References

Anfinsen CB (1973) Principles that govern the folding of protein chains. Science 181: 223–230

Bergman LW, Kuehl WM (1979) Formation of an intrachain disulfide bond on nascent immunoglobulin light chains. J Biol Chem 254: 8869–8876

Buchberger A, Schröder H, Hesterkamp T, Schönfeld H-J, Bukau B (1996) Substrate shuttling between the DnaK and GroEL systems indicates a chaperone network promoting protein folding. J Mol Biol 261: 328–333

Chen W, Helenius J, Braakman I, Helenius A (1995) Cotranslational folding and calnexin binding during glycoprotein synthesis. Proc Natl Acad Sci USA 92: 6229–6233

Ellis RG (1994) Role of molecular chaperones in protein folding. Curr Opin Struct Biol 4: 117–122

Fedorov AN, Baldwin TO (1995) Contribution of cotranslational folding to the rate of formation of native protein structure. Proc Natl Acad Sci USA 92: 1227–1231

Fink AL (1999) Chaperone-mediated protein folding. Physiol Rev 79: 425–449

Frydman J, Nimmesgern E, Ohtsuka K, Hartl FU (1994) Folding of nascent polypeptide chains in a high molecular mass assembly with molecular chaperones. Nature 370: 111–117

Frydman J, Erdjument-Bromage H, Tempst R, Hartl FU (1999) Co-translational domain folding as the structural basis for the rapid de novo folding of firefly luciferase. Nature Struct Biol, 6: 697–705

Georgopoulos C, Welch WJ (1993) Role of the major heat shock proteins as molecular chaperones. Ann Rev Cell Biol 9: 601–634

Gething M-J, Sambrook J (1992) Protein folding in the cell. Nature 355: 33–45

Gilmore R, Coffey MC, Leone G, McLure K, Lee PWK (1996) Co-translational trimerization of the reovirus cell attachment protein. EMBO J 15: 2651–2658

Hanes J, Plckthun A (1997) In vitro selection and evolution of functional proteins by using ribosome display. Proc Natl Acad Sci USA 94: 4937–4942

Hartl FU (1996) Molecular chaperones in cellular protein folding. Nature 381: 571–580

Kiho Y, Rich A (1964) Induced enzyme formed on bacterial polyribosomes. Proc Natl Acad Sci USA 51: 111–118

Kim J, Klein PG, Mullet JE (1991) Ribosome pause at specific sites during synthesis of membrane-bound chloroplast reaction centre protein D1. J Biol Chem 266: 14931–14938

Kolb VA, Makeyev EV, Spirin AS (1994) Folding of firefly luciferase during translation in a cell-free system. EMBO J 13: 3631–3637

Kolb VA, Makeyev EV, Kommer A, Spirin AS (1995) Cotranslational folding of proteins. Biochem Cell Biol 73: 1217–1220

Kolb VA, Makeyev EV, Spirin AS (2000) Co-translational of an eukaryotic multidomain protein in a prokaryotic translation system. J Biol Chem 275: 16597–16601

Komar AA, Kommer A, Krasheninnikov IA, Spirin AS (1997) Cotranslational folding of globin. J Biol Chem 272: 10646–10651

Kudlicki W, Chirgwin J, Kramer G, Hardesty B (1995) Folding of an enzyme into an active conformation while bound as peptidyl-tRNA to the ribosome. Biochemistry 34: 14284–14287

Lewis SA, Tian G, Vainberg IE, Cowan NJ (1996) Chaperonin-mediated folding of actin and tubulin. J Cell Biol 132: 1–4

Lin L, DeMartino GN, Greene WC (1998) Cotranslational biogenesis of NF-kB p50 by the 26S proteasome. Cell 92: 819–828

Makeyev EV, Kolb VA, Spirin AS (1996) Enzymatic activity of the ribosome-bound nascent polypeptide. FEBS Lett 378: 166–170

Mullet JE, Klein PG, Klein RR (1990) Chlorophyll regulates accumulation of the plastid-encoded chlorophyll apoprotein CP43 and D1 by increasing apoprotein stability. Proc Natl Acad Sci USA 87: 4038–4042

Netzer WJ, Hartl FU (1997) Recombination of protein domains facilitated by co-translational folding in eukaryotes. Nature 388: 343–349

Nicola AV, Chen W, Helenius A (1999) Co-translational folding of an alphavirus capsid protein in the cytosol of living cell. Nat Cell Biol 1: 341–345

Peters T, Davidson LK (1982) The biosynthesis of rat serum albumin. J Biol Chem 257: 8847–8853

Phillips DC (1967) The hen egg-white lysozyme molecule. Proc Natl Acad Sci USA, 57: 484–495

Ryabova LA, Desplancq D, Spirin AS, Plückthun A (1997) Functional antibody production using cell-free translation: effects of protein disulfide isomerase and chaperones. Nature Biotechnol 15: 79–84

Schröder H, Langer T, Hartl FU, Bukau B (1993) DnaK, DnaJ and GrpE form a cellular chaperone machinery capable of repairing heat-induced protein damage. EMBO J 12: 4137–4144

Srikakulam R, Winkelmann D (1999) Myosin II folding is mediated by a molecular chaperonin. J Biol Chem 274: 27265–27273

Vysokanov AV (1995) Synthesis of chloramphenicol acetyltransferase in a coupled transcription-translation in vitro system lacking the chaperones DnaK and DnaJ. FEBS Lett 375: 211–214

CHAPTER 12

Expression of an Aggregation-Prone Protein in the RTS 500 System

12

JEAN-MICHEL BETTON*

Introduction

The most widely used host for high-level production of recombinant proteins is *Escherichia coli*. However, many of these proteins are produced in biologically inactive and aggregated forms or inclusion bodies. The conventional molecular mechanism of this process suggests that nascent polypeptide chains have two alternative and competitive pathways between folding and aggregation (King et al. 1996). Although solubilization/refolding procedures exist to obtain proteins in soluble form, they are lengthy processes, often inefficient and not generalizable (Lilie et al. 1998). Roche Molecular Biochemicals' recent development of an in vitro protein biosynthesis system, the Rapid Translation System or RTS 500, is a very attractive alternative for recombinant protein production. In this paper, the RTS 500 system was evaluated for the expression of the maltose-binding protein (MalE), an exported protein of *E. coli*, which serves as periplasmic receptor for the high-affinity transport of maltodextrins and for a defective MalE folding mutant, MalE31, that forms inclusion bodies when expressed in bacteria (Betton and Hofnung 1996).

Materials and Methods

Expression Vectors

To construct vectors for RTS 500 expression, *malE* alleles, corresponding to mature proteins without signal sequence, were amplified by the polymerase chain reaction (PCR) from plasmids p1H and p31H (Betton et al. 1996) using the forward primer ME5 with *NdeI* site and two different reverse primers (ME3 or ME3B) with *HpaI* site. The primer ME3 was designed to fuse tag-sequences to the C-terminus of MalE, and the 5′ end of ME3B primer containing the natural TAA

* J-M. Betton (✉) (e-mail: jmbetton@pasteur.fr; Tel.: +33–1–45–688959; Fax: +33–1–40–613043)
Unité de Biochimie Cellulaire, Institut Pasteur, CNRS (URA2185), 28 rue du Dr Roux, 75724 Paris Cedex 15, France

stop codon of *malE* and *malE31* was used to express the corresponding proteins without a tag. The primer sequences are as follows:

- ME5 (CTC<u>CATATG</u>AAAATCGAAGAAGGTAAACTGGTAATC)
- ME3 (TCTC<u>GTTAAC</u>CTTGGTGATACGAGTCTGCGCG)
- ME3B (CTC<u>GTTAAC</u>TTACTTGGTGATACGAGTCTGCG)

The PCR products were cloned into the pCR-Script Cam SK vector (Stratagene, La Jolla CA, USA), and nucleotide sequences of inserts were confirmed. Plasmids pIV2.1ME, pIV2.2ME, pIV2.3ME, and pIV2.4ME, carrying the wild-type *malE* gene under the control of T7 promoter, were constructed by subcloning PCR inserts into the corresponding pIVEX2.1MCS, pIVEX2.2bNde, pIVEX2.3MCS, and pIVEX2.4bNde vectors (Roche Molecular Biochemicals, Mannheim, Germany), digested by *NdeI* and *SmaI*. Plasmids pIV3ME and pIV3ME31 were constructed by subcloning *malE* and *malE31*, amplified by the primer pair ME5/ME3B, and inserted into the pIVEX2.3MCS vector as described above.

Protein Synthesis Reaction

In vitro MalE synthesis was performed with the RTS 500 instrument essentially as described in the instruction manual from Roche Molecular Biochemicals. The coupled transcription/translation reaction, initiated by adding 15 μg of pIVEX plasmid DNA, was carried out in 1 ml total volume for 20 h at 30°C. Proteins were then separated by 12% SDS-polyacrylamide gel and stained with Coomassie blue.

Protein Purification

All MalE proteins were purified by affinity chromatography using a cross-linked amylose resin as described (Betton and Hofnung 1996). Whole RTS 500 extracts (0.6 ml) were loaded on amylose columns (1 ml of gel) equilibrated with 20 mM Tris-HCl buffer, pH 7.5 containing 0.1 M NaCl. After a washing step, the proteins, eluted (1 ml) by 10 mM maltose in the same buffer, were analyzed by SDS-gel electrophoresis, and MalE concentrations were determined from the absorbance at 280 nm with an extinction coefficient of 68,750 M^{-1} cm^{-1}.

Results and Discussion

To examine the influence of fusion tags on the production level of MalE in the RTS 500, four plasmids were constructed that produced MalE fused with either a His-tag (pIV2.3ME and pIVME2.4ME) or a Strep-tag (pIV2.1ME and pIV2.2ME), either at the N-terminus (pIV2.2ME and pIV2.4ME) or at the C-terminus (pIV2.1ME and pIV2.3ME). Since tag-sequences could influence the expression level, one plasmid (pIV3ME) that produces MalE without a tag was constructed as a control (Fig. 1). For the same reason, to evaluate more directly the production

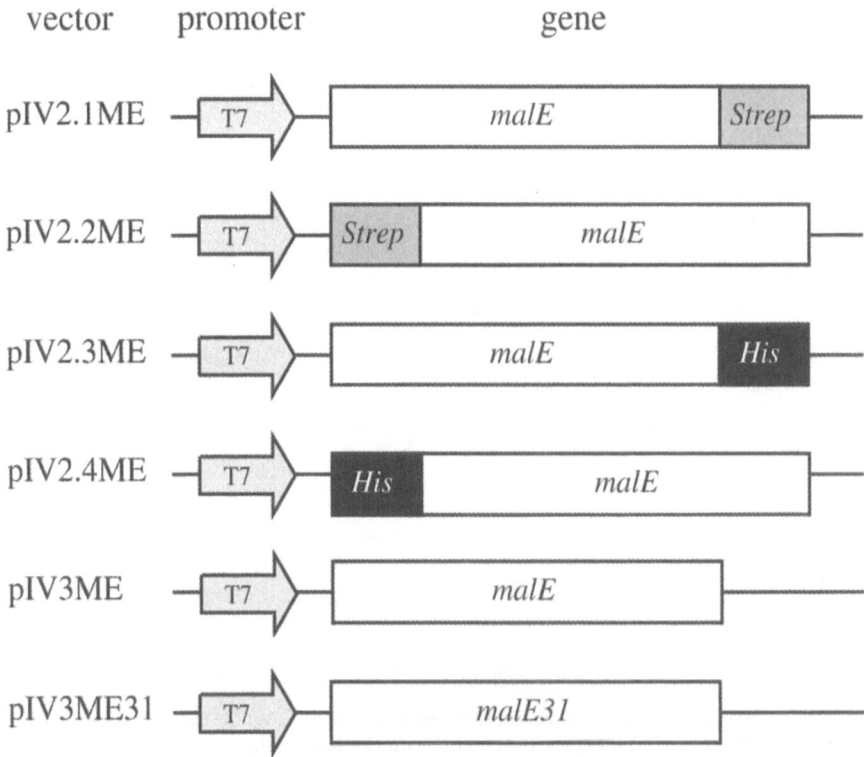

Fig. 1. In vitro expression vectors were constructed by subcloning PCR fragments containing the *malE* and *malE31* genes into the various pIVEX plasmids, digested by *Nde*I and *Sma*I. pIV3ME and pIV3ME31 vectors expressed proteins without a fusion tag

level of MalE31, this defective folding mutant was expressed from pIV3ME31 without a tag.

The study began by comparing the steady-state productions of MalE and MalE31 encoded by the various pIVEX plasmids in the RTS 500 (Fig. 2). When whole extracts were analyzed by SDS-polyacrylamide gel, stained with Coomassie blue, all proteins except one (MalE expressed from pIV2.2ME) were detected at a comparable yield to GFP used for the control reaction. In the case of Strep-tagged MalE to its N-terminus, the protein was synthesized at a very low level. Since His-tagged MalE at the N-terminus was correctly expressed, this result indicated that the Strep-tag sequence rather than the N-terminal position had a negative influence on MalE production. Perhaps an unfavorable secondary structure of the translation initial region was formed in this vector context (Yarchuk et al. 1992). Alternatively, the specific Strep-tag amino acid sequence at the N-terminus could decrease the folding rate or destabilize MalE (Ganesh et al. 1999) and promote its excessive degradation. It will be interesting to explore this phenomenon with other proteins.

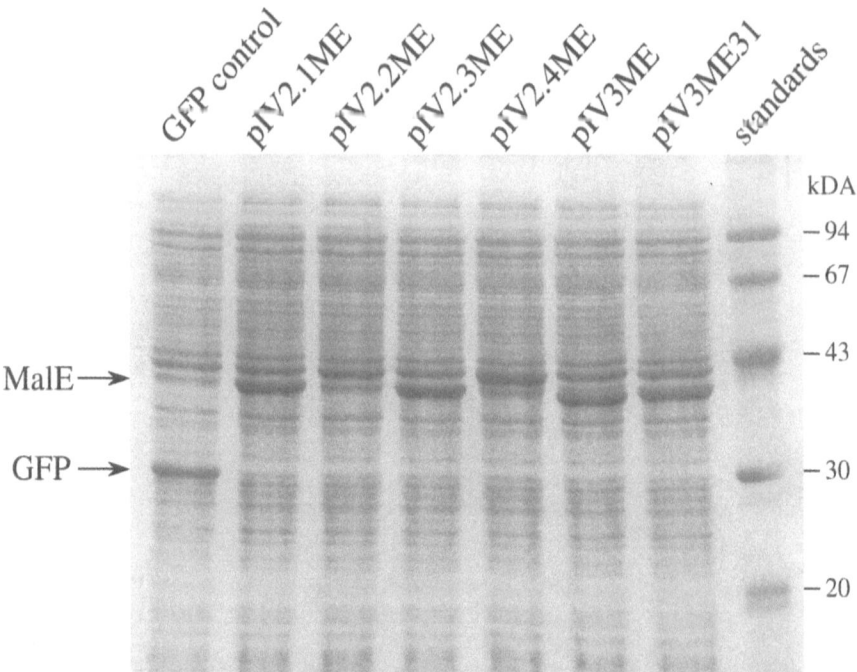

Fig. 2. In vitro production of MalE and MalE31 after 20 h at 30°C. Whole extracts were separated on 12% SDS-polyacrylamide gels and stained with Coomassie blue

Because of the advantage provided by the affinity purification of MalE, the ability of various proteins to bind cross-linked amylose was tested to assess their native conformations (Fig. 3). Furthermore, in comparative studies on the purification yield, amylose columns help to minimize problems of quantification due to differences in fusion tags. All proteins correctly expressed with RTS 500 were successfully purified by this method with yields ranging from 240 to 350 µg/ml (Table 1). Purity and homogeneity of these proteins were further characterized by electrospray mass spectrometry (ESMS), which indicated a single ESMS peak at the expected mass (data not shown). However, the most striking result emerging from these experiments is the yield of native MalE31 that was obtained (280 µg/ml). Indeed, this mutant forms inclusion bodies when expressed in bacteria even at low temperature (Betton and Hofnung 1996). In general, the extent of aggregation is greater at higher temperatures due to the strong temperature-dependence of the hydrophobic interactions that determine these reactions. Hence, the soluble production level of MalE31 at 37°C with RTS 500 was assessed.

Figure 4 shows the comparison between bacterial and RTS 500 expression from pIV3ME31 at different temperatures. At 25°C in bacteria, MalE31 was mainly recovered in the pellet (I) by centrifugation. While this aggregation-prone protein was found to be entirely soluble, even at 37°C, when expressed in vitro. This result

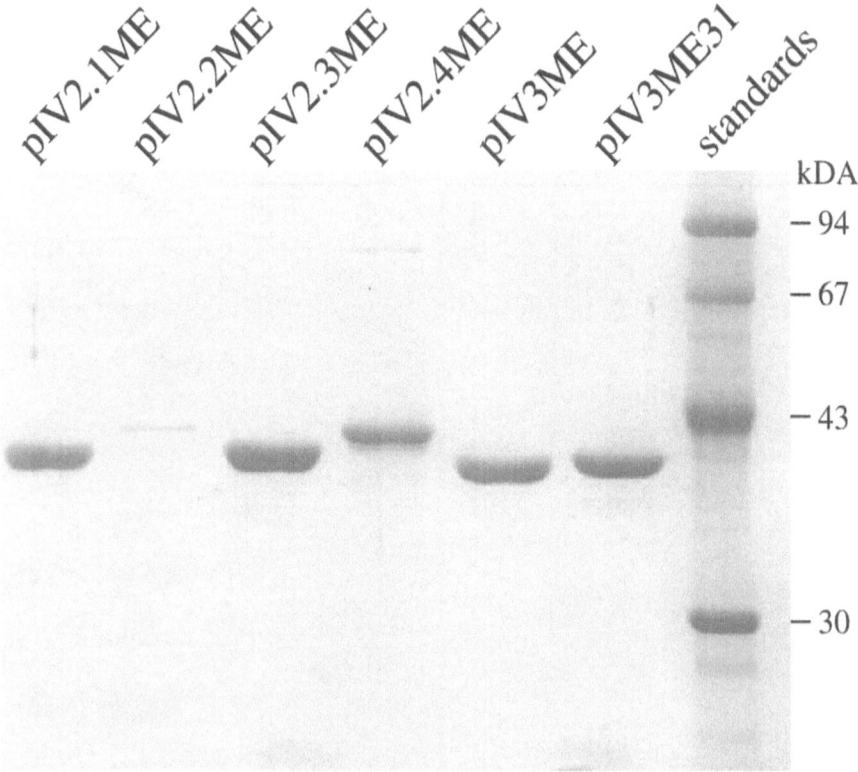

Fig. 3. MalE and MalE31 were purified by amylose affinity chromatography. The maltose-eluted proteins were analyzed on 12% SDS-polyacrylamide gels and stained with Coomassie blue

Table 1. Yields of MalE and MalE31 purified by affinity chromatography

Plasmids	Purification yield (μg ml^{-1})
pIV2.1ME	240
pIV2.2ME	<20
pIV2.3ME	350
pIV2.4ME	240
pIV3ME	320
pIV3ME31	280

indicated that RTS 500 expression could considerably reduce unproductive inter-actions of nascent polypeptide chains during translational folding. The large difference in macromolecular crowding between cellular and RTS 500 environments may account for the productive folding of MalE31 (van den Berg et al. 1999). Further studies with other aggregation-prone proteins should be performed to confirm this property of RTS 500 expression system.

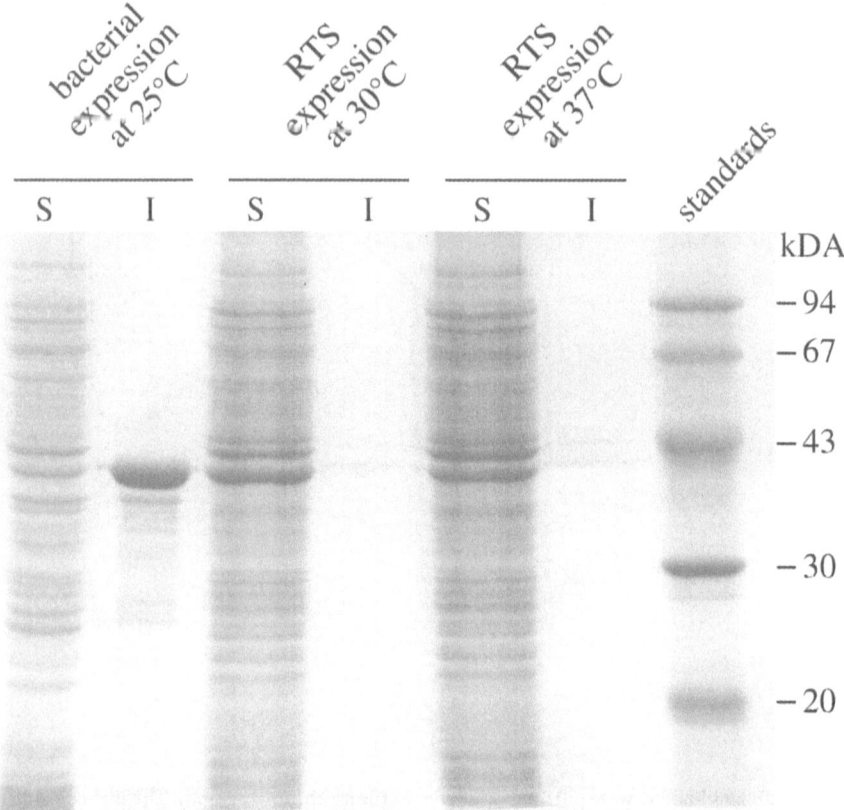

Fig. 4. Comparison of soluble/insoluble production of MalE31 between bacterial (strain BL21-DE3) and RTS 500 expression systems at various temperatures. Soluble (*S*) and insoluble (*I*) fractions were prepared by centrifugation of whole extracts at 15,000 rpm and analyzed on 12% SDS-polyacrylamide gel and stained with Coomassie blue

Conclusion

The RTS 500 could provide a promising alternative for soluble and native proteins, which otherwise form insoluble aggregates in bacterial expression systems. The applications of RTS 500 include not only the quick and simple production of proteins, but also the synthesis of toxic or unstable proteins, the rapid screening of mutants by direct translation of PCR products, and the selective incorporation of isotope-labeled amino acids in proteins for NMR studies.

References

van den Berg B, Ellis RJ, Dobson CM (1999) Effects of macromolecular crowding on protein folding and aggregation. EMBO J 24:6927–6933

Betton J-M, Hofnung M (1996) Folding of a mutant maltose-binding protein of *Escherichia coli*, which forms inclusion bodies. J Biol Chem 271:8046–8052

Betton J-M, Boscus D, Missiakas D, Raina S, Hofnung M (1996) Probing the structural role of an αβ loop of maltose-binding protein by mutagenesis: heat-shock induction by loop variants of the maltose-binding protein that form periplasmic inclusion bodies. J Mol Biol 262:140–150

Ganesh C, Banerjee A, Shah A, Varadarajan R (1999) Disorded N-terminal residues affect the folding thermodynamics and kinetics of maltose-binding protein. FEBS Lett 454:307–311

King J, Haase-Pettingell C, Robinson AS, Speed M, Mitraki A (1996) Thermolabile folding intermediates: inclusion body precursors and chaperonin substrates. FASEB J 10:57–66

Lilie H, Schwarz E, Rudolph R (1998) Advances in refolding of proteins produced in *E. coli*. Curr Opin Biotechnol 9:497–501

Yarchuk O, Jacques N, Guillerez J, Dreyfus M (1992) Interdependence of translation, transcription and mRNA degradation in the *lacZ* gene. J Mol Biol 226:581–596

Chapter 13

Expression of Soluble Murine Endostatin with the RTS System

13

Stacey Traviglia, Monique van Hoek*

Introduction

When recombinant proteins are produced in *Escherichia coli* (*E. coli*) expression systems, they often aggregate due to their inability to fold properly. In some cases, a protein produced as inclusion bodies in *E. coli* can be expressed as a soluble and active protein in an *E. coli*-based cell-free expression system, such as the RTS [1]. However, if aggregation also occurs in cell-free systems, the conditions can easily be adjusted to overcome the problem. For example, the addition of detergents, cofactors, or molecular chaperones can often enhance or achieve protein solubility.

Molecular chaperones are specialized proteins that bind to the non-native states of proteins. Chaperones assist the re-folding of proteins denatured by chemical or environmental stresses and in the folding of some newly translocated proteins [2, 3]. The Hsp70 chaperone system consists of DnaK, DnaJ, and GrpE and can assist folding co-translationally or after the nascent polypeptide chain emerges from the ribosome. The Hsp70 chaperone system can also transfer the unfolded protein to the Hsp60 chaperonin system (GroEL and GroES), which assists protein folding post-translationally. GroEL and GroES form a complex tertiary structure, enabling a protein to fold within its cavity. Overexpression of these chaperones in *E. coli* has been shown to facilitate protein folding in vivo and enhance the production of active proteins [4–6].

We chose murine endostatin as a target protein to test the use of chaperones in the cell-free protein expression system, RTS 500. Endostatin is a therapeutic protein with anti-cancer properties [7] that has recently entered into clinical trials in humans. The endostatin used for these trials is being made in *Pichia pastoris* [8] because endostatin is insoluble when it is expressed in *E. coli* [9, 10]. Here we demonstrate that the addition of chaperones to a cell-free *E. coli*-based lysate achieves production of soluble endostatin.

* Stacey Traviglia and Monique van Hoek (✉) (e-mail: monique.vanhoek@roche.com, stacey.traviglia@diosynth-rtp.com, Tel.: +1–510–883–7956, Fax: +1–510–883–0636)
Roche Molecular Biochemicals, 2929 7th Street, Suite 100, Berkeley, CA 94710, USA

Materials and Methods

Construction of Plasmids

Plasmid pETKH-1 (American Type Culture Collection, Manassas, VA, USA) was used as a template to amplify the sequence encoding mouse endostatin by polymerase chain reaction (PCR) with a proofreading DNA polymerase (Expand High Fidelity PCR System, Roche Molecular Biochemicals, Indianapolis, IN, USA). The forward primer for the 5′ end was GCCGACCGCGGTCATACTCATCAGGACTTTC, and the reverse primer for the 3′ end was CTGGCATCCGGATTATTTGGAGAAA-GAGGTCATG (Operon, Alameda, CA, USA). In addition to the annealing sequences, the forward primer contained a KspI site, and the reverse primer contained an MroI site followed by a stop codon to achieve the in-frame insertion of the construct. After restriction digest, the PCR product was purified by agarose gel separation. The vector pIVEX 2.4b Nde was digested with KspI and XmaCI (Roche Molecular Biochemicals) followed by treatment with shrimp alkaline phosphatase (Roche Molecular Biochemicals). The PCR fragment was ligated into the digested vector with Rapid DNA Ligation Kit (Roche Molecular Biochemicals). After transformation into TOP10F′ cells (Invitrogen, Carlsbad, CA, USA), the clone was found by PCR screening. The sequence of the construct was verified by Davis Sequencing (Davis, Calif., USA).

Protein-Expression Reactions

In vitro protein-expression reactions of murine endostatin (pIVEX2.4bNde-Endostatin) were performed with the RTS 500 kit (Roche Molecular Biochemicals), as described in the manufacturer's instruction manual, except for the following modifications. Each reaction contained 15 µg of purified plasmid DNA prepared using Qiagen Plasmid Kit (Qiagen, Valencia, CA, USA) in a 500 µl reaction volume. For expression with the addition of purified chaperones, final concentrations of 10 µM DnaK, 10 µM DnaJ, 10 µM GrpE, 1 µMGroEL and 2 µMGroES were added to the top reaction chamber. To avoid dilution effects, the chaperone added volume was subtracted from the final volume of reconstitution buffer, which was to be added to the lyophilized reaction mix. For expression with detergents, a final concentration of 0.1% Tween 20 was added to both the top reaction chamber and the bottom feeding chamber. All reactions were carried out overnight at 30°C with efficient stirring (120 rpm) in the RTS 500 instrument.

Protein-Expression Analysis

Samples taken from an RTS reaction were prepared for SDS-PAGE and Western blot in the following manner: 20 µl of the RTS reaction was spun down at maximum speed in an Eppindorf centrifuge at 4°C. The supernatant was aspirated and diluted up to 80 µl with 32 µl water, 20 µl 4× sample application buffer (Invitrogen, Carlsbad, Ca, USA), and 8 μl 500 mM (10×) dithiothreithol (DTT). At least a fourfold dilution is necessary to prevent precipitation of the lithium or sodium dodecyl sulfate (LDS or SDS) from the sample application buffer with the potassium salts from the RTS reaction. The pellet was re-dissolved in 20 μL of 1% SDS, then diluted up to 80 μL in the same manner as the supernatant. Samples analyzed under non-reducing conditions did not contain DTT in the sample application buffer. An amount of 20 μl of each prepared sample was run on a 10% NuPAGE gel (Invitrogen, Carlsbad, Ca, USA) and either stained with Colloidal Blue (Invitrogen, Carlsbad, Ca, USA) or transferred to polyvinylidene difluoride (PVDF) membrane (Roche Molecular Biochemicals). The membrane was blocked with Western Blocking Reagent (Roche Molecular Biochemicals) and then visualized with anti-His antibody (Roche Molecular Biochemicals), anti-mouse IgG (H+L)-AP (Roche Molecular Biochemicals), followed by 5-bromo-4-chloro-3-indolyl-phosphate/nitroblue tetrazolium (BCIP/NBT) (Roche Molecular Biochemicals). The His-tagged endostatin was purified using Ni-NTA resin (Qiagen) and dialyzed into buffer containing 50 mM sodium phosphate (pH 7.4). Protein concentration was determined using a BCA assay (Pierce, Rockford, IL, USA). Reverse phase HPLC analysis was performed with a C18 column (250×4.6 mm, Vydac) on a Dynavex HLPC system with a 5–95% B gradient [A: H_2O+0.1% trifluoroacetic acid (TFA), B: acetonitrile (ACN)+0.085% TFA] over 45 min (1 ml/min flow rate). The circular dichroism (CD) spectrum was obtained at room temperature in a 1.0-cm path length cell.

Results

RTS Expression of Endostatin

When expressed under standard RTS 500 conditions, endostatin was found in an aggregated, insoluble form. Shorter reaction times (1 h, 3 h, 5 h, 7 h, and 23 h) or the addition of detergents (0.1% Tween 20) failed to yield soluble endostatin (data not shown). The addition of chaperones (10 μM DnaK, 10 μM DnaJ, 10 μM GrpE, 1 μMGroEL and 2 μMGroES) to the reaction mix resulted in the expression of soluble endostatin (Fig. 1). The RTS-expressed endostatin remained soluble after purification. The concentration and yield of endostatin after purification was determined to be 200 μg/ml.

Characterization of RTS-Expressed Endostatin

The purified RTS-expressed endostatin has an apparent molecular weight of 20 kDa by reducing SDS-PAGE. Under non-reducing conditions, the RTS-expressed endostatin shifts to a faster mobility, indicating that it is disulfide bonded (Fig. 2). This shift in mobility is consistent with previous reports [9, 11]. In the case of endostatin, it appears that air oxidation is sufficient for disulfide bond formation (Note that the reaction chamber was filled only 50% to ensure good oxygenation of the reaction).

A

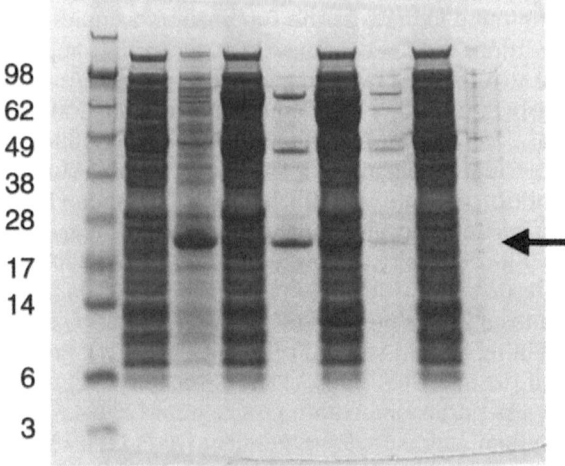

B

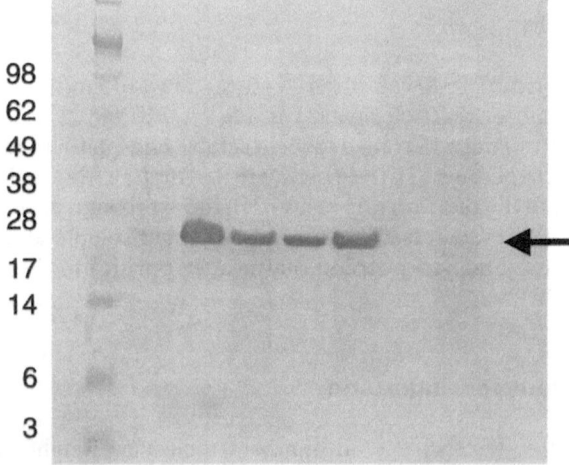

Fig. 1A,B. Addition of chaperones results in expression of soluble endostatin. *Arrows* indicate endostatin. **A** SDS-PAGE stained with Colloidal blue; **B** Western blot with anti-His$_6$ antibody. Lane 1, MW standard; lane 2, supernatant without chaperones; lane 3, pellet of lane 2; lane 4, supernatant with 10 µM DnaK/DnaJ/GrpE; lane 5, pellet of lane 4; lane 6, supernatant with 10 µM DnaK/DnaJ/GrpE and 1 µM GroEL/2 µMGroES; lane 7, pellet of lane 6; lane 8, control reaction without endostatin plasmid

Fig. 2. Reduced and non-reduced SDS-
PAGE of purified endostatin: lane 1, MW
standard; lane 2, reduced endostatin; lane
3, non-reduced endostatin. Under non-
reducing conditions, the disulfide bonds
remain intact, creating a more compact
form of endostatin. The more compact,
non-reduced endostatin migrates faster
through the gel

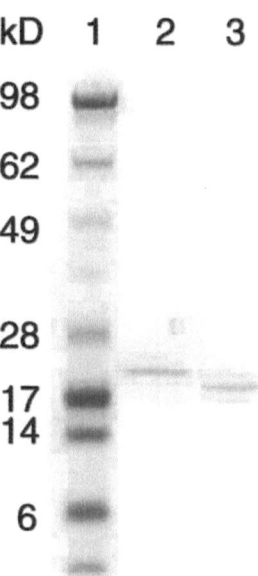

Approximately 1 µg of purified RTS-expressed endostatin was injected onto a
C18 column for reverse phase chromatography. A gradient of 5–95% B [A:
H_2O+0.1% TFA, B: ACN+0.085% TFA] over 45 min resulted in one major peak
with a retention time of 24.6 min. The sharpness of the peak is indicative of the
non-aggregated state of the soluble endostatin (Fig. 3). The two minor peaks present in the analysis are impurities, which do not stain with Coomassie.

The secondary structure of a 50 µg/ml solution of the purified RTS-expressed
endostatin was analyzed by CD analysis in a 1-cm cell at room temperature
(Fig. 4). The CD spectrum was consistent with previous reports [9, 11], having a
characteristic minimum at 205 nm. Estimation of the secondary structure
(Chang's equation) indicated that the RTS-expressed endostatin consisted primarily of β-sheets and had approximately 11% α-helix. These data also correlate to
the determined crystal structure [12].

Discussion

Cell-free protein expression has many advantages over cell-based systems for the
expression of therapeutic proteins. In cell-free systems, there is a more well-
defined profile of contaminants. The purification is more straightforward and not
complicated by cell debris. Furthermore, many potential protein therapeutics are
subject to the same solubility issues in E. coli as endostatin. Shifting to a cell-free
system may solve the problems of insolubility for some proteins. Otherwise, the
cell-free system can also be adjusted in other ways to produce a soluble, folded
target protein.

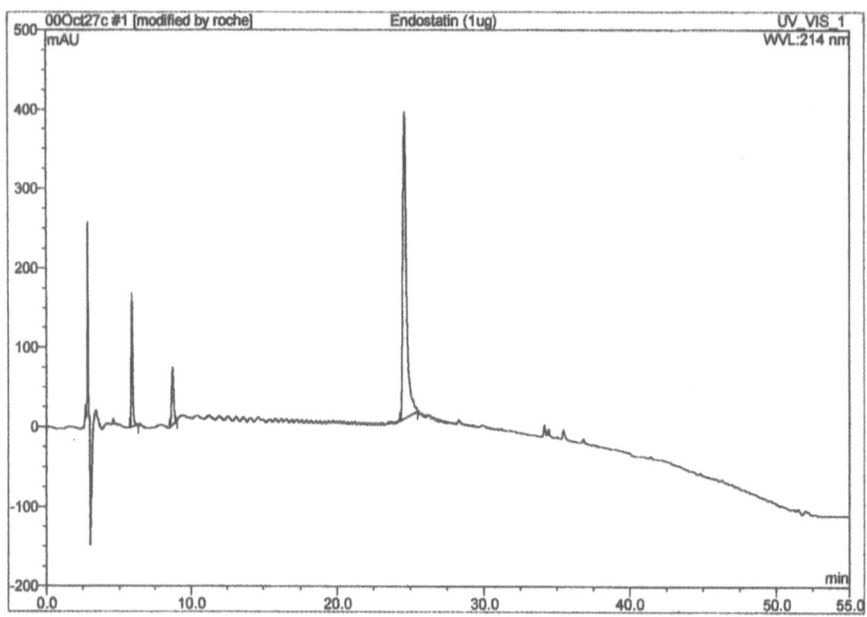

Fig. 3. RP-HPLC analysis of purified endostatin. C18 column with a gradient of 5–95% B [A: H₂O+0.1% TFA, B: ACN+0.085% TFA] over 45 min at a flow rate of 1 ml/min. Peak containing endostatin was eluted at 24.6 min

Conditions to achieve the production of soluble endostatin in an *E. coli* based cell-free system have now been determined. The addition of purified chaperones successfully resulted in >90% solubility of the protein produced. Upon purification, the endostatin remained in solution, demonstrating a stable, folded form. Analytical RP-HPLC resulted in a sharp peak, indicating the protein was not aggregated. A shift in mobility of the non-reduced protein indicates that the endostatin was disulfide bonded. Furthermore, CD analysis demonstrated the correct secondary structure of a properly folded endostatin. Based on these biochemical and biophysical results, the endostatin produced in the presence of chaperones appears to be properly folded into its biologically active confirmation.

While there are reports describing the procedures to purify endostatin from inclusion bodies followed by efficient re-folding protocols [9, 10], it is advantageous to prepare endostatin in a soluble and folded form. Some cell-based *E. coli* expression systems have addressed this problem by co-expression with chaperones. Nishihara et al. have identified that co-overexpression of endostatin with chaperones in *E. coli*, specifically Trigger Factor and GroEL/GroES, results in the production of a soluble product [4]. In that report, Nishihara et al. demonstrated very little effect from overexpression with DnaK/DnaJ/GrpE. In the RTS cell-free system, GroEL/GroES only solubilized a portion of the endostatin being produced (data not shown). Likewise, DnaK/DnaJ/GrpE had the same effect (Fig. 1, lanes 4

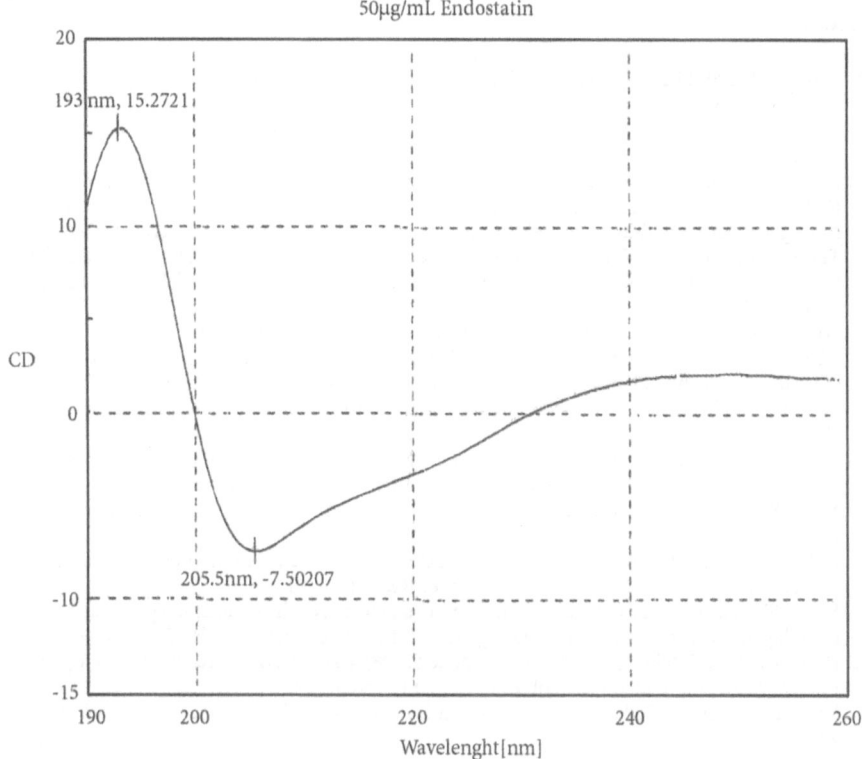

Fig. 4. CD spectra of endostatin in 50 mM sodium phosphate, pH 7.4 at room temperature. Using Chang's equation, the secondary structure of endostatin was estimated to consist primarily of β-sheets and approximately 11% α-helix

and 5). However, both systems combined (DnaK/DnaJ/GrpE and GroEL/GroES) resulted in almost complete solubility (Fig. 1, lanes 6 and 7). The difference in success with GroEL/GroES in cell-based co-expression versus RTS-expression is likely due to a basal level of DnaK/DnaJ/GrpE already present in Nishihara's *E. coli* co-overexpression system.

The results reported here suggest that chaperones may assist the folding of other problematic proteins. While the role of chaperones may be minimal for the expression of a given host's genome [2], chaperones are likely to be more necessary for non-native recombinant proteins. Furthermore, the structural genomics efforts in Japan have shown that a certain percentage of the total protein produced by in vitro protein expression is found to be insoluble [13]. Chaperones may boost the overall yield of in vitro protein expression by reducing the percentage of insoluble products. The combination of chaperones with the speed and high yield of the RTS technology expands the application of in vitro protein expression and provides a valuable tool for high-throughput drug discovery and structural genomics.

References

1. Betton JM (2000) Production of a soluble and active form of a defective folding protein with the RTS 500 system. Roche Biochemica 4–6
2. Bukau B, Deuerling E, Pfund C, Craig EA (2000) Getting newly synthesized proteins into shape. Cell 101:119–22
3. Bukau B and Horwich AL (1998) The Hsp70 and Hsp60 chaperone machines. Cell 92:351–66
4. Nishihara K, Kanemori M, Yanagi H, Yura T (2000) Overexpression of trigger factor prevents aggregation of recombinant proteins in Escherichia coli. Appl Environ Microbiol 66:884–9
5. Thomas JG, Ayling A, Baneyx F (1997) Molecular chaperones, folding catalysts, and the recovery of active recombinant proteins from E. coli. To fold or to refold. Appl Biochem Biotechnol 66:197–238
6. Wall JG and Plückthun A (1995) Effects of overexpressing folding modulators on the in vivo folding of heterologous proteins in Escherichia coli. Curr Opin Biotechnol 6:507–16
7. O'Reilly MS, Boehm T, Shing Y, Fukai N, Vasios G, Lane WS, Flynn E, Birkhead JR, Olsen BR, Folkman J (1997) Endostatin: an endogenous inhibitor of angiogenesis and tumor growth. Cell 88:277–85
8. Dhanabal M, Volk R, Ramchandran R, Simons M, Sukhatme VP (1999) Cloning, expression, and in vitro activity of human endostatin. Biochem Biophys Res Commun 258:345–52
9. You WK et al. (1999) Purification and characterization of recombinant murine endostatin in E. coli. Exp Mol Med 31:197–202
10. Huang X et al. (2001) Soluble recombinant endostatin purified from Escherichia coli: antiangiogenic activity and antitumor effect. Cancer Res 61:478–81
11. Sasaki T et al. (1998) Structure, function and tissue forms of the C-terminal globular domain of collagen XVIII containing the angiogenesis inhibitor endostatin. EMBO J 17:4249–56
12. Hohenester E, Sasaki T, Olsen BR, Timpl R (1998) Crystal structure of the angiogenesis inhibitor endostatin at 1.5 A resolution. EMBO J 17:1656–64
13. Yokoyama S et al. (2000) Structural genomics projects in Japan. Nat Struct Biol [Suppl] 7:943–5

CHAPTER 14

Cell-Free Expression of Soluble Human Erythropoietin

14

CORDULA NEMETZ, STEPHANIE WESSNER, SIMONE KRUPKA,
MANFRED WATZELE, WOLFGANG MUTTER*

Introduction

Erythropoietin (EPO) is a haematopoietic hormone produced by the kidney and secreted into the bloodstream to stimulate self-renewal and differentiation of late erythroid precursor cells toward mature red blood cells (Graber and Krantz 1978). As with many circulating hormones, EPO is highly glycosylated at conserved sites. It contains four complex carbohydrate chains, which have been implicated in the biological activity and stabilization of the protein (Goldwasser et al. 1974, Takeuchi and Kobata 1991). A possible role of these sugar chains for the correct folding and the solubility of EPO has been proposed, but this glycosylation is not necessary for binding of EPO to the specific EPO receptor (Delorme et al. 1992).

To produce human recombinant EPO in a biologically active form, expressions were mainly established in mammalian cells such as CHO (Chinese hamster ovary) cells or COS-1 green monkey kidney cells (Jacob et al. 1985; Lin et al. 1985). All attempts to express recombinant human EPO in *Escherichia coli* resulted in insoluble protein aggregates (Narhi et al. 1991; Boissel et al. 1993). Only with glutathione-*S*-transferase (GST) as a fusion partner, a partial solubility was achieved (Bill et al. 1995).

Here, EPO expression was investigated for the first time using an in vitro system based on *E. coli* lysates, the Rapid Translation System RTS 500. The solubility of unglycosylated EPO produced by cell-free synthesis reactions was analysed in combination with several different N-terminal fusion partners, as well as under solubilizing expression conditions.

* Cordula Nemetz (✉) (e-mail: Cordula.Nemetz@roche.com, Fax: +49–8856–607874),
  Stephanie Wessner, Simone Krupka, Manfred Watzele, Wolfgang Mutter
  Roche Diagnostics GmbH, Nonnenwald 2, 82377, Penzberg, Germany

Materials and Methods

Coupled In Vitro Transcription and Translation Reaction

To perform in vitro high yield protein-synthesis reactions, the Rapid Translation System 500 (RTS 500; Roche Applied Science, RAS, Mannheim, Germany), based on *E. coli* lysates, was used. The reactions for protein expression were performed following the instruction manual but using half-full reaction chambers. The upper chamber of the devices was filled with 0.5 ml of reaction solution. For expression in the presence of detergents, 0.1% of Tween 20 and Chaps were added to the reaction chambers. The chaperone system DnaK/DnaJ/GrpE, kindly provided by Dr. Schönfeld (Roche, Basel, Switzerland), was added in a final concentration of 1 µM.

Cloning

A recombinant DNA clone coding for EPO (accession no. P01588) without the signalling sequence was cloned together with a recognition site for factor Xa protease downstream of the T7 promoter into pIVEX expression vectors supplied within the RTS 500 kit. For N-terminal fusion proteins, the coding sequence for the fusion partner was inserted in frame into the NdeI site downstream of the hexahistidine tag of pIVEX2.4. The sequence of *E. coli* transcription-elongation factor NusA was purchased from Novagen (Madison, Wis., USA), the sequence of the head protein of bacteriophage lambda was kindly obtained from Dr. Eichinger (Roche, Penzberg, Germany), and the MBP sequence was purchased from NEB (Beverly, MA, USA).

SDS-PAGE and Western Blotting

After in vitro expression, the protein solutions were separated in supernatant and pellet fractions by centrifugation for 15 min at 14,000 rpm. We separated 2 µl of protein solution per lane by SDS-PAGE using 10% Bis-Tris gels (Invitrogen Corporation, Carlsbad, Calif., USA) and blotted on nitrocellulose membranes. As an internal standard for quantification, known amounts of a HIS-tagged 10 kDa protein were loaded. After blocking for 1 h with Western Blocking Reagent (Roche Applied Science), the membranes were incubated with horseradish peroxidase conjugated anti-his antibody solution diluted 1:4000 or with a peroxidase conjugated anti-EPO monoclonal antibody diluted 1:1500 (Roche Applied Science). Specific protein signals were visualized with the Lumi Light[Plus] substrate (Roche Applied Science) and quantified with the software program Lumi-Analyst 3.1 belonging to the Lumi-Imager[TM]F1 workstation (Roche Applied Science), as described (Nemetz et al. 2001).

Purification and Proteolysis

The in vitro-expressed fusion protein NusA-EPO was purified from the supernatant fraction by an HiTrap column (His Trap™ Kit, Pharmacia Biotech, Uppsala, Sweden) following the instruction manual. After a dialysis step in 50 mM Tris-HCl, 100 mM NaCl, 0.1 mM $CaCl_2$, pH 8.0, 0.5 µg of purified NusA-EPO protein was digested with 0.1–1.0 µg factor Xa protease (Roche Applied Science) for 3 h at 16°C and analysed by Western blots, as described.

Cell Culture Bioassay

A biological test for EPO activity was established with the UT7 cell line, as described (Komatsu et al. 1991). For that purpose 2×10^5 cells were incubated with different concentrations of the supernatant fraction of in vitro-expressed protein solution for 72 h. After incubation with tetrazolium bromide, the number of living cells was determined via densitometry.

Results

The influence of two tag sequences and several different fusion partners on the amount of in vitro-expressed EPO protein was analysed. Expression of the native N-terminus of the EPO gene without tag sequence did not result in any detectable amount of proteins (Fig. 1A; no tag). After introduction of an N-terminal Strep-tag or a hexa-histidine tag, 300 µg/ml and 900 µg/ml of in vitro-synthesized EPO could be detected by Western blot analysis (Fig. 1A, B; streptag, histag). With the head protein of the bacteriophage lambda (λHead) and the *E. coli* transcription factor NusA as fusion partners, 600 µg/ml and 900 µg/ml were obtained (Fig. 1B). The best expression yields were achieved with N-terminal maltose binding protein (MBP) fusions. Several different RTS expression reactions with a DNA template coding for hexa-histidine tagged MBP-EPO resulted in more than 2 mg/ml synthesized fusion protein (Fig. 1B; His-MBP-EPO). Protein bands corresponding to a shortened version of MBP-EPO represent truncations whose origins could not be clarified.

Analysis of the solubility of unglycosylated EPO via Western blots revealed strong differences depending on the proteins used as fusion partners and depending on the tag sequences introduced (Fig. 1C). With an N-terminal Strep-tag or a hexa-histidine tag, as well as with the λ Head protein, the whole in vitro-expressed EPO was found in the pellet fractions (Fig. 1C). However, the transcription factor NusA as a fusion partner was able to mediate a solubility of 45% to the expressed EPO protein. With N-terminal MBP-EPO fusions, up to 70% of the recombinant proteins were soluble (Fig. 1C). Thus, in vitro expression of human EPO with an N-terminal fusion of MBP resulted in high amounts of soluble recombinant protein.

Next, the solubility of in vitro-expressed MBP-EPO was analysed in more detail using different detergents and chaperones as additives during the synthesis reac-

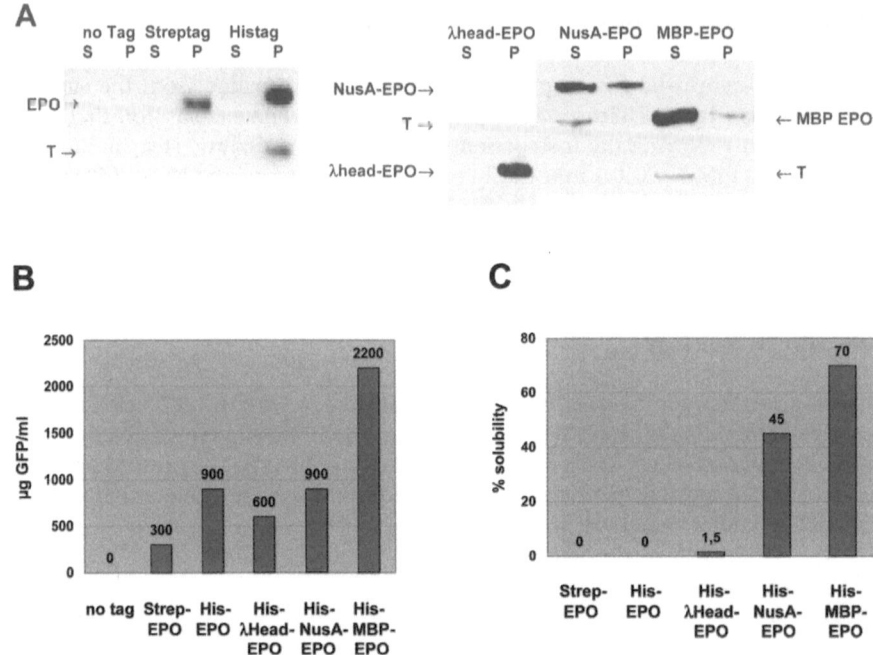

Fig. 1A–C. The amount and the solubility of in vitro-expressed EPO fusion proteins vary strongly. **A** Comparison of different in vitro-expressed EPO fusion proteins by Western blot analysis. We applied 3 μl of RTS 500 reactions on a Bis/Tris-SDS-PAA gel after separation in supernatant (*S*) and pellet (*P*) fractions, and these were analysed by Western blotting using a peroxidase conjugated anti-EPO monoclonal antibody (*left*) and an anti-His$_6$ antibody (*right*). **B** Yield of in vitro-expressed EPO fusion proteins. Determination of the amounts of total protein was performed by Western blot analysis. For densitometric calculation, the samples were spiked with an internal tagged standard protein of known concentration. The concentrations represent an average of at least three independent expressions. **C** Solubility of different EPO fusion proteins. The percentage of soluble EPO fusion proteins regarding the total amount of protein was calculated. Four to five RTS 500 expressions were quantified via Western blot analysis to calculate a mean value

tions to enhance the solubilizing conditions. In the presence of 0.1% of Tween 20, the solubility of MBP-EPO was clearly enhanced (Fig. 2). After expression with 0.1% of Chaps in the reaction chamber of the RTS 500, almost the whole amount of MBP-EPO was found in the supernatant fraction (Fig. 2). As chaperones are known to enhance the correct folding of proteins by binding to partially unfolded proteins, the well-described DnaK/DnaJ/GrpE chaperone system was added to in vitro-synthesis reactions. Compared to the expressions without additives, the chaperone system lead to a higher yield of MBP-EPO in the supernatant fraction (Fig. 2). These results were confirmed with anti-EPO ELISAs as a native test system (data not shown).

Different fusion proteins were then purified by affinity chromatography and subjected to specific proteolysis reactions with factor Xa protease. Whereas the

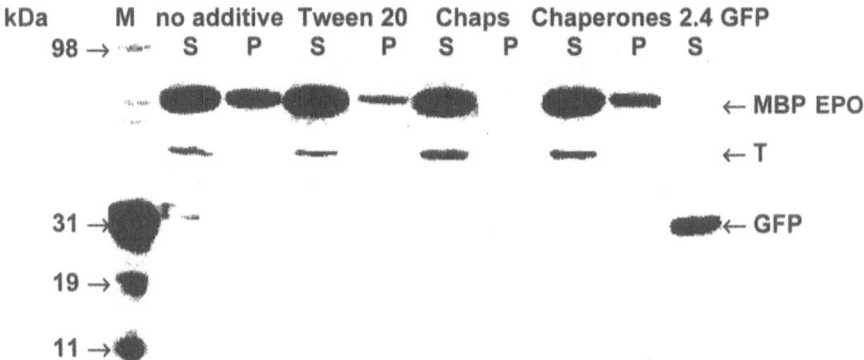

Fig. 2. Addition of detergents and chaperones increases the solubility of MBP-EPO. RTS 500 reactions with a final concentration of 0.1% detergents and 1 μM chaperones were separated in supernatant (*S*) and pellet (*P*) fractions. We applied 0.5 μl on a Bis/Tris-SDS-PAA gel and blotted. For staining, an anti-His$_6$ antibody was used

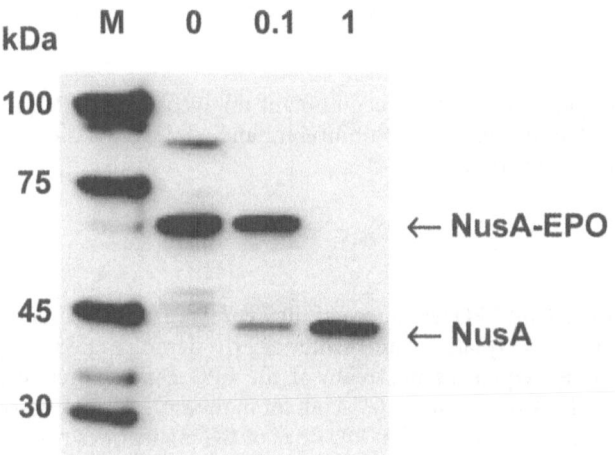

Fig. 3. NusA-EPO is cleaved by factor Xa protease. NusA-EPO was expressed in vitro in RTS 500 reactions. The supernatant fraction was purified, dialyzed and incubated with increasing amounts of factor Xa protease, as indicated (0.1 and 1 μg). The Western blot membrane was reacted with anti-His$_6$ antibody solution. *Arrows* indicate the position of uncleaved fusion protein (*NusA-EPO*) and the proteolytic fragment (*NusA*) after cleavage. *M*, molecular weight marker

MBP-EPO fusion proteins were hardly cleavable even after introduction of a longer linker region between both proteins (data not shown), NusA-EPO was successfully digested after 3 h (Fig. 3).

Finally, the biological activity of soluble in vitro-synthesized MBP-EPO fusion proteins was tested using the human erythroid precursor cell line UT7, as described (Komatsu et al. 1991). Even in the presence of the fusion partner MBP,

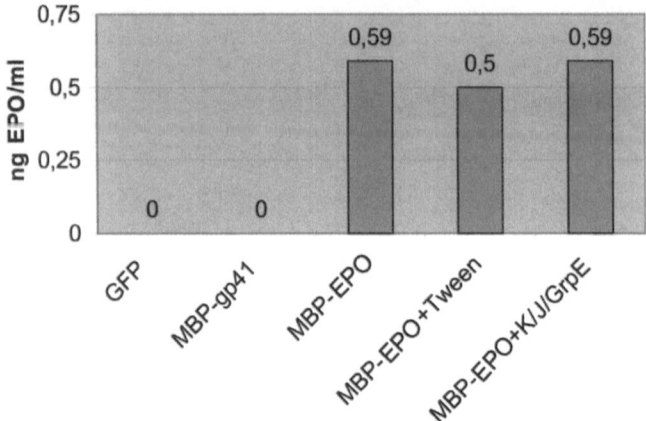

Fig. 4. MBP-EPO fusion protein exhibits biological activity in eukaryotic proliferation assays. For that purpose 2×10⁵ cells of the human leukaemic cell line UT7 were incubated with the MBP-EPO protein solution passed through a sterile filter after expression in RTS 500 reactions. The vitality of the cells was measured after 72 h via photometric determination in comparison to a standard curve. The results are representative for three independent proliferation assays

a specific proliferation enhancing effect of recombinant unglycosylated EPO on UT7 cells was detectable. The presence of solubility-enhancing reagents did not further increase this biological activity (Fig. 4).

Discussion

In this study, human recombinant EPO was successfully expressed with the in vitro RTS system based on *E. coli* lysates. The results clearly demonstrated that depending on the additional sequences upstream of the EPO gene, expression yields varied strongly (Fig. 1). Analysis of the GC-content in the translational start region of corresponding mRNAs revealed that low rates of GC nucleotides resulted in high in vitro-sythesized protein amounts. These data could reflect an inhibitory effect to efficient translation caused by mRNA folding.

In addition to in vitro expression rates, the solubility of EPO differed depending on the N-terminal fusion partner (Fig. 1). The best ratios between supernatant and pellet fractions were obtained with an N-terminal MBP sequence. In vivo, only glutathione-S-transferase (GST) as a fusion partner mediated a partial solubility (Bill et al. 1995). In our in vitro studies, only 10% of the GST-EPO fusion proteins were soluble (data not shown).

Not all mammalian cells produce a form of EPO that is biologically active. As shown, sugar chains of recombinant human EPO expressed in CHO cells relate most closely to urinary human EPO but tend to contain more N-acetyllactosamine-repeating elements than naturally found (Takeuchi et al. 1988). However, glycosylation of EPO seems to be not absolutely required for its biological activity (Delorme et al. 1992). As shown here, even after expression in *E. coli* lysates, where glycosylation is not possible, and under conditions that were not optimized for the formation of disulfide bridges, a biological activity could be clearly demonstrated (Fig. 4). The fact that this EPO-induced proliferation signal was relatively weak may be explained by the presence of the fusion partner MBP. This issue will be examined in the future.

In contrast to the situation in vivo, an in vitro system offers the advantage of expressing proteins in the presence of additives like detergents. In the case of EPO, this possibility helped to increase solubility up to 100% (Fig. 2).

References

Bill RM, Winter PC, McHale CM, Hodges VM, Elder GE, Caley J, Flitsch SL, Bicknell R, Lappin TRJ (1995) Expression and mutagenesis of recombinant human and murine erythropoietins in *Escherichia coli*. Biochim Biophys Acta 1261:35–43

Boissel J-P, Lee W-R, Presnell SR, Cohen FE, Bunn HF (1993) Erythropoietin structure-function relationship: mutant proteins that test a model of tertiary structure. J Biol Chem 268:15983–93

Delorme E, Lorenzini T, Giffin J, Martin F, Jacobsen F, Boone T, Elliott S (1992) Role of glycosylation on the secretion and biological activity of erythropoietin. Biochemistry 31:9871–6

Goldwasser E, Kung CK-H, Eliason J (1974) On the mechanism of erythropoietin-induced differentiation. 13. The role of sialic acid in erythropoietin action. J Biol Chem 249:4202–6

Graber SE and Krantz SB (1978) Erythropoietin and the control of red blood cell production. Annu Rev Med 29:51–66

Jacobs K, Shoemaker C, Rudersdorf R, Neill S. D, Kaufman RJ, Mufson A, Seehra J, Jones SS, Hewick R, Fritsch EF, Kawakita M, Shimizu T, Miyake T (1985) Isolation and characterization of genomic and cDNA clones of human erythropoietin. Nature 313:806–10

Komatsu N, Nakauchi H, Miwa A, Ishihara T, Eguchi M, Moroi M, Okada M, Sato Y, Wada H, Yawata Y (1991) Establishment and characterization of a human leukemic cell line with megakaryotic features: dependency on granulocyte-macrophage colony-stimulating factor, interleukin 3, or erythropoietin for growth and survival. Cancer Res 51:341–8

Lin F-K, Suggs S, Lin C-H, Browne JK, Smalling R, Egrie JC, Chen KK, Fox GM, Martin F, Wasser Z (1985) Cloning and expression of the human erythropoietin gene. Proc Natl Acad Sci USA 82:7850–4

Narhi LO, Arakawa T, Aoki KH, Elmore R, Rohde MF, Boone T, Strickland TW (1991) The effect of carbohydrate on the structure and stability of erythropoietin. J Biol Chem 266:23022–6

Nemetz C, Reichhuber R, Schweizer R, Hloch P, Watzele M (2001) Reliable quantification of in vitro-synthesized green fluorescent protein: comparison of fluorescence activity and total protein levels. Electrophoresis 22:966–969

Takeuchi M, Takasaki S, Miyazaki H, Kato T, Hoshi S. Kochibe N, Kobata A (1988) Comparative study of the asparagine-linked sugar chains of human erythropoietins purified from urine and the culture medium of recombinant Chinese hamster ovary cells. J Biol Chem 263:3657–3663

Takeuchi M and Kobata A (1991) Structures and functional roles of the sugar chains of human erythropoietins. Glycobiology 1:337–46

CHAPTER 15

Cell-Free Expression of a 127 kDa Protein: 15
The Catalytic Subunit of Human Telomerase

ROBIN STEIGERWALD, CORDULA NEMETZ,
BÄRBEL WALCKHOFF, THOMAS EMRICH*

Abbreviations

hTERT   human telomerase reverse transcriptase
hTR   human telomerase RNA
HA   epitope tag from hemagglutinin
RTS   rapid translation system
His   hexa histidine tag

Introduction

Telomeres are highly repetitive sequences at the linear ends of eukaryotic chromosomes (Moyzis et al. 1988). Together with bound protein factors telomeres protect chromosomes from degradation and recombination (Blackburn et al. 1995). Telomeres are maintained by the ribonucleoprotein telomerase. Telomerase also counteracts progressive shortening due to telomeric loss during cellular replication. Cells with inactivated telomerase show a decreased capacity for replication followed by senescence and cell death (Feng et al. 1995).

Telomerase activity cannot be detected in the vast majority of somatic cells but is found in normal germ line cells, stem cells as well as in 85%–90% of human tumour cells (Kim et al. 1994). Therefore, the enzyme is a highly attractive candidate for tumour diagnostics. The catalytic subunit (hTERT, human telomerase reverse transcriptase) of the ribonucleoprotein seems to be strongly regulated, whereas the telomerase RNA component (hTR; Feng et al. 1995) is not restricted to tumour cells, as shown by qualitative analysis.

The gene coding for the catalytic subunit hTERT of the ribonucleoprotein has recently been identified (Nakamura et al. 1997). So far, using conventional cell-based expression techniques, it is not possible to generate large quantities of active telomerase for biochemical characterization or inhibitor screening. Suffi-

* R. Steigerwald, C. Nemetz, B. Walckhoff, T. Emrich
  (✉) (e-mail: thomas.emrich@roche.com, Tel.: +49–8856–603122, Fax: +49–8856–603631)
  Roche Applied Science, Nonnenwald 2, 82372 Penzberg, Germany

cient amounts of soluble full-length protein to generate antibodies can not be produced in a cost-effective manner. The cell-free RTS system was used in the following study to overcome this limitation.

Materials and Methods

Construction of Expression Plasmids

pGRN282, full-length *hTERT*-cDNA with adapted *E. coli* codon usage (J.B. Trager and G.B. Morin, Geron Corp.) was modified for expression in the RTS system.

The sequence coding for the 9-aa *influenza virus* hemagglutinin (HA) epitope (YPYDVPDYA) was fused to the 3'-end of *hTERT* by PCR-mutagenesis. The resulting HA-tagged hTERT-cDNA (hTERT-HA) was fused in frame into the *NdeI* and *SacI* restriction sites of the pIVEX-vectors 2.0, 2.2, and 2.4b (Roche Applied Science, Mannheim, Germany). To test the influence of various fusion partners on expression efficiency and solubility, the corresponding DNA sequences were inserted into the *NdeI* site between the His-tag and the *hTERT*-gene of pIVEX2.4b-hTERT-HA.

In Vitro Expression

Cell-free *in vitro* expression, based on *E. coli* lysates, was performed with the RTS 500 kit (Roche Applied Science, Mannheim, Germany) under the conditions detailed below. Purified chaperones DnaK and DnaJ were kindly provided by Dr. Schönfeld (Roche, Basel, Switzerland) and added to the *E. coli* lysate prior to expression as indicated.

SDS-PAGE and Western Blotting

After SDS-PAGE (Bis-Tris 4–12% polyacrylamide gel; Invitrogen, Carlsbad, CA), proteins were transferred to PVDF membranes and membranes were blocked with 1% Western blocking solution in TBS-T. A high-affinity, anti-HA antibody was used to detect the HA-tagged hTERT (Anti-HA, High Affinity (3F10) and Anti-HA-Peroxidase, High Affinity; Roche Applied Science, Mannheim, Germany). A purified HA-tagged 30-kDa protein was used as a reference protein to quantify the HA-tagged hTERT extracts. An Sf21-extract from baculovirus-infected cells expressing hTERT was run as a positive control (Mikuni et al. 2001). The Multi-Tag-Marker (Roche Applied Science, Mannheim, Germany) or the Full-Range-Rainbow-Marker (Amersham Pharmacia, Uppsala, Sweden) was used to indicate molecular weight. Specific protein bands were detected using the Lumi Light Plus substrate on a Lumi-Imager F1 Workstation and quantified with the LumiAnalyst software package (all products: Roche Applied Science, Mannheim, Germany).

Reconstitution of Telomerase Complexes

The telomerase RNA component hTR was synthesized *in vitro* using pGRN164 linearized with *FspI* in the Ribomax RNA transcription system (Promega) or the SP6/T7 Transcription Kit (Roche Applied Science, Mannheim, Germany). After incubation at 37^0C for 2 hours, synthesized hTR was treated with DNaseI and ethanol precipitated. Extracts of hTERT produced *in vitro* or from Sf21 cells (baculovirus control) were mixed with hTR and incubated at 30^0C for 90 min to allow for complex formation. These telomerase complexes were diluted 16-fold in CHAPS lysis buffer (10mM Tris-CL, pH 7.5, 1mM $MgCl_2$, 1mM EGTA, 0.5% CHAPS, 5mM 2-mercaptoethanol, 10% glycerol) and used for detection of telomerase activity as described below.

Measurement of Telomerase Activity

Telomeric repeat amplification protocol (TRAP) was performed with the *TeloTAGGG* PCR ELISA *Plus* (Roche Applied Science, Mannheim, Germany) or according to standard protocols (Kim et al. 1994) using a radiolabeled TS primer. Detection and analysis of PCR products separated by gel electrophoresis were performed on a phosphoimager.

Results

During heterologous expression of hTERT in different prokaryotic and eukaryotic expression systems, we found either very low levels of hTERT or high levels of insoluble full-length hTERT or hTERT fragments. Expression of full-length hTERT often appeared to be toxic for the cells. Thus, we used the cell-free in vito expression system RTS 500, based on the *E. coli* S30 lysate. For the convenient and sensitive detection of limited amounts of full-length hTERT, an HA-tag (8-aa hemagglutinin epitope) was added to the C-terminus of hTERT. The fusion of a HA-tag had no effect on expression of hTERT in human cell lines or in the baculovirus system (data not shown). Similarly, in the cell free expression system RTS 500, we found equal production of untagged hTERT and hTERT-HA protein (Fig. 1B, His-hTERT and His-hTERT-HA). Thus, in the following experiments the HA epitope was used for detecting and indirectly quantifying hTERT-HA.

Initial expression of full-length hTERT in the RTS 500 resulted in a protein level lower than 100 ng/ml (Fig. 1A, hTERT-HA). The nature of N-terminal tag sequences and fusion partners strongly influences the level of transcriptional initiation and protein expression in the RTS 500 system (Graentzdoerffer et al. this volume). Therefore, we checked the ability of several N-terminal tags and fusion partners to increase the level of hTERT expression. While the addition of a Strep-tag (Skerra and Schmidt 1999; StrepTag II; WSHPQFEK) increased the expression level of Strep-hTERT-HA by a factor of only 1.5-fold relative to hTERT-HA

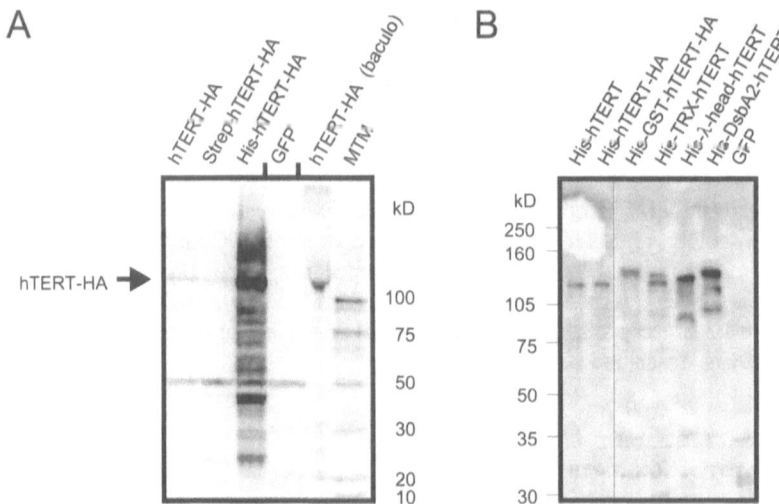

Fig. 1. Influence of N-terminal tags and fusion partners on cell-free expression of hTERT in *E. coli* S30 lysate. **A** hTERT was cloned in pIVEX vectors, resulting in the expression of hTERT-HA (pIVEX2.0) and the N-terminally tagged Strep-hTERT-HA (pIVEX2.2) and His-hTERT-HA (pIVEX2.4b). All constructs have a C-terminal HA-tag. Expression was performed with the RTS 500 kit for 16 h at 30°C. Aliquots of the reaction mixture were separated by SDS-PAGE, and blotted. Blots were probed with a HA-specific antibody. The Multi-Tag-Marker (*MTM*) was used to estimate molecular weight. **B** Different gene fragments were inserted between the His-tag and hTERT-HA in pIVEX2.4-hTERT-HA (His-hTERT-HA) and compared to a non-HA-tagged hTERT construct in a modified RTS System. Aliquots of the reaction mixture were separated on SDS-PAGE, and blotted. The blot was probed with an hTERT-specific antibody, raised from a synthetic peptide. The Rainbow Marker was used for estimation of molecular weight

(Fig. 1A), the fusion of a N-terminal His-tag to the hTERT protein strongly increased the expression level of hTERT by about 150-fold. Typically 5–10 µg/ml of hTERT were found in the reaction mixture. Therefore, His-hTERT-HA was used in further experiments.

We next examined the influence on hTERT expression of protein fusion partners, which have been observed to increase the expression or solubility of many recombinant proteins (Autexier and Bachand 1999; La Vallie et al. 1993; Forrer and Jaussi 1998; Zhang et al. 1998). When glutatione-S-transferase- and thioredoxin fragments were introduced between His-tag and hTERT as fusion partners, expression of hTERT was only slightly higher than the level of His-hTERT-HA (Fig. 1B, His-GST-hTERT, factor 1.5; His-TRX-hTERT-HA, factor 1.2). In contrast, inserting the headprotein of phage lambda and the disulfide bridge oxidoreductase showed enhanced expression with factors of 3.5 and 4.4, respectively (Fig. 1B, His-λ-head-hTERT and His-DsbA2-hTERT).

In addition to high-level expression of hTERT, we focused on generating soluble protein. Thus we analyzed the solubility of hTERT generated with the *E. coli* lysate-based cell-free expression system; this was done by comparing the pellet and supernatant fractions after centrifugation at 20,000 *g* for 30 min. Using the

RTS 500 kit, less than 30% of hTERT were found in the supernatant (data not shown). In order to enhance the solubility of hTERT in the reaction mixture, we investigated the influence of the four fusion partners described above. In addition to enhancing expression level, these four protein fragments are known to improve the solubility of many proteins (GST: Bill et al. 1995; TRX: La Vallie et al. 1993; λ-head: Forrer and Jaussi 1998; DsbA2: Zhang et al. 1998). As discussed above, GST and TRX as fusion partners showed unaltered expression efficiency of hTERT, while λ-head and DsbA2 enhanced expression. Interestingly, after expression in the RTS system, these fusion partners did not influence the solubility of hTERT. The relative amounts of hTERT found in pellet and supernatant fractions after 30 min of centrifugation at 20,000 g did not change (data not shown). Lowering the incubation temperature from 37°C to 25°C in a modified expression protocol increases the solubility of the produced hTERT to 50% (Fig. 2) without influencing the expression level.

Chaperones are well known as cellular tools providing proper protein folding co-translationally and in many cases efficient renaturation after transcription. We tested the effect of purified chaperones on the generation of soluble hTERT. Different amounts of purified E. coli chaperones DnaK and DnaJ were added to the reaction mixture before the expression of hTERT. In the presence of 2 µM or more of DnaK and DnaJ, 96%–98% of the synthesized hTERT was found in the supernatant (Fig. 3). Further centrifugation of the supernatant at 100,000 g did not

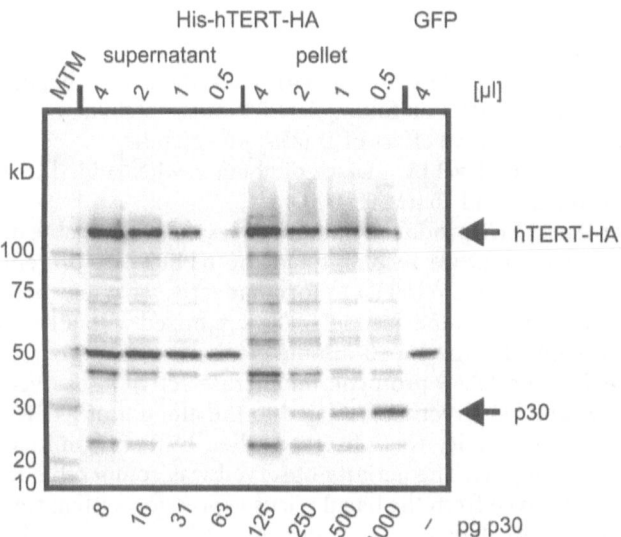

Fig. 2. Solubility of hTERT expressed in *E. coli* lysate. After centrifugation the reaction mixture for 30 min at 20,000 g, the pellet was resuspended in the original volume of buffer. As indicated, decreasing amounts of supernatant and pellet fractions were separated on SDS-PAGE, and blotted. The blot was probed with a HA-specific antibody. To quantify the indicated amounts of p30, a HA-tagged 30-kDa marker protein was spiked into the reaction mixture prior to PAGE. The Multi-Tag-Marker (*MTM*) was used to estimate the molecular weight

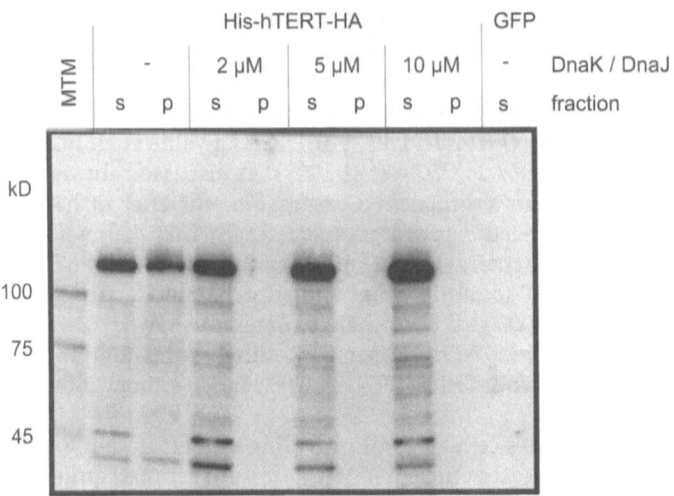

Fig. 3. Influence of chaperones on hTERT expression in *E. coli* lysate. Expression of hTERT was performed in the presence of DnaK and DnaJ at the final concentrations indicated for 16 h at 25°C. Following centrifugation of the reaction mixture for 30 min at 20,000 g, the pellet was resuspended in the original volume of buffer. Aliquots of the supernatant (*s*) and pellet (*p*) fractions were separated on SDS-PAGE, and blotted. Blots were probed with a HA-specific antibody. The Multi-Tag-Marker (*MTM*) was used to estimate the molecular weight

result in new precipitation of hTERT (data not shown). Thus, the *E. coli* chaperones DnaK and DnaJ had a strong effect on folding hTERT and/or stabilizing the soluble form. In addition to the positive effect of DnaK/J on solubility, the chaperones also elevated the expression level by a factor of about 2, when added to a final concentration between 2 μM and 10 μM (Fig. 3).

To determine whether the hTERT produced in the RTS system was capable of folding into a native conformation, we reconstituted the hTERT-HA protein with the RNA component of telomerase (hTR) to form an active enzyme complex. Different amounts of *in vitro* synthesized hTERT were mixed with a fixed amount of *in vitro* transcribed hTR. After a 90-minute incubation, the reconstituted enzyme was assayed by the TRAP protocol. Telomerase activity observed in all of the reconstituted samples directly correlated to the amount of hTERT used for the assay (Fig. 4). No activity was observed when hTR was omitted from the reconstitution mix (Fig. 4). The activity observed was comparable to that obtained when hTERT derived from the baculovirus expression system was used as a protein source (Fig. 4). This demonstrates that at least some of the hTERT produced in the RTS 500 system is capable of forming a native conformation.

For producing hTERT in RTS 500, we found that lower temperatures and N-terminal tags or fusion partners as well as the addition of chaperones positively influenced the level of hTERT production and solubility. This soluble full-length hTERT now can be used for biochemical studies and antibody screening.

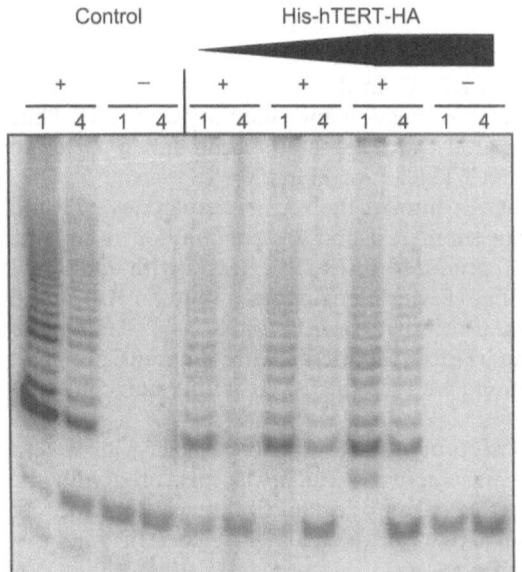

Control   His-hTERT-HA

| + | – | + | + | + | – | addition of hTR |
| 1 | 4 | 1 | 4 | 1 | 4 | 1 | 4 | 1 | 4 | 1 | 4 | fold dilution for TRAP |

Fig. 4. Reconstitution of hTERT expressed in *E. coli* lysate. Reconstitution of active telomerase was performed by incubating extracts containing hTERT protein with *in vitro* transcribed hTR for 90 minutes at 30°C. Following reconstitution, the reaction mixtures were diluted 16-fold in CHAPS lysis buffer. One and four-fold dilutions were used to determine telomerase activity by the TRAP assay. hTERT expressed in the baculovirus system was used as a positive control for reconstitution. hTR was omitted from the reconstitution mixture as a negative control for telomerase activity

Discussion

The RTS 500 System enables many proteins to be produced that could not be expressed in various other prokaryotic and eukaryotic systems. Cell-free expression of hTERT led to what is presumed to be a native full-length protein. The addition of an HA-tag to the C-terminus of hTERT allowed easy and sensitive detection of hTERT produced in various expression systems. This epitope tag seems to influence neither the expression efficiency nor the solubility of the protein. Expression efficiency could be enhanced by adding N-terminal tags (Fig. 1A). The underlying mechanism could be a stable folding of the mRNA after the fast transcription with T7-polymerase, leading to sequential transcription and translation rather than to the coupled transcription/translation reaction typical for expression in *E. coli* cells (Graentzdoerffer et al. and citations therein, this volume). A computer-aided structure prediction (mfold; GCG-software-package) suggests rather stable mRNA structures with easily accessible ribosomal binding sites and AUG start codon when the N-terminal His-tag is used for hTERT expression. In contrast the Strep-tag construct showed unfavourable structures and energy levels. Thus far the predicted mRNA-structures go together with the expression results. Only the N-terminally non-tagged hTERT expression with a fairly well-predicted structure and energy level showed low-level expression (Fig. 1A). This indicates that there seem to be factors other than the mRNA-structure that influence the translation of hTERT in the RTS 500. We found that the chaperones DnaK and DnaJ enhanced expression and solubility of hTERT. Thus it appears that processes occurring after translational initiation have a profound influence on the production of native hTERT protein in the RTS system.

Holt et al. (1999) reported that the human molecular chaperones p23 and Hsp90 directly interact with human telomerase and are essential for its activity. The enhanced solubility of hTERT produced in *E. coli* extracts with additional bacterial chaperones DnaK and DnaJ (*E. coli* homologues of human Hsp70 and Hsp40) indicates that they may make a similar contribution to hTERT folding to a native state. Whether DnaK/J and hTERT still interact after the translation and guided folding is unclear and should be investigated in immunoprecipitation experiments *in vitro*.

In the first reconstitution experiments of the RTS-produced hTERT with *in vitro* transcribed hTR, we found telomerase activity, which was measured with the PCR-based TRAP assay. This shows that eukaryotic modifications or cofactors are not necessary for at least the basal activity of the telomerase core enzyme produced in a cell-free bacterial system.

The hTERT material produced in RTS 500-*E. coli* lysates can now be used in biochemical studies to investigate how eukaryotic chaperones and other cofactors contribute to the activity of a telomerase holoenzyme.

References

Bill RM, Winter PC, McHale CM, Hodges VM, Elder GE, Caley J, Flitsch SL, Bicknell R, Lappin TR (1995) Expression and mutagenesis of recombinant human and murine erythropoietins in *Escherichia coli*. Biochimica and Biophysica Acta 1261:35–43

Blackburn EH, Greider CW (1995) Telomeres. Cold Spring Harbor Laboratory Press, Cold Spring Harbor, NY

Feng J, Funk WD, Wang S-S, Weinrich SL, Avilion AA, Chiu C-P, Adams RR, Chang E, Allsopp RC, Yu J, Le S, West MD, Harley CB, Andrews WH, Greider CW, Villeponteau B (1995) The RNA Component of Human Telomerase. Science 269:1236–1241

Forrer P, Jaussi R (1998) High-level expression of soluble heterologous proteins in the cytoplasm of *Escherichia coli* by fusion of the bacteriophage lambda head protein D. Gene 224: 45–52

Holt SE, Aisner DL, Baur J, Tesmer VM, Dy M, Ouelette M, Trager JB, Morin GB, Toft DO, Shay JW, Wright WE, White MA (1999) Functional requirement of p23 and Hsp90 in telomerase complexes. Genes and Development 13:817–826

Kim NW, Piatyszek MA, Prowse KR, Harley CB, West MD, Ho PL, Coviello GM, Wright WE, Weinrich SL, Shay JW (1994) Specific association of human telomerase activity with immortal cells and cancer. Science 266:2011–2015

La Vallie ER, DiBlasio EA, Kovacic S, Grant KL, Schendel PF, McCoy JM (1993) A thioredoxin gene fusion expression system that circumvents inclusion body formation in the *E. coli* cytoplasm. Bio/Technology 11:187–193

Mikuni O, Trager JB, Ackerly A, Weintrich SL, Akira A, Yamashita Y, Mizukami T, Anazawa H (2001) Reconstitution of telomerase activity utilizing human catalytic subunit expressed in insect cells. Biochemical and Biophysical Research Communications, submitted

Moyzis RK, Buckingham JM, Cram LS, Dani M, Deaven LL, Jones MD, Meyne J, Ratliff RL, Wu JR (1988) A highly conserved repetitive DNA sequence, (TTAGGG)n, present at the telomeres of human chromosomes. Proceedings of the National Academy of Sciences of the United States of America 85:6622–6662

Nakamura TM, Morin GB, Chapman KB, Weinrich SL, Andrews WH, Lingner J, Harley CB, Cech TR (1997) Telomerase catalytic subunit homologs from fission yeast and human. Science 277:955–959

Skerra A, Schmidt TGM (1999) Applications of a peptide ligand for streptavidin: the Strep-tag. Biomolecular Engineering 16:79–186

Zhang Y, Olsen DR, Nguyen KB, Olson PS, Rhodes ET, Mascarenhas D. Expression of eucaryotic proteins in soluble form in Escherichia coli. Protein Expression and Purification 12:159–165

The Expression of Disulfide Bonded Proteins in Cell-Free Protein Expression   16

ERHARD FERNHOLZ*, KATRIN ZAISS, HÜSEYIN BESIR,
WOLFGANG MUTTER

Short description

Due to the reducing conditions in common *E.coli* based cell-free protein Expression systems the formation of disulfide bonds is not possible. By changing the redox environment and by the addition of disulfide bond-forming catalysts it could be demonstrated that significant percentages of active disulfide-bonded proteins can be produced.

Key words

Disulfide bond, cell-free protein expression, *E.coli* lysate, folding.

Introduction

The expression of active disulfide bonded proteins is still a matter of trial and error. The reason for this dilemma is that in nature the particular proteins get folded in the endoplasmatic reticulum (ER) as part of the secretory pathway and it is poorly understood how all the folding catalysts and chaperones in this compartment cooperate with each other [1]. Furthermore, it is very hard to manipulate the composition in the ER making it almost impossible to quickly evaluate the participating factors of each folding process.

Cell-free protein expression allows easy manipulation of the reaction conditions by just adding necessary factors or chemicals. Recently, the problem of productivity could be solved pushing this technology towards a standard procedure. Here we show a convenient method to improve the yield of properly folded disulfide bonded protein using a cell-free protein expression system.

* Erhard Fernholz, (✉) (e-mail: erhard.fernholz@roche.com)
  Katrin Zaiss, Hüseyin Besir, Wolfgang Mutter
  Roche Diagnostics GmbH, Nonnenwald 2, D-82372 Penzberg

Materials and Methods

Materials:

All materials if not stated differently had been made available by Roche Diagnostics.
PDI was kindly provided by J. Buchner (Technische Universität Munich).
The gene of rPA was cloned into pIVEX 2.4 using conventional techniques.
For cell-free protein expression, the RTS 500 *E.coli*, circular template had been used as it is or after substitution of the reducing agent against glutathione (oxidized and reduced). All reactions were stirred at 30 C for at least 18 hours using the device (continuous exchange technology) supplied in the kit.

Activity assay for rPA:

The activity of expressed rPA was assayed using the substrate S2288 (Chomogenix) at 37 C. The release of p-nitroanilin was detected spectroscopically at 405 nm. The amounts of functional product were calculated by comparison with rPA standard protein.

Results

The recombinant plasminogen activator (rPA) is a 39 kDa fragment of the tissue plasminogen activator (tPA) and has 9 disulfide bonds. Since standard protein is available and the activity can easily be quantified it appeared to be a good test protein to study the influences on disulfide bond formation in a cell-free protein expression setup.

The minimal requirement for oxidative folding is to use a redox system, usually reduced and oxidized glutathione (GSH/GSSG). Initial results comparing strictly reducing conditions (by using the RTS 500 *E.coli*, circular template kit as it is) with redox conditions in which the reducing agent had been completely substituted by GSH/GSSG showed that active product had been formed. However, even optimized redox conditions (GSH:GSSG = 10:1) gave only 8% active product at maximum (Fig. 1). The remaining 92% were found in the pellet.

Since incorrect disulfide bonds could have been formed under these conditions, we added different protein disulfide isomerases (PDI, bovine or human) as a disulfide bond repair system but ended up with only limited increase (15%, 30μg/ml) of active product (Fig 1). Interestingly, neither higher amounts of PDI nor a higher concentration of GSH/GSSG further increased the yield. Also the addition of DsbA, a periplasmatic disulfide oxidase of *E.coli* showed no effect (data not shown). We concluded that the process of getting functional rPA was not only dependant on oxidative conditions and on a disulfide bond repair system. In fact, it is known that numerous chaperones are present in the oxidizing compartment of the cells (periplasm, ER). We therefore screened for additional factors and detected a significant increase in activity and solubility if GroEL/GroES (GroE) were added (Fig. 2).

These results are remarkable since GroEL/GroES are cytosolic chaperones of *E.coli* and do not participate in the folding process of disulfide bonded proteins in nature. Furthermore, they seem to work equally well under these redox conditions. Finally, further addition of a peptidyl prolyl cis/trans isomerase led to almost 120 µg/ml active rPA or 65% of total expressed protein (Fig. 3 and 4).

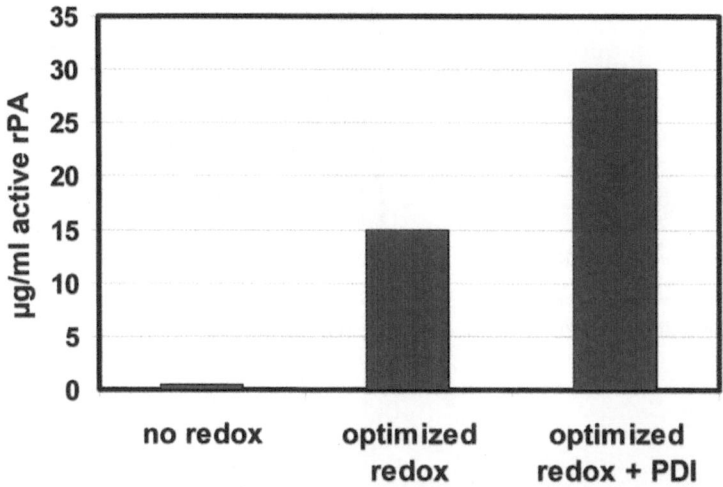

Fig. 1. Cell-free expression of rPA under different redox conditions with and without PDI shows a clear increase of the yield of functional rPA for the combination of PDI and optimized redox conditions

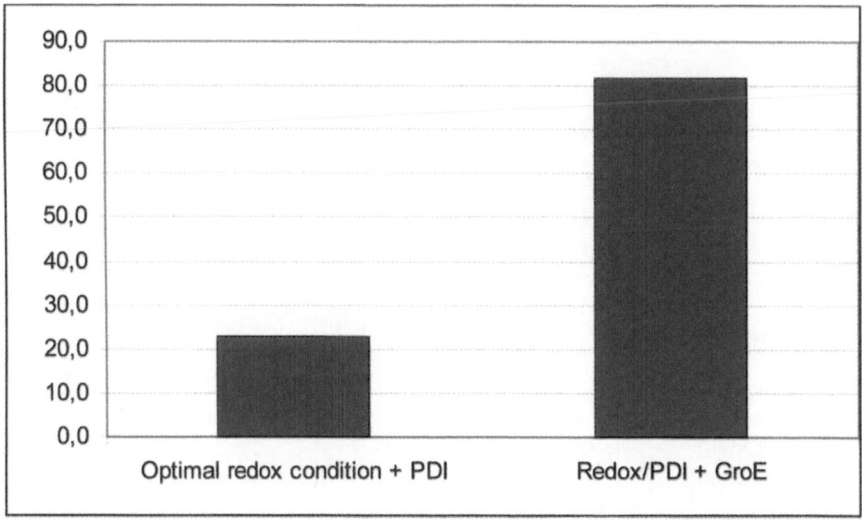

Fig. 2. Cell-free expression of rPA using optimized redox conditions with and without GroEL/ES supplement (yields in µg/ml active rPA)

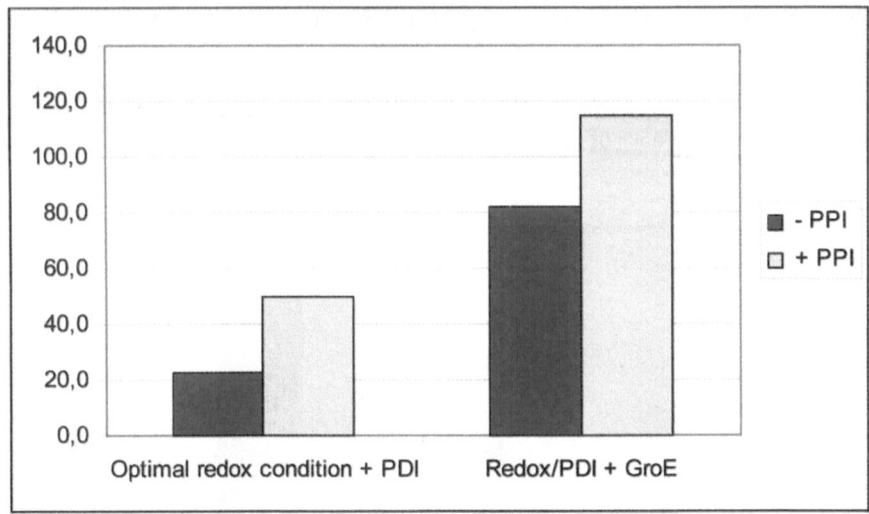

Fig. 3. Effect of the addition of PPI on the yield of active rPA (μg/ml)

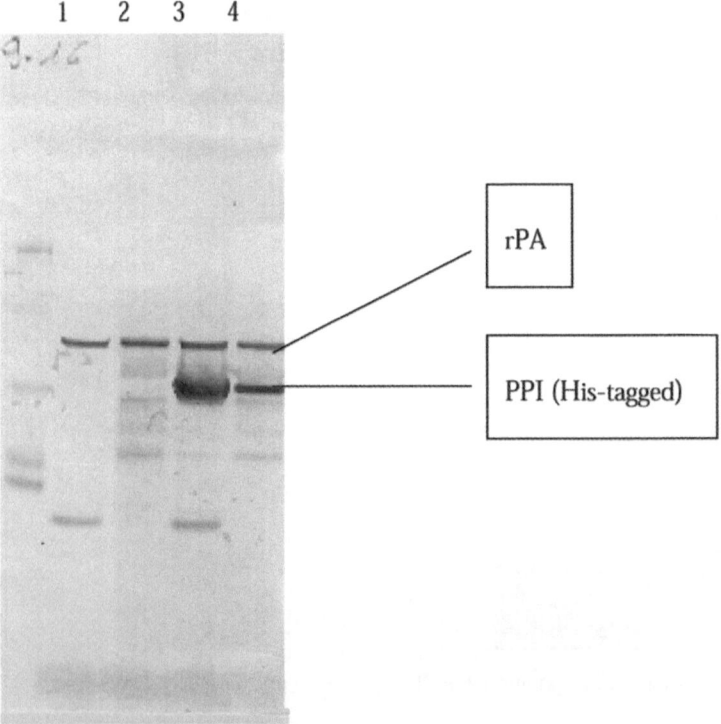

Fig. 4. Effect of the addition of PPI (His-tagged) on the solubility of rPA analyzed via Western blot (lanes 1 and 3 contain the soluble fraction; lanes 2 and 4 show the insoluble fraction)

Discussion and outlook

In this paper we could demonstrate that the conditions of a cell-free protein expression system can be adapted for the expression of active recombinant plasminogen activator as an example of a disulfide bonded protein. Because cell-free expression is basically considered as an open system which can easily be manipulated we could rapidly screen for the most efficient factors and found that even cytosolic chaperones can be used to express active disulfide bonded proteins.

References

1. A.R. Frand, I.W. Cuozzo, C.A. Kaiser (2000) Trends in Cell Biology, Vol. 10, p203

Special Topics and Short Communications V

Structural and Functional Compensation by Proteins for the RNA Deficit of Animal Mitochondrial Translation Systems

17

TSUTOMU SUZUKI[1,2], TAKASHI OHTSUKI[2], YOH-ICHI WATANABE[3],
MAKI TERASAKI[1], TAKAO HANADA[2] and KIMITSUNA WATANABE[1,2]

Abstract

Compared to the more usual bacterial translation systems, animal mitochondrial translation systems are unique in their use of extremely simplified RNAs. It seems that the RNA deficit of these mitochondrial systems is compensated by the presence of novel proteins and/or bacterial homologues that have longer N- or C-terminal regions, certain of which may compensate for missing regions in animal mitochondrial RNA. On the basis of two distinct cases we discuss the possibility that certain structures and functions of bacterial RNAs were transferred to proteins during the evolutionary process from ancestral to extant organisms. Our analyses may provide appropriate models for an experimental verification of the concept of transition from the "RNA world" to the "RNP world" in the early process of the evolution of life.

Introduction

The mitochondrion is a eukaryotic cellular organelle that serves as an ATP supplier. It possesses its own unique and independent genetic information and translation systems. It is believed that mitochondria originated through the symbiosis of an aerobic eubacterium within an ancestral eukaryote. The mitochondrial (mt) translation system thus originated from a eubacterium and has since evolved separately.

Animal mt translation systems possess a characteristic feature in that the RNA is extremely simple. In addition, only 22 different species of tRNA are required for the translation of about 60 different codons in animal mitochondria. This number of tRNA species is considered to be the minimal to enable decoding of the genetic code. In comparison, more than 50 species of tRNA are required in other

[1] Department of Integrated Biosciences, Graduate School of Frontier Sciences,
University of Tokyo, Kashiwanoha 5-1-5, Kashiwa-shi, Chiba, 277-8562, Japan
[2] Department of Chemistry and Biotechnology, Graduate School of Engineering,
University of Tokyo, Hongo 7-3-1, Bunkyo-ku, Tokyo 113-8656, Japan
[3] Department of Biomedical Chemistry, Graduate School of Engineering,
University of Tokyo, Hongo 7-3-1, Bunkyo-ku, Tokyo 113-0033, Japan.
e-mail: kw@kwl.u-tokyo.ac.jp, telephone:+81(0)471363601, fax:+81(0)471363600

translation systems, such as in prokaryotic and eukaryotic cytoplasmic systems (Sprinzl *et al.* 1998).

The mt translation system of the nematode is different in that it possesses Tarm-lacking tRNAs (Keddie *et al.* 1998; Okimoto *et al.* 1992; Okimoto and Wolstenholme 1990; Wolstenholme *et al.* 1987). These RNAs are not found in any other system except for those of some mollusca (Yamazaki *et al.* 1997) and arthropoda (Masta 2000). Of the full complement of 22 tRNA species, 20 are Tarm lacking. The other two are specific for serine, which lack the D arm and have a truncated Tarm [tRNA$^{Ser}_{GCU}$ specific for AGY (Y = U and C) codons and tRNA$^{Ser}_{UGA}$ for UCN (N = A, U, G and C) codons]. The nematode *Ascaris suum* mt tRNA$^{Ser}_{GCU}$ is the shortest tRNA of all tRNAs reported so far, with a chain length of 54 nucleotides (Sprinzl *et al.* 1998). We have identified a new type of elongation factor (EF-Tu) gene with an extended C-terminus, which is responsible for Tarm lacking tRNAs (Ohtsuki *et al.* 2001). It seems that the missing Tarm in tRNAs of nematode mitochondria is compensated through a novel EF-Tu with an extended C-terminus, because the usual EF-Tu requires the tRNA Tarm for its recognition (Nissen *et al.* 1995).

Another characteristic feature of mitochondria is the composition of ribosomes. The mammalian mitochondrial ribosome (mitoribosome) is protein-rich, and the protein to RNA ratio is 3 to 1 (Patel *et al.* 2000). The total length of mitoribosomal RNA (mt rRNA) is shortened to about half of that found in bacterial rRNA. However, the increased protein content of the mitoribosome ensures that the overall sizes of the ribosomes from both organisms are approximately equivalent.

We have systematically analyzed ribosomal proteins from bovine mitoribosomes using a combination of 2D gel electrophoresis and LC/MS analysis (Suzuki *et al.* 2001a; Suzuki *et al.* 2001b). We found that mitochondrial ribosomal proteins whose binding sites on mt rRNA are shortened or lost carry extended peptide chains. In contrast, ribosomal proteins with conserved rRNA-binding sites have a similar molecular mass to their *E. coli* counterparts. From this information, we have hypothesized that the enlarged proteins found in the mitoribosome might compensate for the reduced rRNA portion to ensure maintenance of normal ribosomal function.

In this report, we summarize how mitochondrial proteins may compensate for the deficit in RNA. We refer to typical examples in animal mt translation systems, and discuss the possibility that transfer of some structural and functional roles from RNA to protein has occurred during evolution from ancestral to extant organisms.

Nematoda mitochondrial elongation factor Tu

Recent crystallographic analysis (Nissen *et al.* 1995; Nissen *et al.* 1999) and many biochemical data (Clark *et al.* 1995) have shown that bacterial EF-Tu binds mainly to two regions in tRNA; the terminal region of the acceptor stem and the Tstem helix. However, mt DNAs in some nematodes code for tRNAs that are predicted to lack the Tarm (Keddie *et al.* 1998; Okimoto *et al.* 1992; Okimoto and Wolstenholme 1990; Wolstenholme *et al.* 1987). Many of these tRNAs have actually been isolated (Ohtsuki *et al.* 1996; Watanabe *et al.* 1994; Watanabe *et al.* 1997). There is therefore a question as to how nematode mitochondrial EF-Tu binds to T stem-lacking tRNA. The cDNA sequence of bacterial EF-Tu homologue for *C. elegans* has been determined (Ohtsuki *et al.* 2001). As the corresponding protein was actually found in the nematode mt extract by Western blotting and the recombinant protein appreciably stimulated the poly(U)-dependent poly(Phe) synthesis in a bovine mt *in vitro* translation system, it could be regarded as a nematode mt EF-Tu (Ohtsuki *et al.* 2001). The amino acid residues in contact with the T stem – located in domain 3 (Nissen *et al.* 1999) – are quite different in the nematode mt EF-Tu compared with other EF-Tu's. In addition, the nematode mt EF-Tu possesses a C-terminal extension of 57 amino acids (called domain 3'). Nematode mt EF-Tu specifically binds to the nematode mt aminoacyl-tRNA lacking the Tarm, but bacterial EF-Tu does not. To investigate the function of this extension, we constructed a series of chimeric variants of the nematode mt EF-Tu, where domains were exchanged with bovine mt EF-Tu counterparts that are known to bind to cloverleaf-shaped aminoacyl-tRNAs (Kumazawa *et al.* 1991). This work demonstrated that the extended C-terminal region (domains 3 and 3') of nematode mt EF-Tu is essential for binding to Tarm-lacking aminoacyl-tRNA, and the N-terminal domains 1 and 2 of the mammalian and nematode mt EF-Tu's could be functionally equivalent in binding to aminoacyl-tRNA (Fig. 1). The C-termi-

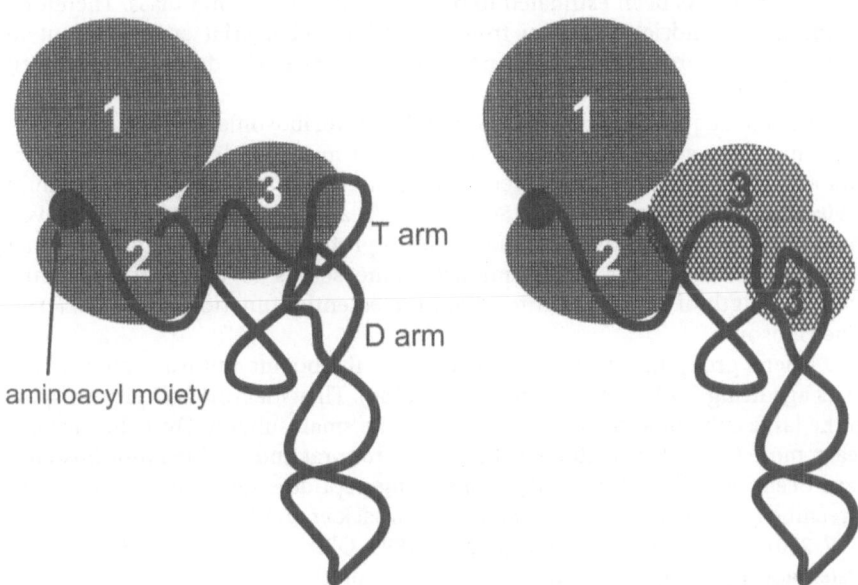

Fig. 1. A schematic representation of bacterial (Nissen *et al.* 1995) or mammalian mt (Andersen *et al.* 2000) ternary complex (left), and a predicted model of its nematode mt counterpart (Ohtsuki *et al.* 2001) (right). GTP is omitted in both. The EF-Tu is shown as a set of hatched circles (domains are numbered). The tRNAs are portrayed as simplified backbones with the aminoacyl moiety at the 3' terminus depicted by filled circles. Domains 3 and 3' of the nematode mt EF-Tu are unique regions for Tarm-lacking tRNAs

nus of bacterial EF-Tu in the ternary complex is located near the connector region of tRNA (Nissen *et al.* 1995), and the short C-terminal extension of bovine mt EF-Tu has been suggested to interact with the connector region of tRNA (Andersen *et al.* 2000). Therefore, domain 3' of the nematode mt EF-Tu may play a compensatory role for the loss of interaction between the tRNA Tarm and domain 3 that is normally found between tRNAs and mammalian mt and bacterial EF-Tu's.

Mammalian mitochondrial ribosome

The mammalian mitochondrial ribosome (mitoribosome) is a highly protein-rich particle in which almost half of rRNA contained in the bacterial ribosome is replaced with proteins. The mitoribosome has a smaller sedimentation coefficient of 55S compared with that of the bacterial ribosome. It consists of a large 39S and a small 28S subunit (O'Brien, 1971). The 39S subunit contains 16S rRNA and the 28S subunit contains 12S rRNAs. In mammals, no counterpart to the bacterial 5S rRNA has been found. A recent physiochemical study has revealed that the rat mitoribosome has a large molecular mass (3.57 MDa) compared with the *E. coli* ribosome (2.49 MDa) (Patel *et al.* 2000), and that the protein to RNA ratio is completely reversed between these two ribosomes. The protein composition of the mitoribosome has been estimated to be about 75% of the total mass. Therefore, during mitochondrial evolution from a possible eubacterial endosymbiont in eukaryotic cell progenitors, large parts of bacterial rRNA domains have been replaced by protein components.

Functional equivalency of the mammalian mitoribosome and bacterial ribosome has been suggested because mammalian mitochondrial translation factors are exchangeable with bacterial ribosome factors (Chung and Spremulli 1990; Liao and Spremulli 1991; Schwartzbach and Spremulli 1989). However, these ribosomes each have a distinct RNA to protein ratio, which indicates that the protein component of the mitoribosome serves to compensate for the reduced length of rRNA chain to retain the essential function of the bacteria-type ribosome.

O'Brien's group first isolated proteins from the bovine mitoribosome twenty years ago using 2D-PAGE (Matthews *et al.* 1982). They identified 52 protein spots in the large subunit and 33 protein spots in the small subunit. Over the past few years, more than 20 mitoribosomal proteins from rat and bovine mitoribosomes have been identified by several groups using peptide sequencing and database screening (Goldschmidt-Reisin *et al.* 1998; Graack *et al.* 1999; Koc *et al.* 1999; Koc *et al.* 2001a; Koc *et al.* 2000; O'Brien *et al.* 1999; O'Brien *et al.* 2000). Recently, we systematically analyzed protein components of the bovine mitoribosome by LC/MS using ESI/iontrap mass spectrometry and identified 55 proteins, including 29 new proteins (Suzuki *et al.* 2001a; Suzuki *et al.* 2001b) (Table 1); 31 proteins from the 39S large subunit, 21 proteins from the 28S small subunit and three proteins from the 55S intact mitoribosome. A total of 37 proteins were identified as orthologues of bacterial ribosomal proteins. These data provide a new landmark for proteomic analysis of the mammalian mitoribosome, genetic information on mt protein synthesis and candidate genes for respiratory defect of mitochondrial human diseases.

A striking result from this work was the finding that mt proteins whose binding sites on rRNA are shortened or lost carry N- or C-terminal extensions, while the proteins with conserved rRNA-binding sites have a similar molecular weight to their *E. coli* counterparts. The enlarged proteins, together with several extra proteins specifically found in the mitoribosome, obviously provide both structural and functional compensation for the deficit of rRNA in the mitoribosome, such that the whole molecular architecture of the ribosome is conserved between bacteria and mitochondria.

39S large subunit

Among the 31 protein components in the large 39S subunit, 24 proteins have been identified as prokaryotic homologues of ribosomal proteins (Suzuki *et al.* 2001a) (Table 1). Significant homology of mitoribosomal proteins to bacterial ribosome proteins throughout the sequences indicates similar topological orientations of these proteins. Partial conservation of the specific interactions between rRNA and protein components in mitoribosome is suggested. The mitoribosomal proteins have characteristic N- or C-terminal extensions, resulting in an increased molecular mass when compared with *E. coli* counterparts. The calculated average molecular masses of the 24 ribosomal proteins (*E. coli*) and the human homologues are 13.8 kDa and 21.1 kDa, respectively. Mitochondrial ribosomal proteins therefore have an average 1.5-fold increase in their molecular masses compared with those of their *E. coli* counterparts. In addition to these enlarged protein homologues, there are specific extra proteins in the mitoribosome that are predicted to play a role in the compensation for the decrease of the rRNA component in the mitoribosome.

Figure 2 maps the mitoribosomal proteins on to the secondary structure of rRNA (Fig. 2) to make visualization of the entire mitoribosome easier. The secondary structure of human mt rRNA large subunit (gray line) is superimposed on the corresponding *E. coli* 23S rRNA (black line) (Gutell *et al.* 1990). Mitoribosomal proteins with well-conserved binding sites on mt rRNA have similar molecular masses to those of their prokaryotic counterparts. In contrast, the proteins having small (reduced) or no distinct binding sites on mt rRNA showed increased molecular masses to the N- or C-terminal extensions. The crystal structure of the large subunit has been determined (Ban *et al.* 2000), and helices 66, 61, 53, 43/44 and 95 in rRNA are the main binding sites for L2, L22, L23, L11 and L14, respectively. These helices are well-conserved in mt rRNA, and the protein homologues in the mitoribosome have similar molecular masses to their *E. coli* counterparts (Fig. 2, Table 1). L24, L4 and L15 have many contacts with domains I and II of 50S subunits in *Haloarcula marismortui* (Ban *et al.* 2000), whereas their binding sites are almost lost in mt rRNA. The molecular masses of L24mt, L4mt and L15mt are increased by over 15 kDa when compared with the bacterial counterparts. Similar compensation was also observed in the L19mt homologue where most of the binding sites have been lost in mt rRNA. Lack of the helices 77 and 78 in mt rRNA

Table 1. Protein components of the human mitoribosome. 31 proteins from the 39S large subunit, 21 proteins from the 28S small subunit and three proteins from the intact 55S mitoribosome are summarized. The molecular weight of each protein was calculated from the amino acid sequence of the mature protein. The apparent molecular weight value, estimated by SDS-PAGE, is indicated with an asterisk. The ratio and difference in molecular weights between each protein for human mitochondria and *E. coli* were calculated (the columns Ratio (II/E) and Diff (H-E), respectively). Orthologs from *S. cerevisiae* (S. cere.), *C. elegans* (C. eleg.) and *D. melanogaster* (D. mela.) were obtained through BLAST searches of databases using human mitoribosomal proteins as queries. Reference numbers are indicated: 1 for (Goldschmidt-Reisin *et al.* 1998), 2 for (O'Brien *et al.* 1999), 3 for (Graack *et al.* 1999), 4 for (O'Brien *et al.* 2000), 5 for (Ou *et al.* 1987), 6 for (Tsang *et al.* 1995), 7 for (Mao *et al.* 1998), 8 for (Koc *et al.* 1999), 9 for (Koc *et al.* 2000), 10 for (Koc *et al.* 2001a), 11 for (Koc *et al.* 2001b) and 12 for (Walker *et al.* 1992)

Spot name	Family name	Mr (KDa) Human mt.	E. coli	Ratio (H/E)	Diff. (H-E)	Ortholog S. cere.	C. eleg.	D. mela.	Reference
bMRP-16	L1	36.2*	24.6	147	11.6	S52681	F33D4.5	CG7494	Our study
bMRP-26	L2	27.1	29.7	91	-2.7	Rml2p	AF045646	CG7636	2
bMRP-15	L3	34.4	22.2	155	12.2	YmL9	U13875	CG8288	Our study, (5)
bMRP-18	L4	38.9*	22.1	176	16.8	P51998	T23B12.2	AAF45007	Our study
bMRP-31	L7/12A	16.5	12.2	135	4.3	YGL068W	W09D10.3	CG5012	3
bMRP-34	L7/12B	16.0	12.2						
bMRP-28	L9	24.8	15.8	157	9.0	–	–	CG4923	Our study
bMRP-24	L10	26.4	17.6	150	8.8	YmL11	K01C8.6	CG11488	1
bMRP-32	L11	20.7*	14.7	141	6.0	Mrpl19p	B0303.15	CG3351	Our study
bMRP-33	L13	21.2*	16.0	132	5.2	Mrpl23p	F13G3.7	CG10603	Our study
bMRP-53	L14	12.7	13.5	94	-0.9	YmL38	F45E12.4	CAB63504	1
bMRP-17	L15	31.3	15.0	209	16.3	YmL10	F37F2.1	CG5219	2
bMRP-25	L16	25.3	15.3	166	10.1	YmL47	T04A8.11	CAA15945	Our study
bMRP-39	L17	17.8*	14.4	124	3.4	YmL8	AC024810	CG13880	2
bMRP-21	L19	34.7*	13.0	267	21.7	Img1p	(CAB61204)	CG8039	3
bMRP-50	L20	16.4	13.4	122	3.0	–	Y48C3A.1	CG11258	Our study
bMRP-43	L22	19.4	12.2	159	7.2	Yml177cp	Y39A1A.6	CG4742	1
bMRP-38	L23	17.8*	11.2	159	6.6	YmL41	T08B2.8	CG1320	Our study, (6)
bMRP-30	L24	26.5	11.2	237	15.3	–	F59A3.3	CG8849	4
bMRP-55	L27	13.0	9.0	144	4.0	YmL2	–	–	Our study
bMRP-52	L30	14.7	6.4	230	8.3	YmL33	W04B5.4	CG7038	1
bMRP-59b	L32	12.8	6.3	203	6.5	YmL32	C30C11.1	CG12220	Our study
bMRP-66	L33	6.7	6.2	108	0.5	YmL39	–	CG3712	Our study, (7)
bMRP-68	L34	5.4	5.4	99	0.0	Ydr155wp	–	–	Our study
bMRP-69	L36	4.9	4.7	104	0.2	Ypl183bw	–	CG7528	Our study
bMRP-8	MRP-L2	49.2	–	–	–	–	Y48E1B.5	CG6547	3
bMRP-11	MRP-L3	46.0	–	–	–	Yml35	Y34D9A.f	CG15871	3
bMRP-13	MRP-L5	40.9	–	–	–	–	–	CG17166	3
bMRP-41	MRPL22(24)	19.2	–	–	–	TmL28	F54C4.1	CG5242	1
bMRP-51	YmL27	14.1	–	–	–	YmL27	B0432.3	(CG12954)	1
bMRP-54	MRPL31	11.3	–	–	–	–	ZC410.7	CG12921	1
bMRP-36a	(CI-B8)	15.8*	–	–	–	Ypr100wp	C25A1.13	CG5479	Our study

Table header spanning row: **Protein components in the 39S large ribosomal subunit from human mitochondria**

Table 1. *Continued.*

Spot name	Family name	Mr (KDa) Human mt.	E. coli	Ratio (H/E)	Diff. (H-E)	Ortholog S. cere.	C. eleg.	D. mela.	Reference
Protein components in the 28S small ribosomal subunit from human mitochondria									
bMRP-S21	S2	33.0*	26.7	124	6.3	Mrp4p	T23B12.3	CG2937	Our study
bMRP-20	S5	35.0*	17.5	200	17.5	MRP17	Q93425	–	10
bMRP-48a	S6	14.1	15.7	90	-1.6	MRPS7	R12E2.12	CG15016	Our study
bMRP-27a	S7	24.5	19.9	123	4.6	–	Y57G11C.4	CG5108	8
bMRP-42	S11	17.8*	13.7	130	4.1	YNR036C	W04D2.5	CG5184	Our study
bMRP-62	S12	12.8*	13.6	89	-1.5	MRP2	T03D.2	P10735	Our study*
bMRP-59a	S14	13.2*	11.4	116	1.8	Mrps28	D49391	CG12211	9
bMRP-35	S15	24.0	10.1	237	13.9	Mrps24	–	CG4207	Our study
bMRP-48b	S16	14.4*	9.2	154	5.2	Ymr188c	F561.3	CG8338	Our study
bMRP-46	S17	15.1*	9.6	158	5.5	–	C05D11.10	CG4326	Our study
bMRP-22	S18a	29.3*	8.9	329	20.4	Mrp21p	T13H5.5	CG10757	Our study
bMRP-45	S18b	18.8	8.9	211	9.9		T14B4.2	CG11744	Our study
bMRP-61	S21	10.5*	8.4	125	2.1	–	F29B9.10	–	Our study
bMRP-12	S22	38.0				–	C14A4.14	CG12261	9
bMRP-29	S23	21.8*				–	ZK1098.7	–	9
bMRP-47	S24	14.0*				–	–	CG13608	9
bMRP-36b	S25	17.7*				–	Y55F3AM.1	CG14413	9
bMRP-37	S26 (MRP-S13)	21.3				–	–	CG7354	1
bMRP-27b	S34 (MRP-S12)	27.5				–	T23838	CG13037	4
bMRP-44	S35 (MRP-S28)	13.0				–	Y43F8C.8	CG5497	4
bMRP-10	DAP3	48.0				YGL129c	CAB60994	CG3633	Our study,11

Spot name	Family name	Mr (KDa)	Ortholog C. eleg.	D. mela.	Reference
Protein components in the 55S intact mitoribosome (unassigned localization)					
bMRP-63	new	11.3	–	CAA15941	Our study
bMRP-64	new	11.7	T12G3.5	CG13098	Our study
bMRP-49	CI-B14	15.0	Y57G11C.12	CG7712	Our study, 12

may be substituted for by the extended termini of L1mt and L9mt. The L3 binding sites in domain VI (helices 94 and 100) are also missing in mt rRNA; L3mt shows long N- and C-terminal extensions. Taken together, these observations strongly suggest that enlarged mitoribosomal proteins may compensate for sequence losses of rRNAs in the mitoribosome.

We constructed 3D-models for mitochondrial rRNAs from the human and *C. elegans* (Fig. 3) based on the 50S crystal structure of *H. marismortui* (Ban *et al.* 2000).

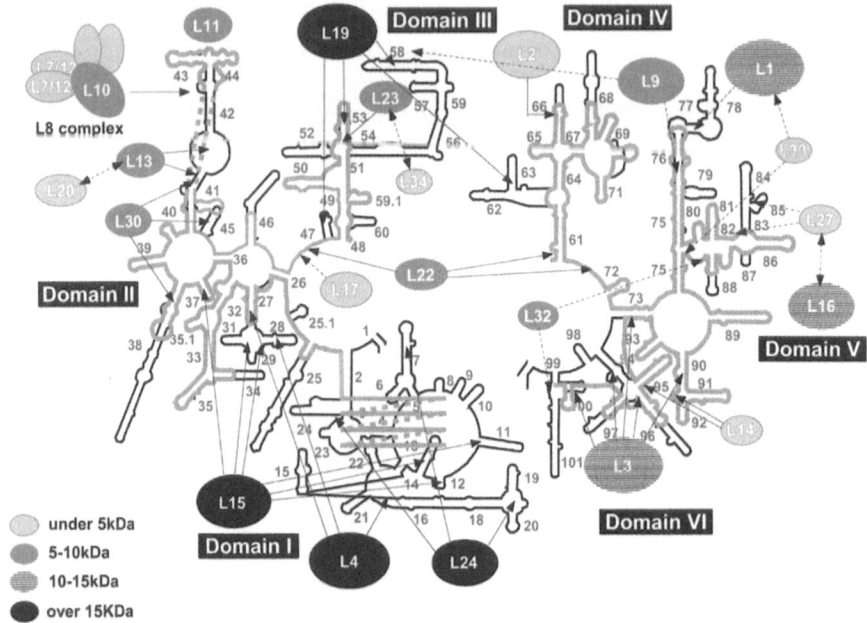

Fig. 2. Secondary structures for large ribosomal RNAs of bacterial and mammalian mitochondria, along with interacting ribosomal proteins. The secondary structure of human mitochondrial 16S rRNA (gray line) is superimposed on the secondary structure of *E. coli* 23S rRNA (black line) (Gutell *et al.* 1990) according to the format described by Ban et al. (Ban *et al.* 2000). The 5' region of the mitochondrial 16S rRNA (about 160 bases) could not be aligned with domain I of the bacterial 23S rRNA. Ovals representing large ribosomal proteins are mapped on the secondary structure of rRNA and their interaction is indicated by arrows. Solid arrows show interactions that have been identified through the crystal structure of the 50S subunit (Ban *et al.* 2000). Broken arrows indicate interactions derived from biochemical studies (Baranov *et al.* 1999; Egebjerg *et al.* 1991). The size of the oval for each protein represents its relative molecular weight. The size of the compensatory protein enlargement in the mitoribosome compared with the *E. coli* counterpart (Table 1) is indicated through the use of four different patterns

The shortened mt rRNA forms a spherical cluster at the center of the peptidyl transferase, while several lost rRNA portions, which are localized discontinuously in the secondary structure (Fig. 2), form large missing domains. These domains are located at the bottom, the back, the central protuberance, and the left side when viewing the structure from the subunit interface. In the mammalian mitoribosome (Fig. 3B), the missing bottom domain was composed of H47 (domain II), H54–58 (domain III), H62/63 (domain IV) and H99–101 (domain VI). Helices 83–85 (domain V), H38 (domain II) and 5S rRNA necessary for forming the central protuberance have all been lost. The missing back and the left domains are composed mostly of domain I, H25/28–31/41/42/45/46 (domain II), H52 (domain III), H66/68 (domain IV), H77–79/88 (domain V), and H97 (domain VI). A number of ribosomal proteins are localized in rRNA portions that are not present in the mitoribosome. The portions of rRNA that bind ribosomal protein sites appear to be shortened, and are replaced

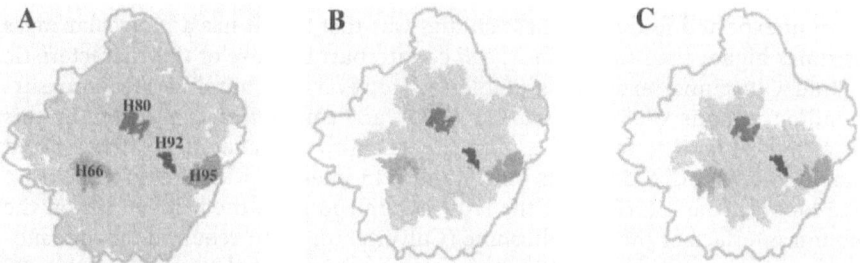

Fig. 3. The 3-D models for large mitoribosomal RNA based on the crystal structure of the 50S subunit. All atomic coordinates for the *H. marismortui* 50S subunit (A) were obtained from the Protein Data Bank [accession number 1FFK; (Ban *et al.* 2000)]. Model structures for two mitoribosomes were constructed by removing non-conserved rRNA portions from the atomic coordinates of 1FFK, based on the secondary structure of large mitoribosomal RNA from mammalian (B) and *C. elegans* (C) (Gutell *et al.* 1990). The 3D structures were displayed using Rasmol Ver. 2.6 (Sayle and Milner-White 1995). The outer edge line represents the crystal structure of 50S subunit when viewing from the subunit interface. Some functional rRNA domains are highlighted: helix 80 (P loop), helix 92 (A loop), helix 95 (S/R loop) and helix 66 (L2 binding helix)

by other lengthened ribosomal proteins. Thus, ribosomal proteins fill up the empty space on the rRNA with extended N- or C-termini.

28S small subunit

In a previous study on the 28S small subunit, we identified 21 proteins, including 11 that were newly identified (Suzuki *et al.* 2001b)(Table 1). Of these, 13 were homologues of bacterial ribosomal proteins and eight were extra proteins found only in the mitoribosome. The results and comparison between the bacterial ribosome and the mitoribosome demonstrate that there are characteristic features of small subunit proteins, and that there is a similar structural compensation for the reduced rRNA content in the mitoribosome by the inclusion of larger proteins, as observed for the 39S subunit.

The S12 protein is a key component of the ribosomal small subunit, and controls decoding fidelity and is susceptible to aminoglycoside antibiotics (Brakier-Gingras and Phoenix 1984). The mammalian mitochondrial S12 (S12mt) has a similar size (Table 1) and has strong sequence homology to bacterial S12. This is consistent with the complete conservation of the RNA-binding site and the peripheral region beside the decoding center. A genetic study in *Drosophila* has demonstrated that a single point mutation in the S12mt gene causes a bang senseless mutant called *tko*. The phenotype of this mutant is similar to sensorineural hearing loss in humans. The cause of this hearing loss in humans has been related to mitochondrial dysfunction (Royden *et al.* 1987). The human autosomal dominant deafness gene, DFNA4, has been mapped to Chr 19q13 (Chen *et al.* 1995), which is the same location that S12mt gene has been mapped to (RPMS12) (Johnson *et al.* 1998; Shah *et al.* 1998). It has been suggested that mutations in the S12mt gene result in translational misreading and mitochondrial dysfunction. Thus, S12mt is a candidate gene for DFNA4.

An unexpected finding in these studies was that S15mt has a molecular mass 2.4 times higher than that of its *E. coli* counterpart because of the characteristic N- and C-terminal extensions (Table 1). However, the binding sites and surrounding regions of S15mt on 12S rRNA are completely conserved. In this instance it is hard to forward a plausible scientific explanation for the 2.4 fold increase in mass of S15mt. It is known that S15 interacts with the tip of helix 34 (715 loop) in the 23S rRNA of the large subunit to form the bridge (B4) on the subunit interface of the 70S ribosome (Culver *et al.* 1999) required for intersubunit association (Champney 1980). According to the crystal structure of the 70S ribosome and the S15-rRNA complex (Cate *et al.* 1999; Culver *et al.* 1999; Nikulin *et al.* 2000), helix α4 of S15 is a likely candidate for binding to the 715 loop in the large subunit. Since S15mt has significant similarity to bacterial S15 throughout its sequence, including the helix α4 region, the intersubunit interaction of helix 34-S15mt should also be conserved in the 55S mitoribosome. Helix 34 of mitochondrial 16S lrRNA in the 39S subunit is shortened compared with the bacterial helix 34 (Table 1) (Gutell *et al.* 1990). The large size of S15mt may enable it to access the short mitoribosomal helix 34 through an extended protein chain. This remains the most plausible reason for why the 28S subunit contains a much larger S15 homologue. No S15 homologue can be found in the *C. elegans* mitoribosome (Table 1), which does not have helix 34 (Gutell *et al.* 1990). Thus, helix 34 and S15 might have coevolved to conserve the interaction.

The mechanism of translation initiation in mammalian mitochondria is one of the most characteristic features of mitochondrial protein synthesis. Mitochondrial mRNAs encoded by mt DNA do not have significant leader sequences 5' to the initiation codon. There is no Shine-Dalgarno sequence, nor is there a cap structure at the 5' end (Attardi 1985; Ojala *et al.* 1980). It is considered that mitochondrial protein synthesis starts directly at initiation codons located at the 5' end of these mRNAs. One notable feature of the 28S subunit is the presence of a GTP-binding protein that may be involved in this characteristic initiation step (Denslow *et al.* 1991). We have found a potential GTP-binding protein, DAP3 (death-associated protein 3), that is an apoptosis-related protein in the 28S subunit. It has been revealed that DAP3 is a positive mediator of TNF-α and Fas-induced cell death (Kissil *et al.* 1999; Kissil *et al.* 1995; Levy-Strumpf and Kimchi 1998). Further study is needed to clarify how this extra 28S subunit protein compensates the reduced rRNA content of the mitoribosome both structurally and functionally, and to clarify the role of the GTP binding activity of DAP3 in protein synthesis and/or programmed cell death. Recently, Spremulli's group has also identified two apoptosis-related proteins from the bovine mitochondrial 28S subunit, one of which is DAP3 (Koc *et al.* 2001b).

Summary and Perspective

In this report we have provided two examples where mt RNAs are shortened, and where ribosomal proteins compensate for this deficit both structurally and functionally. Questions that arise from this finding are why have mitochondria RNAs shortened during evolution, and why are proteins imported from the cytoplasm into the mitochondria. A plausible explanation may be that in the course of mt evolution, mt genome sizes have been shortened by discarding spacer regions and transferring most of genes coding for mt proteins to the host genome (Lang *et al.* 1997). Eventually, through this process, the vertebrate mitochondrion now only has 37 genes, including all of the RNA genes necessary for translation (2 rRNAs and 22 tRNAs), and 13 subunit proteins of the enzyme complex for respiration.

Economizing the mt genome would have some advantages for maintenance and/or breeding of mitochondria, such as for rapid and efficient replication of the mt genome. Vertebrate mt genomes are known to replicate in a σ–replacement manner, where one strand is kept in a single-stranded form for a certain period of time. Since the mitochondrial matrix is a highly oxidative environment, single-stranded DNA would be easily oxidized, resulting in mutations. Thus, it would be advantageous to make the mt genome size as small as possible to ensure accurate replication. As part of this process, the number of rRNA genes was reduced to two, corresponding to large and small subunit rRNAs, while the equivalent bacterial 5S RNA gene was discarded. In addition, the number of tRNA genes was reduced to 22 species, which are the minimal set of tRNAs capable of decoding 60 sense codons. Moreover, the sizes of both rRNAs and tRNAs were shortened; rRNAs became half the size of their *E. coli* counterparts, and the average size of mt tRNAs was decreased. In particular, some species of tRNAs lost the D arm or Tarm (Sprinzl *et al.* 1998). In order to compensate for such losses in rRNAs and tRNAs, proteins from the cytoplasm would have been imported into mitochondria. This process would be through the placement of signal sequences upstream of the relevant genes so that proteins are transported through the mitochondrial membranes.

Investigations into systems where proteins are used to compensate for deficits in RNA may provide some useful insights. One insight might include information on how the structures of RNAs are required for proper function and what is the minimal prerequisite RNA structure. A second insight might be into how there was a transition from the "RNA world" to the "RNP world" in the early process of the evolution of life. Further examples of protein compensation systems should be obtained. More extensive studies on such systems may help determine a more complete and accurate explanation for the reasons such why protein compensations have occured in nature.

References

Andersen GR, Thirup S, Spremulli LL and Nyborg J (2000) High resolution crystal structure of bovine mitochondrial EF-Tu in complex with GDP. *J Mol Biol*, 297:421–36.

Attardi G (1985) Animal mitochondrial DNA: an extreme example of genetic economy. *Int Rev Cytol*, 93:93–145.

Ban N, Nissen P, Hansen J, Moore PB and Steitz TA (2000) The complete atomic structure of the large ribosomal subunit at 2.4 A resolution. *Science*, 289:905–20.

Baranov PV, Kubarenko AV, Gurvich OL, Shamolina, T.A. and Brimacombe, R. (1999) The Database of Ribosomal Cross-links: an update. *Nucleic Acids Res*, 27:184–5.

Brakier-Gingras L and Phoenix P (1984) The control of accuracy during protein synthesis in Escherichia coli and perturbations of this control by streptomycin, neomycin, or ribosomal mutations. *Can J Biochem Cell Biol*, 62:231–44.

Cate JH, Yusupov MM, Yusupova GZ, Earnest TN and Noller HF (1999) X-ray crystal structures of 70S ribosome functional complexes. *Science*, 285:2095–104.

Champney WS (1980) Protein synthesis defects in temperature-sensitive mutants of Escherichia coli with altered ribosomal proteins. *Biochim Biophys Acta*, 609:464–74.

Chen AH, Ni L, Fukushima K, Marietta J, O'Neill M, Coucke P, Willems P and Smith RJ (1995) Linkage of a gene for dominant non-syndromic deafness to chromosome 19. *Hum Mol Genet*, 4:1073–6.

Chung HK and Spremulli LL (1990) Purification and characterization of elongation factor G from bovine liver mitochondria. *J Biol Chem*, 265:21000–4.

Clark BFC, Kjeldgaard M, Barciszewski J and Sprinzl M (1995) *Recognition of aminoacyl-tRNAs by protein elongation factors*. American Society for Microbiology, Washington, D. C.

Culver GM, Cate JH, Yusupova GZ, Yusupov MM and Noller HF (1999) Identification of an RNA-protein bridge spanning the ribosomal subunit interface. *Science*, 285:2133–6.

Denslow ND, Anders JC and O'Brien TW (1991) Bovine mitochondrial ribosomes possess a high affinity binding site for guanine nucleotides. *J Biol Chem*, 266:9586–90.

Egebjerg J, Christiansen J and Garrett RA (1991) Attachment sites of primary binding proteins L1, L2 and L23 on 23 S ribosomal RNA of Escherichia coli. *J Mol Biol*, 222:251–64.

Goldschmidt-Reisin S, Kitakawa M, Herfurth E, Wittmann-Liebold B, Grohmann L and Graack HR (1998) Mammalian mitochondrial ribosomal proteins. N-terminal amino acid sequencing, characterization, and identification of corresponding gene sequences. *J Biol Chem*, 273:34828–36.

Graack HR, Bryant ML and O'Brien TW (1999) Identification of mammalian mitochondrial ribosomal proteins (MRPs) by N-terminal sequencing of purified bovine MRPs and comparison to data bank sequences: the large subribosomal particle. *Biochemistry*, 38: 16569–77.

Gutell RR, Schnare MN and Gray MW (1990) A compilation of large subunit (23S-like) ribosomal RNA sequences presented in a secondary structure format. *Nucleic Acids Res*, 18: 2319–30.

Johnson DF, Hamon M and Fischel-Ghodsian N (1998) Characterization of the human mitochondrial ribosomal S12 gene. *Genomics*, 52:363–8.

Keddie EM, Higazi T and Unnasch TR (1998) The mitochondrial genome of Onchocerca volvulus: sequence, structure and phylogenetic analysis. *Mol Biochem Parasitol*, 95:111–27.

Kissil JL, Cohen O, Raveh T and Kimchi A (1999) Structure-function analysis of an evolutionary conserved protein, DAP3, which mediates TNF-alpha- and Fas-induced cell death. *Embo J*, 18:353–62.

Kissil JL, Deiss LP, Bayewitch M, Raveh T, Khaspekov G and Kimchi A (1995) Isolation of DAP3, a novel mediator of interferon-gamma-induced cell death. *J Biol Chem*, 270:27932–6.

Koc EC, Blackburn K, Burkhart W and Spremulli LL (1999) Identification of a mammalian mitochondrial homolog of ribosomal protein S7. *Biochem Biophys Res Commun*, 266: 141–6.

Koc EC, Burkhart W, Blackburn K, Koc H, Moseley A and Spremulli LL (2001a) Identification of four proteins from the small subunit of the mammalian mitochondrial ribosome using a proteomics approach. *Protein Sci*, **in press.**

Koc EC, Burkhart W, Blackburn K, Moseley A, Koc H and Spremulli LL (2000) A Proteomics Approach to the Identification of Mammalian Mitochondrial Small Subunit Ribosomal Proteins. *J Biol Chem*.

Koc EC, Ranasinghe A, Burkhart W, Blackburn K, Koc H, Moseley A and Spremulli LL (2001b) A new face on apoptosis: death-associated protein 3 and PDCD9 are mitochondrial ribosomal proteins. *FEBS Lett*, 492:166–70.

Kumazawa Y, Schwartzbach CJ, Liao HX, Mizumoto K, Kaziro Y, Miura K, Watanabe K and Spremulli LL (1991) Interactions of bovine mitochondrial phenylalanyl-tRNA with ribosomes and elongation factors from mitochondria and bacteria. *Biochim Biophys Acta*, 1090:167–72.

Lang BF, Burger G, O'Kelly CJ, Cedergren R, Golding GB, Lemieux C, Sankoff D, Turmel M and Gray MW (1997) An ancestral mitochondrial DNA resembling a eubacterial genome in miniature. *Nature*, 387:493–7.

Levy-Strumpf N and Kimchi A (1998) Death associated proteins (DAPs): from gene identification to the analysis of their apoptotic and tumor suppressive functions. *Oncogene*, 17:3331–40.

Liao HX and Spremulli LL (1991) Initiation of protein synthesis in animal mitochondria. Purification and characterization of translational initiation factor 2. *J Biol Chem*, 266:20714–9.

Mao M, Fu G, Wu JS, Zhang QH, Zhou J, Kan LX, Huang QH, He KL, Gu BW, Han ZG, Shen Y, Gu J, Yu YP, Xu SH, Wang YX, Chen SJ and Chen Z (1998) Identification of genes expressed in human CD34(+) hematopoietic stem/progenitor cells by expressed sequence tags and efficient full-length cDNA cloning. *Proc Natl Acad Sci U S A*, 95:8175–80.

Masta SE (2000) Mitochondrial sequence evolution in spiders: intraspecific variation in tRNAs lacking the TPsiC Arm. *Mol Biol Evol*, 17:1091–100.

Matthews DE, Hessler RA, Denslow ND, Edwards JS and O'Brien TW (1982) Protein composition of the bovine mitochondrial ribosome. *J Biol Chem*, 257:8788–94.

Nikulin A, Serganov A, Ennifar E, Tishchenko S, Nevskaya N, Shepard W, Portier C, Garber M, Ehresmann B, Ehresmann C, Nikonov S and Dumas P (2000) Crystal structure of the S15-rRNA complex. *Nat Struct Biol*, 7:273–7.

Nissen P, Kjeldgaard M, Thirup S, Polekhina G, Reshetnikova L, Clark BF and Nyborg J (1995) Crystal structure of the ternary complex of Phe-tRNAPhe, EF-Tu, and a GTP analog. *Science*, 270:1464–72.

Nissen P, Thirup S, Kjeldgaard M and Nyborg J (1999) The crystal structure of Cys-tRNACys-EF-Tu-GDPNP reveals general and specific features in the ternary complex and in tRNA. *Structure Fold Des*, 7:143–56.

O'Brien TW (1971) The general occurrence of 55 S ribosomes in mammalian liver mitochondria. *J Biol Chem*, 246:3409–17.

O'Brien TW, Fiesler SE, Denslow ND, Thiede B, Wittmann-Liebold B, Mougey EB, Sylvester JE and Graack HR (1999) Mammalian mitochondrial ribosomal proteins (2). Amino acid sequencing, characterization, and identification of corresponding gene sequences. *J Biol Chem*, 274:36043–51.

O'Brien TW, Liu J, Sylvester JE, Mougey EB, Fischel-Ghodsian N, Thiede B, Wittmann-Liebold B and Graack HR (2000) Mammalian mitochondrial ribosomal proteins (4). Amino acid sequencing, characterization, and identification of corresponding gene sequences. *J Biol Chem*, 275:18153–9.

Ohtsuki T, Kawai G, Watanabe Y, Kita K, Nishikawa K and Watanabe K (1996) Preparation of biologically active Ascaris suum mitochondrial tRNAMet with a TV-replacement loop by ligation of chemically synthesized RNA fragments. *Nucleic Acids Res*, 24:662–7.

Ohtsuki T, Watanabe Yi Y, Takemoto C, Kawai G, Ueda T, Kita K, Kojima S, Kaziro Y, Nyborg J and Watanabe K (2001) An "Elongated" translation elongation factor Tu for truncated tRNAs In nematode mitochondria. *J Biol Chem*, in press.

Ojala DK, Montoya J and Attardi G (1980) The putative mRNA for subunit II of human cytochrome c oxidase starts directly at the translation initiator codon. *Nature*, 287:79–82.

Okimoto R, Macfarlane JL, Clary DO and Wolstenholme DR (1992) The mitochondrial genomes of two nematodes, Caenorhabditis elegans and Ascaris suum. *Genetics*, 130:471–98.

Okimoto R and Wolstenholme DR (1990) A set of tRNAs that lack either the T psi C arm or the dihydrouridine arm: towards a minimal tRNA adaptor. *Embo J*, 9:3405–11.

Ou JH, Yen TS, Wang YF, Kam WK and Rutter WJ (1987) Cloning and characterization of a human ribosomal protein gene with enhanced expression in fetal and neoplastic cells. *Nucleic Acids Res*, 15:8919–34.

Patel VB, Cunningham CC and Hantgan RR (2001) Physiochemical properties of rat liver mitochondrial ribosomes. *J Biol Chem* 276, 6739–6746.

Royden CS, Pirrotta V and Jan LY (1987) The tko locus, site of a behavioral mutation in D. melanogaster, codes for a protein homologous to prokaryotic ribosomal protein S12. *Cell*, 51:165–73.

Sayle RA and Milner-White EJ (1995) RASMOL: biomolecular graphics for all. *Trends Biochem Sci*, 20:374.

Schwartzbach CJ and Spremulli LL (1989) Bovine mitochondrial protein synthesis elongation factors. Identification and initial characterization of an elongation factor Tu-elongation factor Ts complex. *J Biol Chem*, 264:19125–31.

Shah ZH, Migliosi V, Miller SC, Wang A, Friedman TB and Jacobs HT (1998) Chromosomal locations of three human nuclear genes (RPSM12, TUFM, and AFG3L1) specifying putative components of the mitochondrial gene expression apparatus. *Genomics*, **48**:384–8.

Sprinzl M, Horn C, Brown M, Ioudovitch A and Steinberg S (1998) Compilation of tRNA sequences and sequences of tRNA genes. *Nucleic Acids Res*, **26**:148–53.

Suzuki T, Terasaki M, Takemoto-Hori C, Hanada T, Ueda T, Wada A and Watanabe K (2001a) Proteomic analysis of mammalian mitochondrial ribosome; Identification of the 28S small subunit proteins. *J Biol Chem*, submitted.

Suzuki T, Terasaki M, Takemoto-Hori C, Hanada T, Ueda T, Wada A and Watanabe K (2001b) Structural compensation for deficit of rRNA with proteins in mammalian mitochondrial ribosome; systematic analysis of protein components of the large ribosomal subunit from mammalian mitochondria. *J Biol Chem*, **in press**.

Tsang P, Gilles F, Yuan L, Kuo YH, Lupu F, Samara G, Moosikasuwan J, Goye A, Zelenetz AD, Selleri L et al. (1995) A novel L23-related gene 40 kb downstream of the imprinted H19 gene is biallelically expressed in mid-fetal and adult human tissues. *Hum Mol Genet*, **4**:1499–507.

Walker JE, Arizmendi JM, Dupuis A, Fearnley IM, Finel M, Medd SM, Pilkington SJ, Runswick MJ and Skehel JM (1992) Sequences of 20 subunits of NADH:ubiquinone oxidoreductase from bovine heart mitochondria. Application of a novel strategy for sequencing proteins using the polymerase chain reaction. *J Mol Biol*, **226**:1051–72.

Watanabe Y, Tsurui H, Ueda T, Furushima R, Takamiya S, Kita K, Nishikawa K and Watanabe K (1994) Primary and higher order structures of nematode (Ascaris suum) mitochondrial tRNAs lacking either the T or D stem. *J Biol Chem*, **269**:22902–6.

Watanabe Y, Tsurui H, Ueda T, Furusihima-Shimogawara R, Takamiya S, Kita K, Nishikawa K and Watanabe K (1997) Primary sequence of mitochondrial tRNA(Arg) of a nematode Ascaris suum: occurrence of unmodified adenosine at the first position of the anticodon. *Biochim Biophys Acta*, **1350**:119–22.

Wolstenholme DR, Macfarlane JL, Okimoto R, Clary DO and Wahleithner JA (1987) Bizarre tRNAs inferred from DNA sequences of mitochondrial genomes of nematode worms. *Proc Natl Acad Sci U S A*, **84**:1324–8.

Yamazaki N, Ueshima R, Terrett JA, Yokobori S, Kaifu M, Segawa R, Kobayashi T, Numachi K, Ueda T, Nishikawa K, Watanabe K and Thomas RH (1997) Evolution of pulmonate gastropod mitochondrial genomes: comparisons of gene organizations of Euhadra, Cepaea and Albinaria and implications of unusual tRNA secondary structures. *Genetics*, **145**:749–58.

CHAPTER 18

Proficient Target Selection in Structural Genomics by In Vitro Protein Expression on Gateway Recombination Plasmids

18

VINCENT MONCHOIS*, RENAUD VINCENTELLI, CÉLINE DEREGNAUCOURT, CHANTAL ABERGEL, JEAN-MICHEL CLAVERIE

Introduction

The numerous whole genome-sequencing projects of the recent genomic era resulted in identifying a huge number of genes with unknown functions, which are now waiting to be characterized. The bioinformatic annotation of prokaryotic genomes, for instance, identifies most of the genes, but leaves 30%–50% of them without any functional attributes [1]. To face this new challenge, many "functional genomics" and "structural genomics" initiatives have been launched throughout the world. One of the goals of structural genomics is to accelerate the discovery of original protein folds as the basis to better understand the protein folding mechanisms, and, thus, improve the performances of computer programs for *ab initio* 3-D structure modeling or "threading" of proteins. The 3-D structure determination of proteins of known function usually provides detailed insights into their mechanisms of action at the molecular level. It is now expected that the 3-D structure of proteins of unknown function will, in many cases, reveal their similarity to previously described protein families and will immediately provide functional hints that can then be tested experimentally.

Structural genomics programs centrally rely on the systematic expression of the proteins encoded by a large selection of candidate genes. It is now clear that a major bottleneck is the high-throughput production of soluble proteins pure enough for nuclear magnetic resonance or crystallographic studies [2]. *Escherichia coli* is the most favoured host for protein over-expression, especially for prokaryotic genes. It is generally admitted that only half of the candidate genes (not including those encoding membrane proteins) can be expressed using standard procedures due to the unpredictable behavior of protein products (i.e. toxicity, formation of inclusion bodies, etc.). Statistically, only 20% of the selected genes quickly lead to usable protein crystals [3]. Thus, structure determination at the genome level requires additional time-consuming stages, such as optimizing

* V. Monchois, C. Abergel, C. Deregnaucourt, J.-M. Claverie
 (✉) (e-mail: vincent.monchois@igs.cnrs-mrs.fr, Tel: +33-0491–164548, Fax: +33-0491–164549)
 Information Génétique et Structurale, UMR1889 CNRS-AVENTIS, 31 Chemin Joseph Aiguier, 13402 Marseille, CEDEX 20, France (http://igs-server.cnrs-mrs.fr)
** R. Vincentelli, present adress: A.F.M.B, UMR 6098 CNRS, 31 chemin Joseph Aiguier, 13402 Marseille CEDEX 20, France

expression conditions, using alternative expression systems, the separate expression of protein sub-domains or the expression of orthologous genes from hyperthermophilic microorganisms [4]. Another possibility is to use cell-free protein synthesis, an approach that has already proven efficient for parallel protein expression straight from the sub-cloned DNA and for optimizing expression conditions, while overcoming the toxicity problems encountered in vivo [5, 6]. While the continuous-flow cell-free (CFCF) system [7] is an expensive strategy for protein expression, it was recently chosen by the RIKEN structural genomics consortium [8] to produce proteins in the mg/ml range [9].

In this article, we present the results of a pilot project involving 29 *E. coli* K12 ORFs of unknown function encoding non-membrane proteins for which we designed an original approach combining recombination cloning and in vitro expression to allow for the proficient selection of soluble proteins and the immediate scale-up of in vivo or in vitro expression using *E. coli* extracts formulated by Roche Molecular Biochemicals. Those highly efficient extracts are, to our knowledge, the only commercially available cell-free extracts allowing the visualization by regular Coomassie Blue staining of SDS-PAGE gels.

Material and Methods

Plasmid Construction

Oligonucleotide primers were designed for 29 full-length *E. coli* ORFs of unknown function that were predicted to be non-membrane proteins. The specific Lambda phage recombination sites, attB1 and attB2, were added at their 5′ end to promote directional cloning by using the GATEWAY system (Invitrogen, [10]). Each ORF was polymerase chain reaction (PCR) amplified from *E. coli* K-12 MG1655 genomic DNA using Pwo DNA polymerase (Roche Molecular Biochemicals). PCR products were inserted by homologous recombination into the pDONR201 donor vector to be transferred into pDEST17 (Invitrogen) in order to express these ORFs in phase with an N-terminal His$_6$ tag under the control of a T7 promoter. After being transformed into *E. coli* DH5α cells, the different purified plasmids were used for the over-expression of the fusion proteins both in a cell-free system and in *E. coli* BL21(DE3) cells.

Cell-Free Protein Expression

The in vitro screening of protein expression and solubility were carried out using the *E. coli* extract RTS 100 HY that was kindly provided by Roche Molecular Biochemicals. In vitro protein expression was carried out in 50 μl extract according to the manufacturer's protocol, except for the recommended plasmid quantity of 500 ng that we reduced to about 200 ng. The positive control used 500 ng of pIVEX that carried the green fluorescence protein (GFP) encoding gene. After 6 hour reaction at 30°C, the total proteins, as well as the soluble proteins recovered from the supernatant after 5 min of centrifugation, were analysed by SDS-PAGE electrophoresis with the standard Coomassie blue staining. When ambiguous results were obtained, we confirmed the expression of the target as a soluble protein by running a batch affinity purification using Ni-NTA resins (Qiagen). The eluted fractions were then loaded on the SDS-PAGE gel to reveal the specific target expression using standard Coomassie blue staining.

In Vivo Protein Expression

Proteins were expressed in *E. coli* BL21(DE3) carrying the relevant plasmids in 4 ml LB+Amp medium at 37°C. Protein expression was induced when OD_{600} reached 0.5 by adding 0.5 mM IPTG. After 3 h of induction, cells were harvested and lysed in SDS-PAGE loading buffer that contained 2 M urea (for total proteins) or in Bug-Buster buffer (Novagen) to recover the soluble protein fraction in the supernatant after centrifugation. Both total and soluble proteins were then analysed by SDS-PAGE electrophoresis revealed by standard Coomassie blue staining.

Results and Discussion

In Vitro Results

Of the 29 ORFs tested, 19 have been shown to be expressed in vitro, 17 as soluble proteins and two as inclusion bodies (Fig. 1). The sizes of the encoded proteins ranged from 21.0 to 74.0 kDa, with an average molecular weight of 37.2 kDa (Fig. 1). The results obtained in vitro suggest a relationship between the size of the targets and its successful expression since 84% of the expressed ORFs have a molecular weight lower than 37.2 kDa. Most of the ORFs exhibited expression intensities that were comparable to the GFP control (Fig. 2, GFP predicted yield 0.4 mg/ml extract) when analysed on electrophoresis gels and revealed by Coomassie blue staining. The exceptions were ORF#24 and #26 (Fig. 1), which displayed markedly higher levels of expression, and ORF#1, which was only detectable by affinity purification.

In Vivo Results

The in vivo over-expression of the 29 ORFs using transformed *E coli* BL21 was tested using a standard procedure. Out of the 29 ORFs, 15 exhibited detectable in vivo expression. Seven of them resulted in soluble proteins but eight were observed in inclusion bodies (Fig. 1).

The overall comparison between the two expression protocols therefore led to 66% of the proteins expressed in vitro versus 52% in *E. coli* (Fig. 1). There is no significant difference between these two systems for protein expression (Fisher's exact test $P=0.2$). In addition, most of the expressed proteins (87%) have, as was the case with in vitro, a molecular weight lower than 37.2 kDa. An interesting finding is that 90% of the expressed proteins are obtained as soluble in vitro versus 47% (seven ORFs) in *E. coli*. The solubility is significantly improved in vitro

A

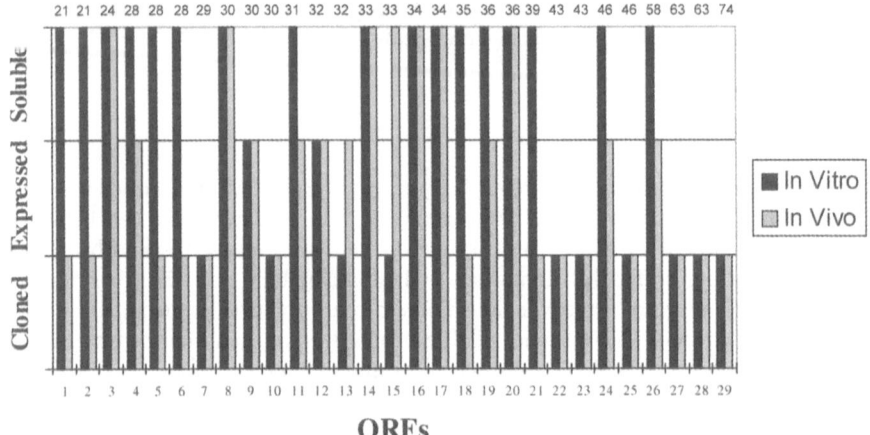

B

ORFs	*In vitro* system	*In vivo* system
Expressed	**19 (66%)**	**15 (52%)**
Soluble	17 (90%)	7 (47%)
Insoluble	2 (10%)	8 (53%)
Non expressed	**10 (33%)**	**14 (48%)**

Fig. 1a–b. Comparison of protein expression and solubility in vitro and in vivo of 29 *E. coli* ORFs of unknown function. The histogram (a) displays the results (summarized in **b**) of expression and solubility of each ORFs in vitro (*black bars*) and in vivo (*grey bars*). ORFs are ranking from the lowest (21 kDa) to the highest (74 kDa) in molecular weight

(Fisher's exact test *P*=0.009). Moreover, most of the proteins not expressed in vitro were not detected with *E. coli*. This suggests that the cell-free system could serve as a reliable screen for "soluble expression" in the context of large structural genomics projects for which it is extremely important to quickly identify potentially soluble targets. The results of our study estimate a 7% error rate (false negative) for such a screening protocol, given that two ORFs (#13 and #15) that were not detected in the cell-free system were found to be expressed in vivo. The 7% false negatives estimated from this study, although a satisfactory percentage, should be diminished by optimizing the in vitro expression protocol (i.e. plasmid concentration, temperature, coupled transcription/translation reaction time, etc.) in order to obtain an even closer correlation between in vitro and in vivo results. On the other hand, 20% of the ORFs were expressed as soluble proteins in vitro

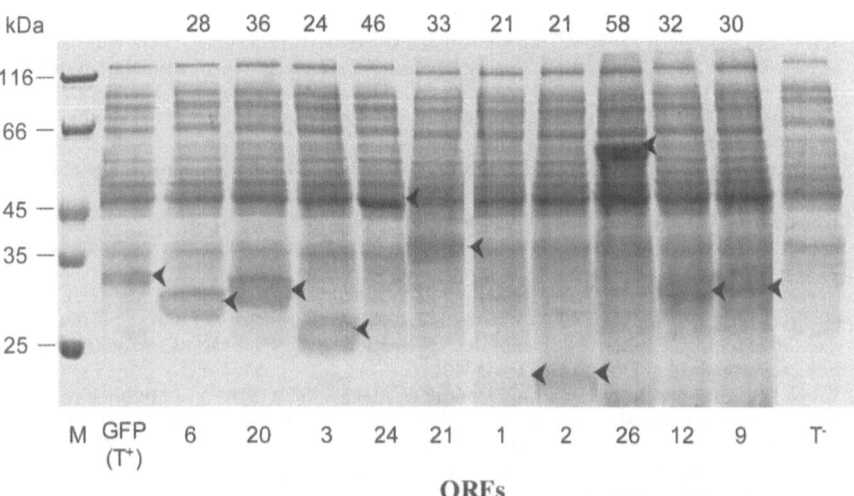

Fig. 2. Examples of total protein expression in the *E. coli* RTS 100 HY extract. The ORFs #1, #2, #3, #6, #9, #12, #21, #24 and #26 were over-expressed as well as GFP (positive control) at 30°C as described in the Materials and Methods section. We analysed 4 μl on SDS-PAGE gel followed by Coomassie blue staining. Each *arrow* corresponds to the location of the expressed protein on the gel. ORFs molecular weight is indicated at the *top* of the gel and corresponds to the ORFs theoretical molecular weight with the extra 3-kDa attB1 N-terminal linker

(#1, #2, #5, #6 #18, #21, Fig. 1) but were not detected in vivo, suggesting that further optimization of in vivo expression conditions could succeed in producing soluble proteins.

Conclusion

For this pilot project, we associated recombination cloning with in vitro expression screening. Recombination cloning was chosen as an efficient strategy to uniformly clone all PCR amplified targets in a standardized expression vector and speedup the process by allowing all targets to be treated in parallel. Our results showed that this Gateway vector, as well as other vectors carrying a T7 promoter [11], can be used for both in vivo and in vitro protein expression. Using the RTS-100 *E. coli* HY extract fits the high-throughput screening of recombinant proteins expression and solubility required by structural genomics. In a first pass, once soluble proteins have been quickly screened by in vitro expression, these targets will be prioritized for quantitative protein production and the less expensive in vivo system will be preferred, optimizing expression conditions when needed. In a second pass, targets that have failed the first screen will be re-cloned in other contexts for subsequent in vitro trials. Thus, cell-free expression should speedup target selection at a genomic scale.

References

1. Stover CK, Pham XQ, Erwin AL, Mizoguchi SD, Warrener P, Hickey MJ, Brinkman FS, Hufnagle WO, Kowalik DJ, Lagrou M, Garber RL, Goltry L, Tolentino E, Westbrock-Wadman S, Yuan Y, Brody LL, Coulter SN, Folger KR, Kas A, Larbig K, Lim R, Smith K, Spencer D, Wong GK, Wu Z, Paulsen IT (2000) Complete genome sequence of *Pseudomonas aeruginosa* PA01 an opportunistic pathogen. Nature 406:959–964
2. Christendat D, Turnbull J (1996) Identification of active site residues of chorismate mutase-prephenate dehydrogenase from *Escherichia coli*. Biochemistry 35:4468–4479
3. Thornton J (2001) Structural genomics takes off. Trends Biochem Sci 26:88–89
4. Baneyx F (1999) Recombinant protein expression in *Escherichia coli*. Curr Opin Biotechnol 10:411–421
5. Nakano H, Yamane T (1998) Cell-free protein synthesis systems. Biotechnol Adv 16:367–384
6. Martemyanov KA, Shirokov VA, Kurnasov OV, Gudkov AT, Spirin AS (2001) Cell-free production of biologically active polypeptides: application to the synthesis of antibacterial peptide cecropin. Protein Expr Purif 21: 456–461
7. Spirin AS, Baranov VI, Ryabova LA, Ovodov SY, Alakhov YB (1988) A continuous cell-free translation system capable of producing polypeptides in high yield. Science 242:1162–1164
8. Yokoyama S, Hirota H, Kigawa T, Yabuki T, Shirouzu M, Terada T, Ito Y, Matsuo Y, Kuroda Y, Nishimura S, Kyogoku Y, Miki K, Masui R, Kuramitsu S (2000) Structural genomics projects in Japan. Nat Struct Biol 7 Suppl: 943–945
9. Kigawa T, Yabuki T, Yoshida Y, Tsutsui M, Ito Y, Shibata T, Yokoyama S (1999) Cell-free production and stable-isotope labeling of milligram quantities of proteins. FEBS Lett 442:15–19
10. Hartley JL, Temple GF, Brasch MA (2000) DNA cloning using in vitro site-specific recombination. Genome Res 10:1788–1795
11. Waldo GS, Standish BM, Berendzen J, Terwilliger TC (1999) Rapid protein-folding assay using green fluorescent protein. Nat Biotechnol 17:691–695

CHAPTER 19

High-Level Cell-Free Protein Expression   **19**
from PCR-Generated DNA Templates

THOMAS HOFFMANN, CORDULA NEMETZ, REGINA SCHWEIZER,
WOLFGANG MUTTER, MANFRED WATZELE*

Introduction

Cell-free DNA-dependent in vitro transcription/translation is a well-established procedure when working with the expression of circular closed DNA and with long linear DNA. Attempts of expression from short pieces of linear DNA were only partially successful. The smaller the DNA used, the more difficult it was to produce relevant amounts of protein. It was shown that these difficulties mainly resulted from the presence of exonucleases. During in vitro transcription and translation with S30 lysates from *Escherichia coli*, it was demonstrated that exonuclease V was responsible for the degradation of linear DNA. Exonuclease V consists of three subunits (the gene products of *recB*, *recC*, *recD*). This exonuclease cleaves linear DNA from its 3'-ends.

It was tried to circumvent these problems by mutating the subunits of this exonuclease to destroy its lytic activity. Yang et al. (1980) describe an improved protein synthesis from a linear DNA template with lysates from the *E. coli* strain CF300 containing deletions of exonuclease V (elimination of genes *recB* and *recC*; in a recB21 strain).

Basset et al. (1983) introduced RNase- and polynucleotide-phosphorylase mutant genes (*rna-19 pnp*-7) into this strain, which resulted in strain CLB7. With this strain, a significantly higher protein expression with linear DNA templates was obtained after a 1-h incubation period.

Lesley et al. (1991) worked with an exonuclease V-deficient *recD* strain, SL 119, and were the first to describe protein synthesis from PCR-generated expression constructs using *E. coli* lysates.

The disadvantage of these methods is that these mutant strains usually have a lower growth rate, and lysates from these strains also have a significantly decreased protein-synthesis activity.

Another possible method to protect nucleic acids from degradation by exonucleases is to modify the ends of the nucleic acid by introducing modified nucleotide units, as described in the literature for antisense strategies (Pandolfi

* Thomas Hoffmann, Cordula Nemetz, Regina Schweizer, Wolfgang Mutter, Manfred Watzele
(✉) (e-mail: Manfred.Watzele@roche.com, Fax: +49–8856–607609)
Roche Diagnostics GmbH, Nonnenwald 2, 82377 Penzberg, Germany

et al. 1999; Verheijen et al. 2000; Kandimalla et al. 1997; Tohda et al. 1994; Tang et al. 1993; Hirao et al. 1993; Yoshizawa et al. 1994). These examples, however, deal mostly with single-stranded DNA or RNA and were only applied for in vivo systems. In this paper, we show strategies to protect double-stranded DNA that is derived from a PCR reaction and can be used directly for in vitro protein synthesis.

Materials and Methods

Preparation of Template DNA

Plasmid pIVEX2.1 GFP, containing the cycle 3 mutant (Crameri et al. 1996) sequence for green fluorescent protein (GFP) from *Aequorea victoria*, together with T7-promoter, ribosome binding site and T7-terminator, was used as a template for the PCR reaction.

A 1,115-bp fragment was amplified using the Expand High Fidelity PCR Kit (Roche Diagnostics, Mannheim, Germany). The PCR product started 30 bp upstream of the T7-promoter and stopped with the T7-terminator. The following primers were used:

1. Sense Primer:
   5′-gcttagatcgagatctcgatcccgcgaaattaatacgactcactatagggagaccacaacggtttc
2. Antisense Primer:
   5′ggaagctttcagcaaaaaacccctcaagacccgtttagaggccccaagg

We used 5 ng pIVEX2.1 GFP as a template with the following PCR-cycles: 1 min 94°C, 1 min 65°C, 1 min 72°C with 30 cycles in total. The concentration of the PCR product was determined over agarose gels. PCR product was purified, in some cases using the High Pure PCR Purification Kit (Roche Diagnostics).

Coupled In Vitro Transcription/Translation Reaction

Trancription/translation reactions were performed in a total volume of 50 μl for 2 h at 30°C. The reaction solution contained 80.5 mM potassium acetate, 10 mM magnesium acetate, 35 mM ammonium chloride, 4 mM magnesium chloride, 4% polyethylene glycol 8000, 1 mM ATP, 0.5 mM CTP, 1 mM GTP, 0.5 mM UTP, 30 mM phosphoenol pyruvate, 8 μg/ml pyruvate kinase, 400 μM of each amino acid, 0.1 mM folinic acid, 0.1 mM EDTA, 50 mM HEPES-KOH, pH 7.6, 20 μg/ml Rifampicin, 0.03% sodium azide, 2 μg/ml Aprotinin, 1 mg/ml Leupeptin, 1 μg/ml Pepstatin A, 10 mM acetyl phosphate, 100 μg/ml tRNA from *E. coli* MRE600, 8 mM Dithiothreitol, 100 U/ml RNase-inhibitor, 15 μl *E. coli* lysate, 0.5 U/μl T7-RNA polymerase. The *E. coli* lysate was prepared from the A19 strain, according to the method of Zubay (1973). If not specified otherwise, 1 μg of template DNA was added to each reaction.

Exonuclease Assay

From a coupled in vitro transcription/translation reaction (200 µl total volume), 13 µl samples were withdrawn after the time points indicated and immediately heated at 65°C for 15 min. After cooling on ice for 15 min, 107 µl H_2O and 3 µl RNAse (Roche Diagnostics) were added and incubated for 30 min at 37°C. Then 12 µl of SDS 5% and 3 µl Proteinase K (Roche Diagnostics) were added and incubated for 30 min at 37°C. After precipitation with 13 µl 3 M NaAc (pH 4.8) in 400 µl of ice-cold ETOH for 30 min at –20°C and washing with 200 µl of cold 70% ETOH, the pellet was dried and applied in total on a 1% TBE-agarose gel.

Modification of 3′ Ends with Dideoxy-ATP with Terminal Transferase

We incubated 45 µg of PCR product for 40 min at 37°C with 250 U terminal transferase (Roche Diagnostics) and 30 nmol dideoxy-ATP (Roche Diagnostics) or phosphorothioate-ATP (Adenosin 5′-O-1-thiotriphosphate) (Firma NAPS, Göttingen, Germany) in 500 µl of reaction buffer for terminal transferase (Roche Diagnostics). The reaction product was then purified over mini Quick Spin RNA columns (Roche Diagnostics).

Modification of 3′ Ends by Ligating the 5′ Ends of the Primers with the 3′ Ends of the Opposite DNA Strand

The following PCR primers, in which a pair of abasic linkers replaced two nucleoside monomer units was used for the PCR reaction amplifying the GFP sequence, as described above (as ribose a β-2′-Deoxy-D-ribofuranose was used).
1. Sense Primer:
     5′-agc gca cgc gtt ttc gcg tgc g-ribose-ribose-cgt ccg gcg tag agg atc g-3′
2. Antisense Primer:
     5′-acc gct ccc ggt ttt ccg gga gcg g-ribose-ribose-atc atg gcg acc aca ccc gt-3′

After the PCR reaction, the resulting 5′ overhangs were ligated to the 3′ ends of the opposite strands with T4 ligase (Roche Diagnostics) to form hairpin-formed loops at both ends of the template.

Quantification of Expressed GFP

GFP needs molecular oxygen to form the fluorophor post-translationally (Coxon and Bestor 1995). Therefore, after the expression, the reaction solution was stored in a 2-ml vial for 24 h at 4°C for maturation of GFP.

   The fluorescence of GFP was then measured using a spectral fluorimeter (Kontron, Biotek Neufahrn, Germany) at an excitation wavelength of 395 nm and an emission wavelength of 504 nm and quantified using a GFP standard protein (Roche Diagnostics).

Results and Discussion

The stability of linear DNA in an in vitro transcription/translation reaction containing an S30 extract from the prokaryote *E. coli* is very low. This is shown in Fig. 1a, where a PCR amplified DNA template for expression was analysed at different time intervals during an in vitro protein-synthesis reaction. After 5 min of incubation, linear DNA is no longer detectable, while circular plasmid DNA remains stable for several hours during the whole synthesis reaction (not shown).

When alpha-phosphorothioate-deoxy ATP residues were introduced into the 5′ ends of the oligonucleotide primers used for the PCR reaction, the template stability was only slightly improved (data not shown). This implicates that a 3′-

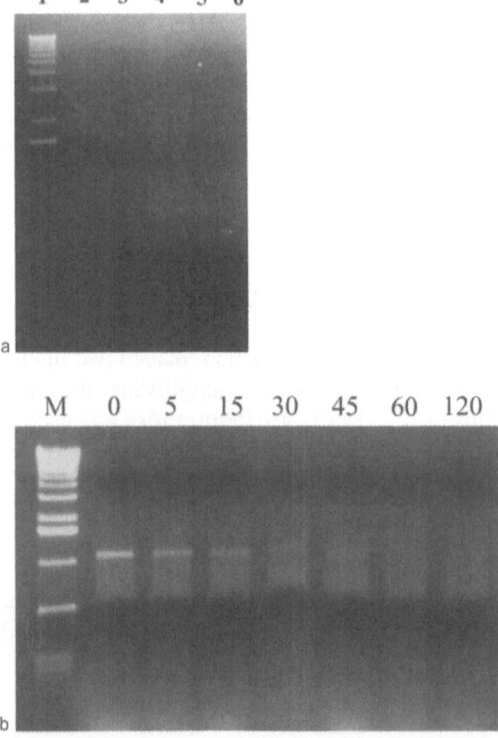

Fig. 1. a The stability of linear DNA is demonstrated by incubation of the templates in the reaction solution. After an RNase and protease treatment, equal aliquots of the reaction were applied onto agarose gels. After 5 min of incubation, linear DNA is no longer detectable. Lane 1, molecular weight markers; lanes 2–6, linear DNA after 5-, 15-, 30-, 45- and 60-min incubation in an in vitro transcription–translation reaction. **b** The stability of a template is shown, where the 3′-ends were modified with terminal transferase using dideoxy-ATP or phosphorothioate-ATP as substrate. After an RNase and protease treatment equal aliquots of the reaction were applied onto agarose gels. Even after 10 min, linear DNA is detectable in this reaction. Lane 1, molecular weight markers; lanes 2–8, linear DNA after 0-, 5-, 15-, 30-, 45-, 60- and 120-min incubation in an in vitro transcription–translation reaction

exonuclease was responsible for the degradation of the template. Therefore, residues like dideoxy-ATP or phosphothioate-ATP, which are more stable against enzymatic hydrolysis were incorporated with terminal transferase. Figure 1b clearly demonstrates that the stability of the template so modified was markedly improved. Even after 10 min, linear DNA is detectable in this reaction.

This end modification also resulted in increased protein-synthesis rates when compared with non-modified templates, as shown in Fig. 2. While unmodified templates, depending on the amount of template used, yielded from 3 to 9 µg of GFP protein per millilitre of reaction, modified templates increased the protein production more than sixfold up to 60 µg.

Since the addition of modified nucleotides at the 3′ ends only leads to a partial stabilization, a totally different approach was planned, in which an internal circularization of the linear template should occur.

PCR primers, in which a pair of abasic linkers replaced two nucleoside monomer units, were used for the PCR reaction described above. In this case, simple deoxyribose units were used as abasic linkers. These linkers would result in stopping the PCR reaction as described by Newton (1990), yielding two single-stranded overhangs at the 5′ ends of the template. These 5′ overhangs were constructed in a way that hairpin-formed loops could be built with their 5′ ends opposing the 3′ ends of the opposite strands.

By the following ligation, the DNA gaps were closed, resulting in a dumbbell-shaped single-stranded ring, as demonstrated in Scheme 1. (Only the end of the sense primer is shown).

Templates containing terminal hairpin loops already showed a slightly increased stability, as can be seen in Fig. 3a. The template DNA could be detected for 10 min.

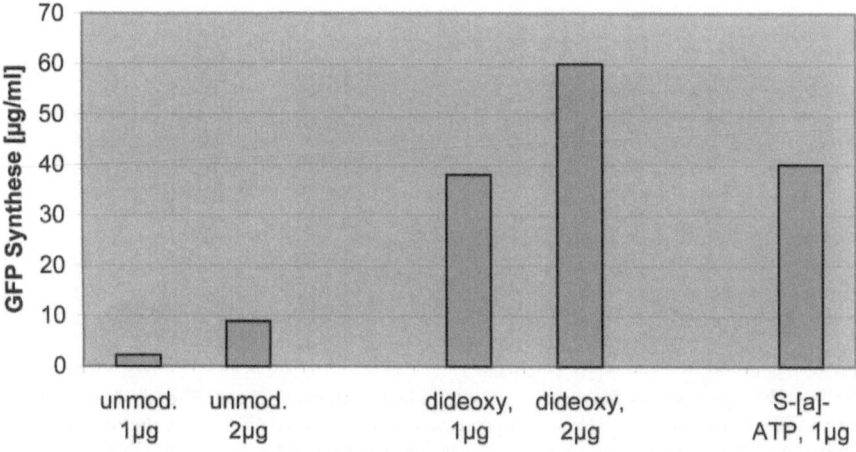

Fig. 2. The figure shows an improved protein synthesis by the modification of the 3′ ends of linear DNA templates with dideoxy-ATP or phosphorothioate-ATP when compared with non-modified templates. Amounts of 1 µg and 2 µg, respectively, of DNA template, as indicated, containing the gene for green fluorescent protein GFP were used in a 100 µl transcription–translation reaction. After the reaction, GFP was quantified using fluorescence reading (395-nm excitation, 508-nm emission)

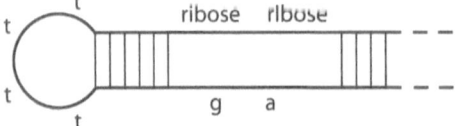

```
ttcgcgtgcg-ribose-ribose-cgtccggcgtagaggatcg...
ttgcgcacgc   g         a       gctggccgcatgtcctagc...
```

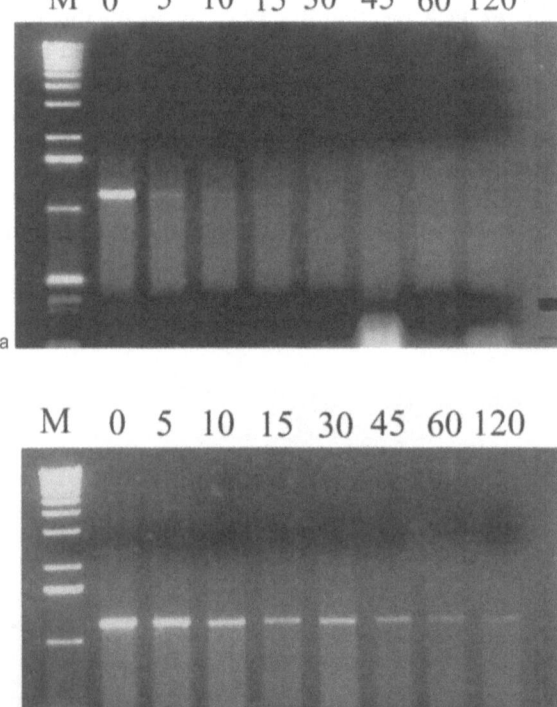

Scheme 1. Sense primer 5'-agc gca cgc gtt ttc gcg tgc g-ribose-ribose-cgt ccg gcg tag agg atc g-3'; after ligation

Fig. 3. a The figure shows the stability of a template, with overhanging 5'-ends that can form hairpin-like loops. After an RNase and protease treatment, equal aliquots of the reaction were applied onto agarose gels. Lane 1, molecular weight markers; lanes 2–9, linear DNA after 0-, 5-, 10-, 15-, 30-, 45-, 60- and 120-min incubation in an in vitro transcription–translation reaction. After more than 10 min, the linear DNA has disappeared. b The same templates as in a, which were ligated using T4 ligase before the incubation in an in vitro transcription–translation reaction. Lane 1, molecular weight markers; lanes 2–9, linear DNA after 0-, 5-, 10-, 15-, 30-, 45-, 60- and 120-min incubation in an in vitro transcription–translation reaction. The template was stable for more than 120 min

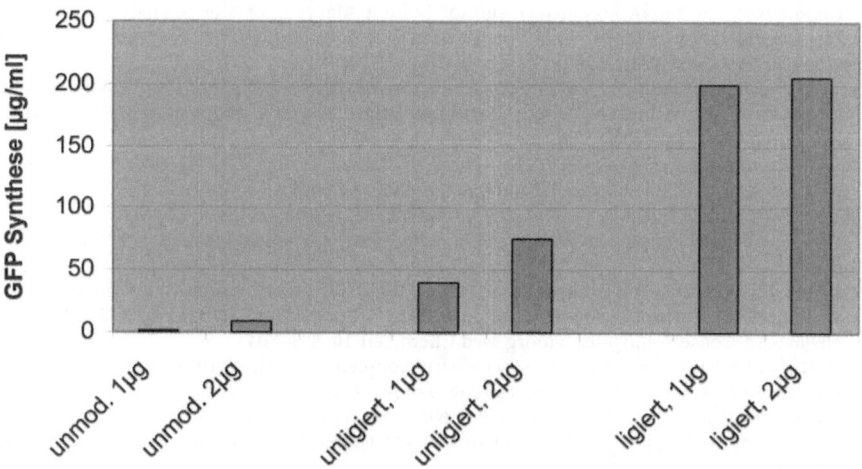

Fig. 4. The figure shows an improved protein synthesis by the modification of the 3′ ends of linear DNA templates by the introduction of hairpin loops. Amounts of 1 μg and 2 μg, respectively, of DNA template, as indicated, containing the gene for green fluorescent protein (*GFP*) were used in a 50-μl transcription–translation reaction. 70 μg/ml of GFP with the unligated ("*unligiert*") template and more than 200 μg/ml with ligated ("*ligiert*") template were obtained

In this case, the 5′ ends were not ligated to the 3′ ends of the opposite strand. After the internal ligation of the two strands, however, the stability of the linear DNA is sufficient to allow protein synthesis during the entire reaction of 120 min as demonstrated in Fig. 3b.

These template modifications also resulted in a much higher protein synthesis, which is shown in Fig. 4. Even without ligation, the templates allow GFP synthesis up to 70 μg/ml. After ligation, the protein production was further increased to more than 200 μg/ml, as was the case with circular plasmid templates (data not shown). It can also be observed, that only 1 μg of PCR-generated template was sufficient for optimum production rates.

References

Coxon A, Bestor TH (1995) Proteins that glow in green and blue. Chem Biol 2: 119

Crameri A, Whitehorn EA, Tate E et al. (1996) Improved Green Fluorescent Protein by Molecular Evolution Using DNA Shuffling. Nat Biotechnol 14: 315–319

Hirao I, Yoshizawa S, Miura K (1993) Stabilization of mRNA in an *Escherichia coli* cell-free translation system. FEBS Lett 321: 169–172

Kandimalla ER, Manning A, Zhao Q et al. (1997) Mixed backbone antisense oligonucleotides: design, biochemical and biological properties of oligonucleotides containing 2′-5′-ribo- and 3′-5′-deoxyribonucleotide segments. Nucleic Acids Res 25: 370–378

Basset CL, Rawson JRY (1983) In Vitro Coupled Transcription-Translation of Linear DNA Fragments in a Lysate Derived from a *recB rna pnp* Strain of *Escherichia coli*. J Bacteriol 156: 1359–1362

Lesley SA, Brow MAD, Burgess RR (1991) Use of *in Vitro* Protein Synthesis from Polymerase Chain Reaction-generated Tempaltes to Study Interaction of *Escherichia coli* Transcription

Factors with Core RNA Polymerase and for Epitope Mapping of Monoclonal Antibodies. J Biol Chem 266: 2632–2638

Newton CR (1990) Amplification processes. EP 0 416817 B1

Pandolfi D, Rauzi F, Capobianco ML (1999) Evaluation of Different Types of End-Capping Modifications on the Stability of Oligonucleotides Toward 3′-and 5′-exonucleases. Nucleosides Nucleotides 18: 2051–2069

Tang J Temsamani J, Agrawal S (1993) Self-stabilized antisense oligodeoxynucleotide phosphorothioates: properties and anti-HIV activity. Nucleic Acids Res 11: 2729–2735

Tohda H, Chikazumi N, Ueda T et al. (1994) Efficient expression of E. coli dihydrofolate reductase gene by an in vitro translation system using phosphorothioate mRNA. J Biotechnol 34: 61–69

Verheijen JC, van Roon AM, Meeuwenoord NJ et al. (2000) Incorporation of a 4-Hydroxy-N-acetylprolinol Nucleotide Analogue Improves the 3′-Exonuclease Stability of 2′-5′-Oligoadenylate-Antisense Sonjugates. Bioorg Med Chem Lett 10: 801–804

Yang HL, Ivashkiv L, Chen HZ et al. (1980) Cell-free coupled transcription-translation system for investigation of linear DNA segments. Proc Natl Acad Sci USA 77: 7029–7033

Yoshizawa S, Ueda T, Ishido Y et al. (1994) Nuclease resistance of an extraordinarily thermostable mini-hairpin DNA fragment, d(GCGAAGC) and ist application to in vitro protein synthesis. Nucleic Acids Res 22: 2217–2221

Zubay G (1973) In vitro synthesis of protein in microbiol systems. Annu Rev Genet 7: 267–287

Optimization of the Translation Initiation Region of Prokaryotic Expression Vectors: High Yield In Vitro Protein Expression and mRNA Folding

20

ANDREA GRAENTZDOERFFER, MANFRED WATZELE,
BERND BUCHBERGER, SABINE WIZEMANN, THOMAS METZLER,
WOLFGANG MUTTER, CORDULA NEMETZ

Introduction

The process of in vitro protein synthesis that is based on T7 RNA polymerases differs strongly from what occurs in *E. coli* (Studier et al. 1990). Since the RNA polymerase of bacteriophage T7 works more than five times faster than endogenous RNA polymerases and the *E. coli* translation machinery, in vitro synthesized mRNAs are less protected by bacterial ribosomes. Consequently, no real coupling of prokaryotic transcription and translation can take place in vitro (Spirin 1999). As unprotected mRNAs easily form secondary structures, double-stranded regions can block the accessibility of important regulatory elements like the ribosomal binding site (RBS or Shine-Dalgarno site) and the start codon (AUG) and, thereby, inhibit the initiation of translation. Here, the initial region of prokaryotic expression vectors containing a T7 promoter and a T7 gene 10 enhancer was investigated and optimized for in vitro protein expression reactions.

Materials and Methods

Cloning Procedures

Green fluorescent protein (GFP, cycle3, Crameri et al. 1996) was cloned into pIVEX2.2 and pIVEX2.4 using the restriction endonucleases *Rca*I and *Sma*I (Roche Applied Sciences, Mannheim, Germany).

To clone the wild-type form of GFP (GFPwt), PCR was performed using oligonucleotides specific for GFPwt (sense primer: TCA TGA CTA AAG GTG AAG AAC TTT TCA CTG G; antisense primer TTA CCC GGG TTG GTA CAG TTC ATC CAT GCC) and a recombinant DNA clone coding for the wild-type form of GFP obtained from IBA (Institute for Bioanalytics GmbH, Goettingen, Germany) as

* C. Nemetz (✉) (e-mail: cordula.nemetz@roche.com, Tel.: +49–8856–603134,
Fax: +49–8856–607874)
A. Graentzdoerffer, M. Watzele, B. Buchberger, S. Wizemann, T. Metzler, W. Mutter,
Roche Diagnostics GmbH, Nonnenwald 2, 82377 Penzberg, Germany

template. The PCR product was cloned with *Rca*I and *Sma*I into pIVEX2.2 and pIVEX2.4, respectively.

Small capsid protein 1049 of cytomegalovirus (CMV 1049) was cloned with *Nco*I and *Sma*I into pIVEX2.1 and pIVEX2.2. The cDNA coding for CMV 1049 was kindly provided by Dr. Martin Messerle (Genzentrum, LMU Munich, Germany).

Coupled In Vitro Transcription and Translation Reaction

To perform in vitro high-yield protein synthesis reactions, the Rapid Translation System 500 (RTS 500; Roche Applied Sciences, Mannheim, Germany) based on *E. coli* lysates was used. The reactions for protein expression were performed following the instruction manual, but half-full reaction chambers for GFP and GFPwt expressions were used. The upper chamber of the devices was filled with 0.5 ml of reaction solution.

Quantification of Expressed GFP

GFP needs molecular oxygen to form the fluorophor post-translationally (Coxon and Bestor 1995). Therefore, the reaction solution was stored in a 2-ml vial for 24 h at 4°C so that the GFP could mature after expression.

The fluorescent activity of GFP was then measured using the fluorescence spectrometer SFM25 (Kontron Instruments, Basel, Switzerland) at an excitation wavelength of 395 nm and an emission range of 430–580 nm. Concentrations were calculated by referring to the fluorescent activity of standard GFP of known concentration.

SDS-PAGE and Western Blotting

After in vitro expression, 2 µl of the protein solution per lane were separated by SDS-PAGE using 12% Bis-Tris gels (Invitrogen Corporation, Carlsbad, CA) and stained with Coomassie blue (SimplyBlue Safe Stain, Invitrogen) or blotted on nitrocellulose membranes. All the reagents needed to detect Strep-tagged proteins are described below and purchased from Roche Applied Sciences. After blocking for 1 h with Western Blocking Reagent, the membranes were washed three times with PBST buffer containing 0.1 % Tween 20 for 5 min followed by an incubation in avidin buffer (2 µg/ml avidin in PBST) for 10 min. A strepavidin-AP conjugate was added and incubated for another 60 min. After washing three times with PBST and two times with PBS for 2 min each, the specific protein signals were visualized using NBT/BCIP ready-to-use tablets.

Prediction of mRNA Secondary Structure

The folding of the mRNA was predicted by using MFold, the program from the GCG Wisconsin Package version 7.2 (Genetics Computer Group, Wisconsin, USA) which is based on Zuker and Jaeger's algorithm (Zuker 1989; Jaeger et al. 1989).

Results and Discussion

First, the transcription level of various prokaryotic expression constructs was compared to in vitro transcription reactions and subsequent Northern blot analysis. On mRNA level no obvious differences were observed between constructs that resulted in a high yield of protein synthesis after in vitro transcription/translation and constructs leading to no detectable expression (data not shown). For that reason we assumed that the initiation of translation might be the critical step for achieving high expression levels. mRNA, especially if not protected by ribosomes, may fold into secondary structures that sometimes force the ribosomal binding site and the AUG start codon into a stable stem-loop structure. This inhibits the ribosomal binding and therefore the translation initiation (Fig. 1). In an attempt to understand the dependence of the expression of proteins on the secondary structure around ribosomal binding site and start codon, we investigated the expression of different proteins in different vector contexts and compared the expression levels to the accessibility of the translation initiation region.

To evaluate a possible role for the first nucleotides downstream of the start codon, GFP (version cycle 3) and GFPwt were investigated in combination with different tags. The tag sequence strongly influenced the levels of expression

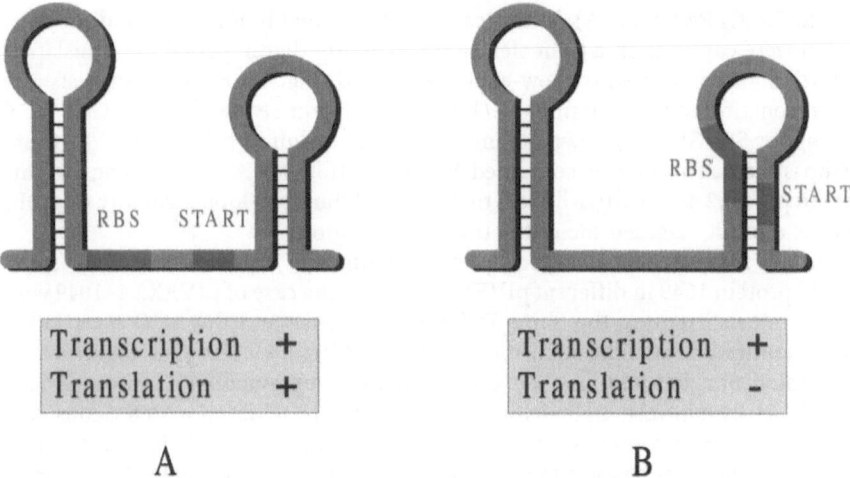

Fig. 1. Model of mRNA folding that enabled (A) or impeded (B) translation initiation

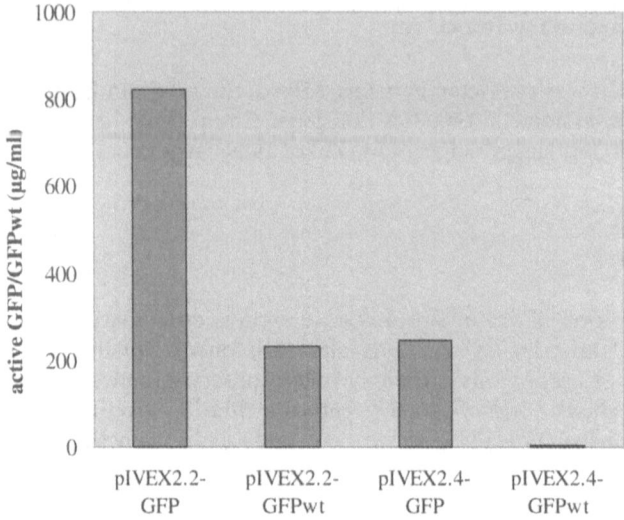

Fig. 2. Comparison of the expression levels of GFP and GFPwt depended on the tag sequence. GFP and GFPwt containing either a Strep-tag (pIVEX2.2) or a His-tag (pIVEX2.4) were expressed in RTS500 reactions. The concentrations were calculated according to the fluorescent activity of standard

(Fig. 2). These results were confirmed on protein level via SDS-PAGE so that a direct influence of the tag on GFP activity could be excluded (data not shown).

Deleting nucleotides 8–10 and mutation at position 15, which represented the differences between GFP and GFPwt at the 5′-end of both cloned genes, decreased the expression levels two to tenfold (Fig. 2).

In the case of GFPwt an unfavourable secondary structure of the translational initial region was formed with the N-terminal His-tagged vector context (pIVEX2.4-GFPwt, Fig. 3A). In this case the ribosomal binding site and the start codon were captured in a tight stem-loop structure that impeded the translation initiation and explained the low-expression yield (Fig. 4A, B). The synthesis rate was reconstituted by changing a G/T rich 9 bp-linker sequence (TCT GGT TCT, coding for Ser-Gly-Ser) downstream of the start codon in pIVEX2.4-GFPwt into a 6 bp-stretch. This mainly contained A/C nucleotides (ACC AGC, coding for Thr-Ser) in pIVEX2.4-GFPwt(Δ-linker) that resolved the stem-loop structure (Fig. 3B) and, as a result, released the initial translation region.

Similar results were obtained by investigating the expression of CMV small capsid protein 1049 in different pIVEX vectors. In the case of pIVEX2.1–1049 with a C-terminal Strep-tag the Shine-Dalgarno sequence and the AUG start codon were captured in a tight stem-loop structure (Fig. 5A). For this construct no expression was detected in a Western blot using streptavidin-AP conjugate (lane 2 in Fig. 6). By contrast, high yields of CMV 1049 were achieved with the construct pIVEX2.2–1049 introducing an N-terminal Strep-tag (Fig. 6, lane 1). The different bases after the start codon provided by the N-terminal Strep-tag led to a secondary structure of the corresponding mRNA where the translation initial sequence

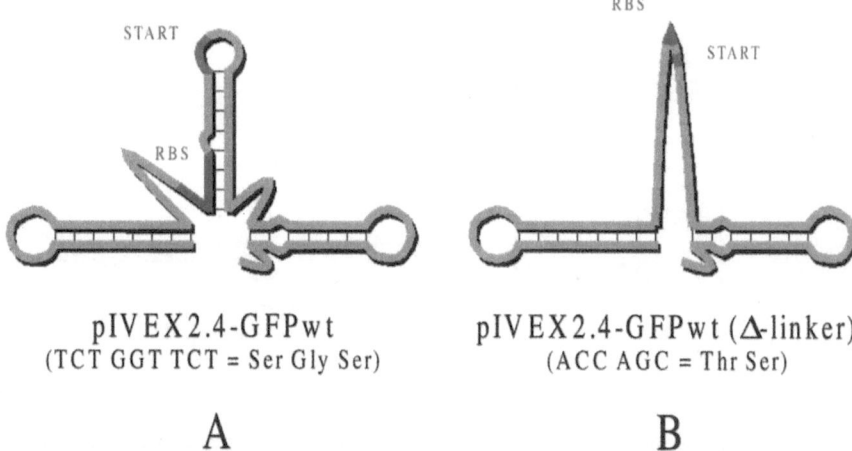

pIVEX2.4-GFPwt
(TCT GGT TCT = Ser Gly Ser)

pIVEX2.4-GFPwt (Δ-linker)
(ACC AGC = Thr Ser)

A   B

Fig. 3. Graphical presentation of predicted secondary structures of the translation initial region in (**A**) pIVEX2.4-GFPwt and (**B**) pIVEX2.4-GFPwt (Δ-linker)

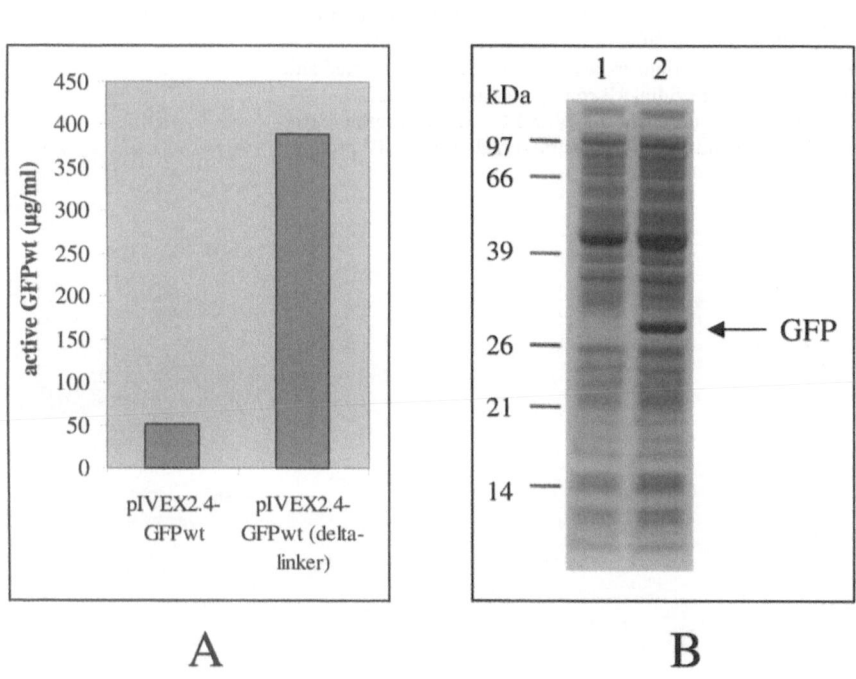

A   B

Fig. 4. Influence of secondary structures in the translation initial region on the expression level of GFPwt. **A** Fluorescent activity of pIVEX2.4-GFPwt and pIVEX2.4-GFPwt(Δ-linker). The GFPwt constructs were expressed in RTS500 reactions. The concentrations were calculated according to the fluorescent activity of standard GFPwt. **B** Aliquots of the in vitro expression reactions with the GFP constructs were applied on a SDS-gel and stained with Coomassie Blue. Lane 1, pIVEX2.4-GFPwt; lane 2, pIVEX2.4-GFPwt(Δ-linker). The *arrow* indicates the position of the GFP protein

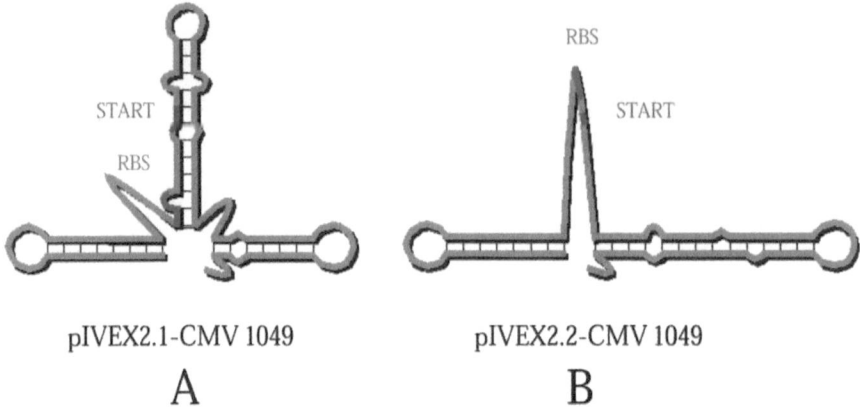

Fig. 5. Predicted secondary structures of 5′-termini of CMV-mRNA transcribed from different expression vectors: (A) pIVEX2.1–1049 and (B) pIVEX2.2–1049

Fig. 6. Western blot analysis of CMV small capsid protein produced with RTS500 in different pIVEX vectors. Aliquots of the in vitro expression reactions were applied on SDS-gels, blotted and detected with strepavidin-AP conjugate. Lane 1, pIVEX2.2–1049 (N-terminal Strep-tag); lane 2, pIVEX2.1–1049 (C-terminal Strep-tag). The *arrow* indicates the specific signal of the CMV protein

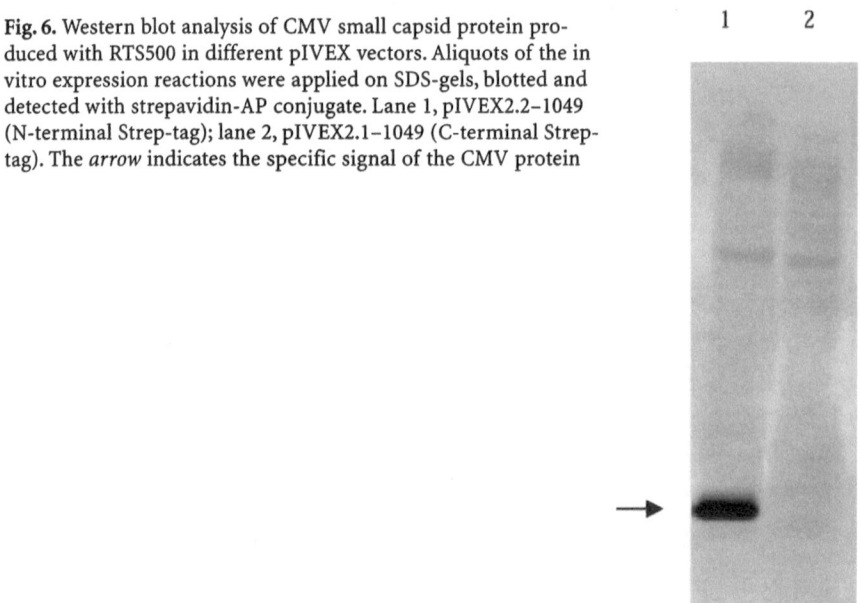

seemed to be more accessible for interaction with ribosomes (Fig. 5B). Similar results were obtained with the respective His-tagged vectors (data not shown).

The protein synthesis rates of more than 100 different structure genes were analysed in combination with C- or N-terminal Strep-tag or His-tag sequences (data not shown). The sequences of these genes in different vector contexts resulted in different intramolecular mRNA foldings and led to a variation in the amount of expressed proteins. The probability of successful expression increased with the number of different vectors tested.

In summary, we could show that the sequence between ribosomal binding site and start codon as well as the initial gene specific nucleotides significantly influences the synthesis rates. As a consequence, the in vitro expression vectors will be optimized to obtain an intramolecular folding of mRNAs in which regulatory elements are well exposed. In addition, a bioinformatical approach is under development in order to predict favourable mRNA secondary structures and to select an optimal vector context for a given nucleotide sequence.

References

Coxon A, Bestor TH (1995) Chem Biol 2:119–121
Crameri A, Whitehorn EA, Tate E, Stemmer WP (1996) Nat Biotechnol 14:315–319
Jaeger JA, Turner DH, Zuker M (1989) Proc Natl Acad Sci USA 86:7706–7710
Spirin AS (1999) Ribosomes. Kluwer Academic/Plenum Publishers, New York Boston Dordrecht London Moscow
Studier FW, Rosenberg AH, Dunn JJ, Dubendorff JW (1990) Methods Enzymol 185:60–89
Zuker M (1989) Science 244:48–52

Chapter 21

Selective Labeling of Proteins in the RTS 500 System

21

Jean-Michel Betton[1], Nicolae Palibroda[2], Abdelkader Namane[3],
Thomas Metzler[4], Octavian Bârzu[3]

Introduction

The benefits of combining heteronuclear, multidimensional NMR with isotope-labeling strategies are well documented by recent studies of protein solution NMR spectroscopy. For example, the use of deuteration in concert with ^{13}C-, ^{15}N-labeling and the development of the TROSY technique have significantly increased the size of proteins that are amenable for study by NMR (Riek et al. 2000).

It has become an almost routine practice to prepare uniformly ^{13}C- or ^{15}N-labeled proteins from high-level expression plasmids by growing transformed *Escherichia coli* cells in a minimal medium containing ^{13}C-glucose or ^{15}N ammonium salts as the sole carbon or nitrogen sources. Although bacterial expression remains an economical system for producing uniformly labeled proteins, selective labeling with one or more ^{13}C- or ^{15}N-enriched amino acids is not always possible because of the amino acid metabolic pathways of *E. coli*.

In this study, we have assessed the new in vitro protein biosynthesis system, RTS 500, from Roche Molecular Biochemicals, for the selective incorporation of ^{13}C/^{15}N-aspartic acid and ^{13}C/^{15}N-arginine into the maltose-binding protein (MalE), the soluble receptor for the high-affinity transport of maltodextrins of *E. coli*. The isotope content of selectively ^{13}C/^{15}N-labeled MalE proteins, produced from RTS 500 and from bacterial expression, was determined by GC/MS spectrometry and compared to assess the respective isotopic enrichments of these two-labeled amino acid residues.

[1] J-M. Betton (✉) (e-mail: jmbetton@pasteur.fr, Tel.: +33–1–45–688959, Fax: +33–1–40–613043)
Unité de Biochimie Cellulaire, Institut Pasteur, CNRS (URA2185), 28 rue du Dr Roux, 75724 Paris Cedex 15, France
[2] N. Palibroda
Institute of Isotopic and Molecular Technology, 3400 Cluj-Napoca, Romania</authorinfo>
[3] A. Namane, O. Bârzu
Laboratoire de Chimie Structurale des Macromolécules, Institut Pasteur, CNRS (URA2185), 28 rue du Dr Roux, 75724 Paris Cedex 15, France
[4] T. Metzler
Roche Diagnostics, Nonnenwald 2, 82372 Penzberg, Germany

Materials and Methods

Materials

Uniformly ^{13}C/^{15}N-labeled L-arginine and ^{13}C/^{15}N-labeled L-aspartic acid were obtained from Euriso-Top (CEA, France). Amino acids with natural isotope content and N-methyl-N-(tert.-butyldimethylsilyl) trifluoroacetamide (MTBSTFA) were obtained from Sigma, St Louis MO, USA.

Preparation of Selectively Labeled MalE Proteins with ^{13}C/^{15}N-Asp or ^{13}C/^{15}N-Arg

Plasmid pIV2.3ME, carrying the wild-type *malE* gene (without its signal sequence) under the control of the T7 promoter (J.-M. Betton, this volume) was used both as DNA template for RTS 500 expression and to transform *E. coli* BL21, DE3 strain for bacterial expression (Studier et al. 1990). Cell-free MalE synthesis was performed with the RTS 500 instrument essentially as described in the instruction manual from Roche Molecular Biochemicals. Specific RTS 500 kits were supplied in which the reaction mix and feeding mix were prepared without Arg and Asp. These kits were reconstituted by adding the desired ^{13}C/^{15}N-labeled amino acid at a final concentration of 250 μM. The coupled transcription/translation reaction, initiated by adding 15 μg of pIV2.3ME, was carried out in 1 ml (total volume) for 20 h at 30°C. For bacterial expression, the culture medium was 10 ml of M9 minimal medium (Miller 1992) supplemented with 0.2% glucose, 2 mM MgSO$_4$, 0.1 mM CaCl$_2$, 0.01 mM thiamine, and 0.1 mg/ml ampicillin. Cells harboring the pIV2.3ME plasmid were grown at 37°C to mid-log phase before addition of 1 mM ^{15}N/^{13}C-arginine or ^{15}N/^{13}C-aspartic acid. Expression was induced with 1 mM isopropyl-β-D-thiogalactopyranoside (IPTG) 15 min after addition of the labeled amino acids. After 3 h, cells were harvested by centrifugation, re-suspended in 2 ml of 50 mM Tris-HCl buffer (pH 8.0), lyzed by French press and pelleted at 15,000 g for 20 min. Protein production was analyzed by SDS-polyacrylamide gel electrophoresis.

The labeled MalE proteins were purified by affinity chromatography using a cross-linked amylose resin as described (Betton and Hofnung 1996). Whole RTS 500 extracts (1 ml) or supernatants of lyzed cells (2 ml) were loaded on amylose columns equilibrated with 20 mM Tris-HCl buffer, pH 7.5 containing 0.1 M NaCl. After a washing step, the proteins that were eluted (1 ml) by 10 mM maltose in the same buffer were pooled, and MalE concentrations were determined from the absorbance at 280 nm with an extinction coefficient of 68,750 M^{-1} cm^{-1}.

Sample Preparation for Isotopic Enrichment Analysis

MalE proteins (50 μg) were hydrolyzed in 6 N HCl at 110°C for 24 h. Phenol, to protect tyrosine, and norleucine, as internal standard, were added prior to hydrolysis. Then amino acid derivatization was performed by adding 50 μl of acetonitrile, 5 µl of ethanethiol, and 20 µl of MTBSTFA to dried hydrolysates. Reaction at 120°C for 15 min gave good signals for the majority of amino acids, while 60 min at 130°C gave a better yield for the basic amino acids Arg, Lys, and His.

Gas Chromatography/Mass Spectrometry

Amino acid separations were performed on a Varian Star 3400CX gas chromatograph (Varian, Palo Alto CA, USA) equipped with a J&W Scientific DB-5MS silica capillary column (J&W Scientific, Folsom CA, USA). Samples were injected in the splitless mode, and the column temperature was programmed from 50°–275°C. Mass spectra were collected on a Varian Saturn 2000 Ion Trap system operated in the electron ionization (EI) mode. Operating parameters were as follows: ion trap temperature 140°C, manifold temperature 50°C, transfer line temperature 200°C, electron emission 20 µA, electron multiplier voltage 1500 V, EI maximum ionization time 25 ms, and prescan ionization time 100 µs.

Results and Discussion

The isotope labeling of proteins expressed either in the cell-free RTS 500 system or in *E. coli* was examined by analyzing the selectivity of biosynthetic incorporation of two $^{13}C/^{15}N$-labeled amino acids, aspartic acid and arginine, representative of the metabolic pathways of *E. coli*. Indeed, since aspartic acid is a major precursor of amino acid biosynthetic reactions, metabolism can scramble the final position of the ^{13}C and ^{15}N atoms introduced by this single amino acid. In contrast, for arginine, which is not a precursor in amino acid biosynthetic pathways, no such metabolic conversions are expected.

The high level expression in RTS 500 (Fig. 1) and the availability of a single-step purification by affinity chromatography make MalE a suitable model to study the selective incorporation of isotopically labeled amino acids. To facilitate the following isotope analysis, we performed, for both RTS 500 and bacterial expression, a blank experiment with unlabeled amino acids. All purified MalE proteins gave single bands on SDS-polyacrylamide gels. Routinely, 0.3 mg of unlabeled or labeled MalE, either with $^{13}C/^{15}N$-aspartic acid or with $^{13}C/^{15}N$-arginine, was purified from 1 ml of RTS 500 reaction. The purified unlabeled and labeled MalE proteins were completely hydrolyzed by HCl and the constituent amino acids derivatized by MTBSTFA. The mixtures of volatile tBDMS amino acid derivatives were then separated by gas chromatography (GC), and their isotope contents were characterized by mass-spectrometry (MS) using an ion trap detector.

All amino acid derivatives eluted as well-separated single peaks, except for arginine, for which three derivatization products were obtained (Fig. 2). Mass spectra were recorded for amino acid standards with natural isotope contents and for $^{13}C/^{15}N$-labeled aspartic acid and $^{13}C/^{15}N$-arginine. The isotope shifts observed in the mass spectra of these labeled amino acids were helpful in elucidating the structure of the tBDMS derivatives of arginine. The ion trap EI mass spectra generally corresponded to those previously reported by Mawhinney et al. (1986) and

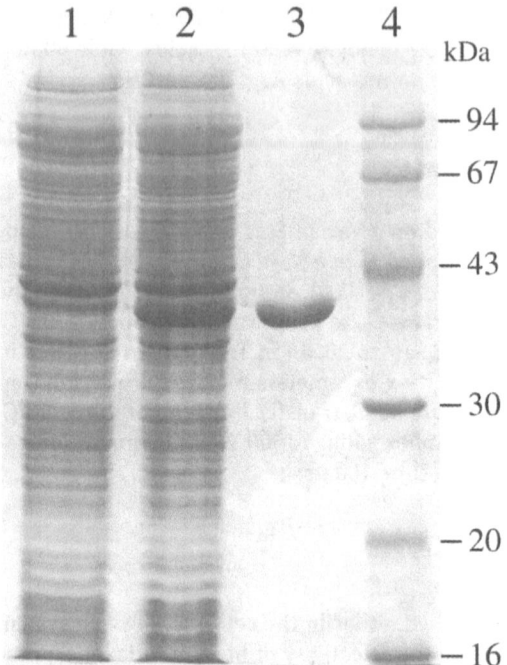

Fig. 1. In vitro production of labeled-MalE in the RTS 500 system. Production and purification of MalE selectively labeled with $^{13}C/^{15}N$-aspartic acid were analyzed by SDS-polyacrylamide gel electrophoresis and stained with Coomassie blue. Lane 1, whole RTS 500 extract containing the control pIVEX2.3MCS plasmid; lane 2, whole RTS 500 extract with the pIV2.3ME expressing MalE plasmid; lane 3, labeled MalE purified by amylose chromatography; lane 4, molecular weight markers

by Chaves Das Neves and Vasconcelos (1987), and present the most intense and characteristic fragment ions, M-57 and M-85 (Fig. 2). Other important ions are the fragment ions M-159, M-43, and M-15. Molecular ions are present as weak peaks with a high tendency to protonate. Exceptions to this rule are lysine and ornithine. Since the loss of 57 mass units came from the fragmentation of a *tert-*butyl group of a silicon atom, the mass spectra of labeled amino acids display the full label on their molecular ion as well as on the fragment ion M-57. The yield of the three arginine derivatives, designated by I, II and III, was about 75:20:5. The mass spectrum of arginine II, dominated by the ions m/z 286 and 474, corresponded to ornithine (N. Palibroda, unpublished results). It carries seven of the parental C and N atoms of arginine. Isotope analysis for the total content of ^{13}C and ^{15}N was performed on the intense fragment ion M-57 and for lysine and ornithine on the molecular ion. True isotope peaks, containing only the contribution of the C and N atoms of the underivatized amino acids, were obtained by subtracting the contributions of the next two ions, M+1 and M+2. These ions correspond to the natural content of silicon isotopes ^{29}Si and ^{30}Si and to carbon and nitrogen isotopes of the tBDMS groups. The intensity ratios of (M+1)/M and

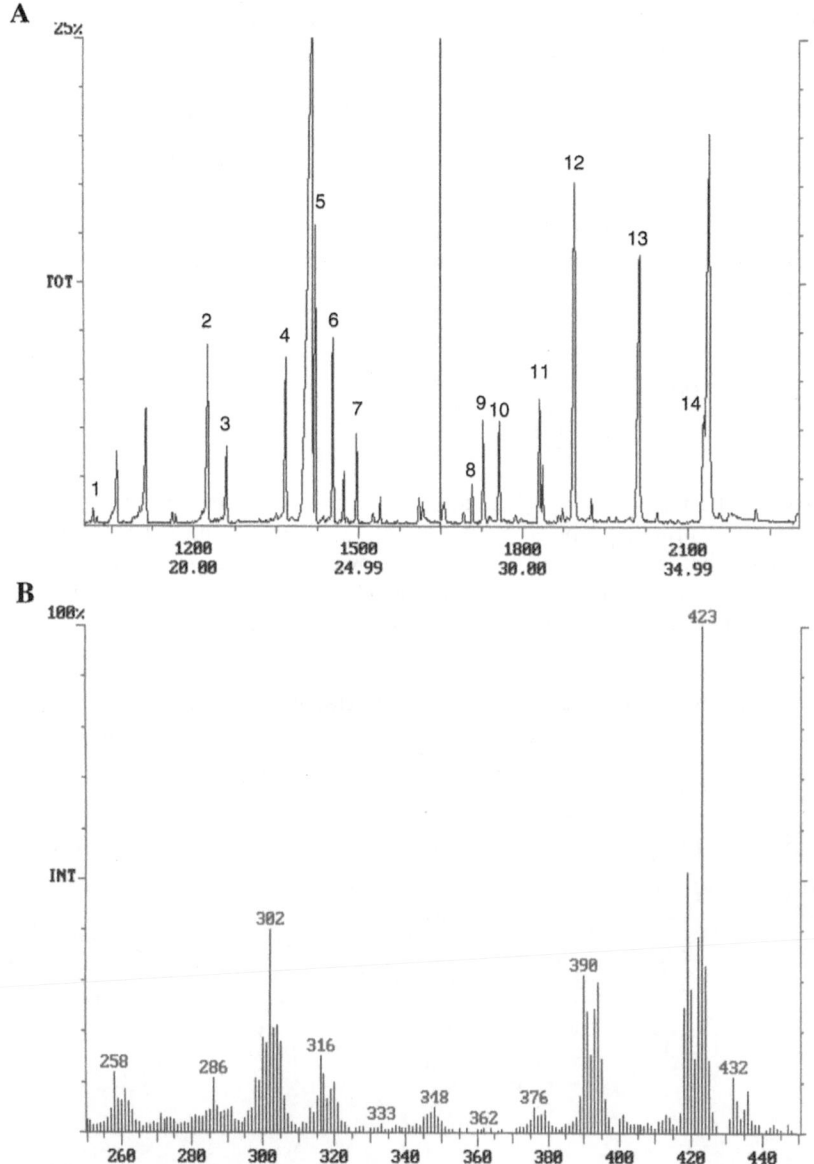

Fig. 2. Typical output from GC/MS spectrophotometer. A Chromatogram showing the separation of tBDMS derivatives of amino acids. The ordinate is the total ion current, while the abscissa is retention time (Rt). The peaks (scans from 1,000 to 2,300) correspond to: ArgI (1; Rt=17.03 min), Ala (2; Rt=20.43), Gly (3; Rt=20.99), Val (4; Rt=22.79), Leu (5; Rt=23.69), Ile (6; Rt=24.23), Pro (7; Rt=24.94), Met (8; Rt=28.46), Ser (9; Rt=28.8), Thr (10; Rt=29.28), Phe (11; 30.48), Asp (12; Rt=31.58), Glu (13; Rt=33.51) and Lys (14; Rt=35.48). **B** Computer-reconstructed mass spectrum of peak 12 (aspartic acid), where relative ion currents are plotted as a function of mass number. The spectrum shows the major M-57 fragment of the tBDMS-aspartic acid derivative (m/z=423) with all of its C and N atoms labeled

(M+2)/M due to the natural isotope content were determined from the spectra of unlabeled amino acids and were assumed to be the same for the labeled amino acids.

The isotope label was defined as the ratio of the total number of ^{13}C and ^{15}N atoms in all isotopic peaks divided by the total number of C and N atoms (A) of the underivatized amino acid molecule. The total number of ^{13}C and ^{15}N atoms in an isotopic peak resulted from the shift number (SN) of the isotopic peak multiplied by its intensity (I). The label value for a given amino acid was calculated using the following equation, where the sum is extended over all isotopic peaks (n):

$$label\,(\%) = 100 \times \frac{\sum_{i=1}^{i=n}(SN_n \times I_n)}{A \times \sum_{i=1}^{i=n} I_n}$$

Table 1 gives the isotope label values for each amino acid obtained from hydrolysates of MalE proteins labeled with aspartic acid or arginine. These results demonstrated that labeling of MalE in bacteria with $^{13}C/^{15}N$-aspartic acid causes isotopic dilution and incorporation of the labels at all amino acid residues on which mass spectra were obtained. As expected, metabolic scrambling did not

Table 1. $^{13}C/^{15}N$ content in amino acids analyzed from labeled-MalE proteins

Amino acid	Number of residues	TBDMS (m/z)	M-57	Bacterial expression		RTS 500 expression	
				$^{13}C/^{15}N$-Asp	$^{13}C/^{15}N$-Arg	$^{13}C/^{15}N$-Asp	$^{13}C/^{15}N$-Arg
Cyanamide (ArgI)		2	213	14	97.5	0	97.8
Ala	44	2	260	10	0	0	0
Gly	32	2	246	20	0	0	0
Val	21	2	288	11	<0.3	1	0
Leu	30	2	302	8	<0.8	1	1
Ile	23	2	302	26	<0.3	0	<0.5
Pro	21	2	286	27	0	0	0
Met	7	2	320	36	0	1	<0.4
Ser	13	3	390	19	0	0	1
Thr	20	3	404	31	0	0	0
Phe	15	2	336	4	0	0	0
Asp (+Asn)	24 (+21)	3	418	46	0	60	0
Glu (+Gln)	27 (+29)	3	432	39	<0.5	1.7	<0.2
Ornithine (ArgII)		3	417	0	97.6	0	98.5
Lys	37	3	431	0	0	0	<0.6
ArgIII	6	3	442	0	99.3	0	98.2
His	9	3	440	0	0	0	0
Tyr	15	3	466	8	0	0	0

occur when ^{13}C/^{15}N-arginine was used for labeling MalE in *E. coli*. In contrast, when MalE was labeled in the RTS 500 system, no detectable ^{13}C or ^{15}N atoms were found in amino acids other than those used for labeling. This result indicates that amino acid metabolism did not occur in the RTS 500 system. The isotopically diluted value of 60% found for aspartic acid in RTS 500 is due to the fact that during acid hydrolysis, the unlabeled asparagine is converted to aspartic acid. The average isotope label expected for the mixing of the 21 asparagine residues of natural isotope content with the 24 highly enriched aspartic acid residues would be about 53%. Therefore, all aspartic acid residues of MalE were completely labeled in the RTS 500 expression system.

Conclusion

We have established that the labeling of MalE with ^{13}C/^{15}N-aspartic acid can be achieved in the RTS 500 system with a high selectivity and without dilution or scrambling. Progress to improve the yield of protein production should help determine the applicability of the method to the preparation of samples for NMR spectroscopy. The clear advantages of selective labeling in the RTS 500 for NMR studies are that resonances may be directly assigned by residue type to facilitate analysis. The applications of RTS 500 include not only the quick and simple production of proteins, but also the amino acid-selective isotope labeling of proteins.

References

Betton J-M, Hofnung M (1996) Folding of a mutant maltose-binding protein of *Escherichia coli*, which forms inclusion bodies. J Biol Chem 271:8046–8052

Chaves Das Neves HJ, Vasconcelos AM (1987) Capillary gas chromatography of amino acids, including asparagine and glutamine: sensitive gas chromatographic-mass spectrometric and selected ion monitoring gas chromatographic-mass spectrometric detection of the N,O(S)-*tert.*-butyldimethylsilyl derivatives. J Chromatogr 392:249–258

Mawhinney TP, Robinet RRS, Atalay A, Madson MA (1986) Analysis of amino acids as their tert.-butyldimethylsilyl derivatives by gas-liquid chromatography and mass spectrometry. J Chromatogr 358:231–242

Miller JH (1992) A short course in bacterial genetics. Cold Spring Harbor Laboratory Press, New York

Riek R, Pervushin K, Wüthrich K (2000) TROSY and CRINEPT: NMR with large molecular and supramolecular structures in solution. Trends Biochem Sci 25:462–468

Studier FW, Rosenberg AH Dunn JJ, Dunbendorf JW (1990) Use of T7 RNA polymerase to direct expression of cloned genes. Methods Enzymol 185:60–89

CHAPTER 22

In Vitro Protein Production for Structure Determination with the RTS System

HO S. CHO*, JEFFREY G. PELTON, WEIRU WANG,
HISAO YOKOTA, DAVID E. WEMMER

Introduction

The goal of the Berkeley Structural Genomics Center is to determine the structures of all proteins encoded in the genomes of *Mycoplasma pneumoniae* and *Mycoplasma genatalium* or structural homologs from other organisms. To achieve this goal, we are working to develop high-throughput methods for protein expression for use in X-ray and NMR structure determination. In collaboration with Roche Molecular Biochemicals (Penzburg, Germany), we are investigating the usefulness of the Rapid Translation System (RTS) of in vitro protein production to generate target proteins in quantities suitable for structure determination. Very encouraging results have been obtained with the test protein phosphoserine phosphatase (PSP) from *Methanococcus jannaschii* (Mj).

Materials and Methods

Preparation of PSP-pIVEX

The gene for phosphoserine phosphatase (Mj) was excised from a pET-21a construct [1] and ligated into pIVEX 2.3-MCS (Roche) using Nde1 and BamH1 restriction sites. The new construct, PSP-pIVEX, was verified by PCR and restriction fragment analysis. The PSP-pIVEX plasmid was amplified in a 1-l LB culture of DH5-α *E. coli* cells and isolated with the Qiafilter Plasmid Maxi Kit (Qiagen, Germantown, Md., USA) Isotech (Miamisburg, Ohio, USA). The final concentration of purified plasmid was approximately 0.37 µg/µl.

* H.S. Cho (✉) (e-mail: ho_s_cho@lbl.gov), J.G. Pelton, H. Yokota, D.E. Wemmer
  Physical Biosciences Division, Lawrence Berkeley National Laboratory, 1 Cyclotron Road,
  Berkeley, CA 94720, USA
  W. Wang, D.E. Wemmer
  Department of Chemistry, University of California, Berkeley, CA 94720, USA

Production of PSP

The lyophilized RTS 500 materials were reconstituted according to Roche Bio-chemicals' instructions. A total of 15 µg of PSP-pIVEX plasmid was used per 1 ml reaction. The 1-ml reaction was continuously dialyzed against 10 ml of a feeding mix, containing additional energy sources and amino acids, in an RTS 500 machine set to 30°C with a stirring speed of 120 rpm. After 20 h of incubation, a 5-µl aliquot was removed and checked for protein production on a 15% SDS-PAGE gel stained with Coomassie blue. After a total of 24 h, the reaction was stopped, and the lysate transferred to a 1.5-ml tube and frozen at –20°C until purification.

Purification of PSP

The 1-ml RTS reaction mixture was diluted with 2 ml of 50 mM Tris at pH 7.5 and separated into three 1.5-ml tubes. The samples were incubated at 70°C for 30 min to precipitate the *E. coli* proteins. The precipitate was pelleted by centrifugation at 13 K rpm for 15 min in a mini-fuge. The supernatant was combined and placed in a dialysis bag with a MWCO of 6–8 kDa and dialyzed against 1 l of 20 mM Tris at pH 8.4 for 4 h. After dialysis, the sample was passed through a 1-ml Hightrap-Q column (Pharmacia, Peapack, NJ, USA) equilibrated with 20 mM Tris, pH 8.4. The column was washed with 5 ml of 20 mM Tris, pH 8.4. The flow-through and wash were collected in 1-ml fractions. We checked 5 µl of each fraction on a 15% SDS-PAGE gel with Coomassie staining. Fractions containing PSP were combined and concentrated to 300 µl in a 4-ml Ultrafree spin concentrator (Millipore) with a 5-kDa MWCO membrane. The buffer was exchanged by twice adding 3 ml of a solution containing 20 mM Tris at pH 7.5, 300 mM NaCl, 1 mM EDTA, and 10 mM DTT and re-concentrating to 300 µl. The sample was then transferred to a 500-µl Ultra-free spin concentrator and further concentrated to 60 µl.

Crystallization and Data Collection

Crystals of the PSP protein, which were generated in the RTS system, were grown using the hanging drop vapor diffusion method with seeding in a buffer, which had previously worked for PSP produced in *E. coli* [1]. We mixed 1 µl of concentrated PSP with 1 µl of a well solution containing 0.1 M sodium acetate buffer at pH 4.5, 0.2 M sodium phosphate dihydrate, 5 mM $MgCl_2$, and 22% polyethylene glycol 2,000 monomethylether (PEG2K MME). Micro-seeding was performed 1 h after the drop was set up. Crystals appeared within 24 h. The concentration of PEG2K MME was then raised to 30% to stabilize the crystals. Crystals from the drop were flash frozen in liquid nitrogen and used directly for cryo-crystallography data collection. X-ray diffraction data were collected at the Advanced Light Source (ALS; Berkeley, Calif., USA) beam line 5.0.2 using an Area Detector System Co. Quantum 4 CCD detector placed 140 mm from the crystal.

"Uniformly" ¹⁵N-labeled PSP

An almost uniformly ¹⁵N-labeled PSP sample was generated by using ¹⁵N algal amino acids (Cambridge Isotope Labs, Andover, Mass., USA) with the RTS 500 labeling kit. A stock solution of ¹⁵N-labeled amino acids was made by dissolving 100 mg of the algal amino acid mixture in 6 ml of reconstitution buffer. Because the algal amino acid mixture is produced from acid hydrolysis of algal proteins, it does not contain Asn, Cys, Gln, or Trp. Except for Trp, which does not occur in PSP, 42 mM solutions of the unlabeled forms of the other three amino acids were used to supplement the ¹⁵N algal amino acids. A total of 2.805 ml of ¹⁵N algal stock solution was combined with 135 µl of Asn, 30 µl of Cys, and 30 µl of Gln to give a final volume of 3.0 ml. This 3.0-ml solution of amino acids was used according to Roche Biochemicals' instructions for the labeling kit. The procedures, described above, for protein production and purification were followed, except for the concentration step. After the Hightrap-Q column, the pooled fractions were concentrated and buffer exchanged to a final volume of 450 µl with 10 mM sodium phosphate at pH 6.5, 10 mM DTT, 20 mM $MgCl_2$, and 0.5 mM EDTA. We added 50 µl of D_2O to the sample and adjusted the pH to 6.5 prior to NMR data collection.

¹⁵N-Gly labeled PSP

A ¹⁵N-Gly labeled PSP sample was produced with the RTS 500 labeling kit by substituting unlabeled Gly with ¹⁵N-Gly obtained from Isotech. The 42 mM stock solutions of each amino acid were combined in amounts depending on how often they occur in PSP (the volumes in microliters and the number of copies of each amino acid are indicated): 240 Ala (17), 105 Arg (8), 135 Asn (9), 195 Asp (13), 30 Cys (2), 30 Gln (1), 375 Glu (27), 225 ¹⁵N-Gly (15), 0 His (0), 300 Ile (22), 270 Leu (20), 450 Lys (33), 45 Met (3), 105 Phe (7), 45 Pro (3), 75 Ser (5), 105 Thr (8), 0 Trp(0), 45 Tyr (3), 225 Val (15). The final 3.0 ml of amino acid solution was used according to Roche Biochemicals' instructions for the RTS 500 labeling kit. The same procedures for protein production, purification, and concentration used for the "uniformly" ¹⁵N-labeled sample was followed.

NMR Data Collection

¹H-¹⁵N HSQC spectra [2] were recorded on Bruker AMX 600 MHz and DRX 500 MHz spectrometers. For each experiment, the sample temperature was set to 25°C, and total recording time was 17 h. The data were processed using the NMR-Pipe software suite [3].

Results and Discussion

Roche Biochemicals' instructions for setting up the RTS 500 reactions proved to be simple and yielded consistent amounts of PSP in each reaction. A 1-ml high yield RTS lysate reaction produced 2–3 mg of protein (Fig. 1). In addition, the purified PSP was active and crystallized in the same conditions as PSP expressed in *E. coli* [1] (Fig. 2). X-ray crystallography data collected with synchrotron radiation at the ALS indicate that the crystals are of high quality, diffracting to a resolution limit of 1.5 Å (Fig. 3). ^{1}H-^{15}N HSQC spectra of almost uniformly ^{15}N and selectively ^{15}N-Gly labeled samples of PSP generated in the RTS system show great promise for NMR structure determination (Figs. 4, 5). Except for the weakest peak, all the peaks in the ^{15}N-Gly spectrum (Fig. 5) can be superimposed with the corresponding peaks in the spectrum of the "uniformly" ^{15}N-labeled sample (Fig. 4). If similar results are obtained for a significant number of target proteins, in vitro protein production with the RTS system should play a major role in structural genomics as well as in structural biology in general.

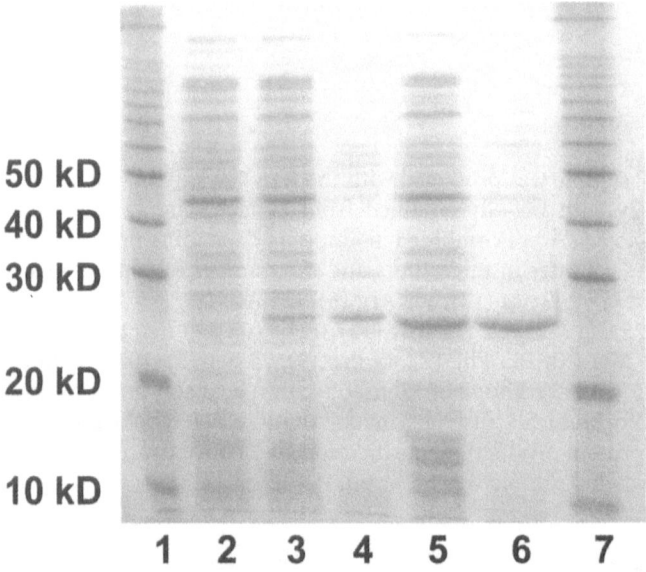

Fig. 1. A Coomassie stained 15% SDS-PAGE gel showing the production of PSP in the RTS system. Lanes 1 and 7, 10 kDa molecular weight markers. Lanes 2–6 each contain approximately 1 μl of reaction material. Lane 2 (control), lysate mixed with protein loading buffer prior to incubation at 30°C. Lane 3, PSP production in the old RTS lysate after incubation at 30°C for 20 h. Lane 4, partially purified lane 3. Lane 5, PSP production in the new high yield RTS lysate after incubation at 30°C for 20 h. Lane 6, partially purified lane 5

A

B

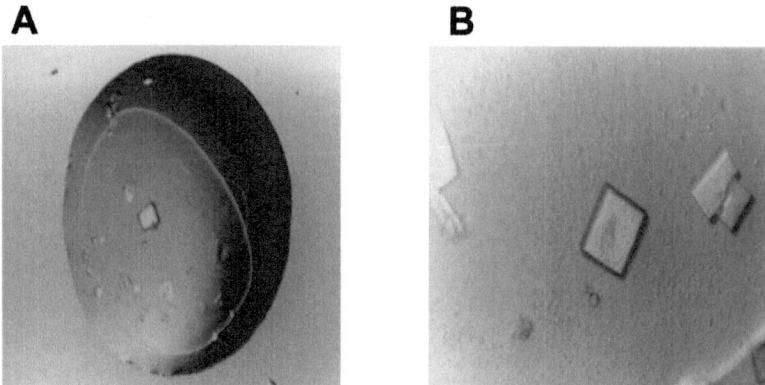

Fig. 2A,B. Crystals of PSP generated with protein isolated from a 1-ml RTS reaction. **A** Picture of the entire crystallization drop. **B** Close-up view of a crystal suitable for X-ray diffraction with dimensions of 100×150×100 µm

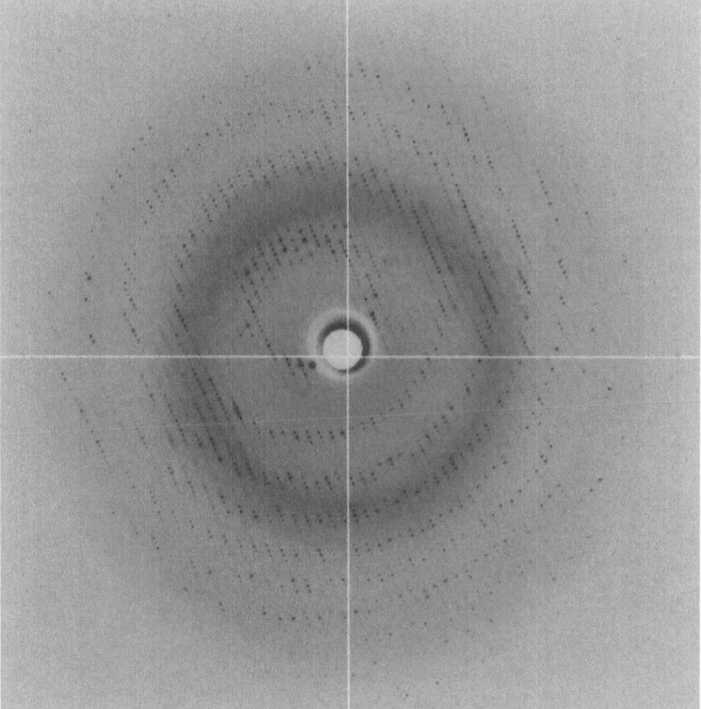

Fig. 3. X-ray diffraction data collected with the crystal in Fig. 2B. The crystal diffracted to a resolution limit of 1.5 Å, comparable to the resolution limit obtained with crystals of PSP expressed in *E. coli*

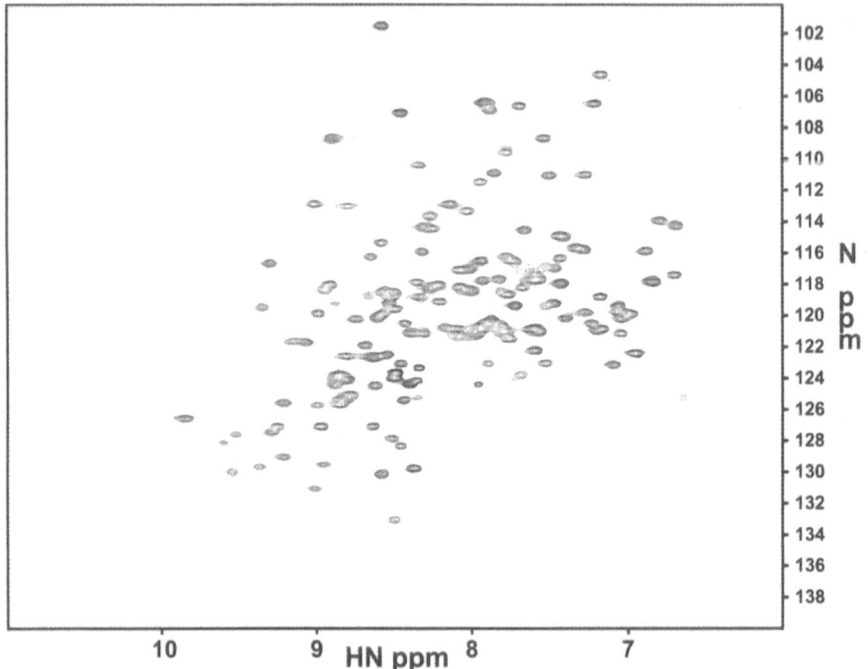

Fig. 4. ^{1}H-^{15}N HSQC spectrum of PSP "uniformly" labeled with ^{15}N-algal amino acids in the RTS system. The spectrum was recorded on a Bruker AMX 600 MHz spectrometer with the sample temperature set to 25°C and a total recording time of 17 h. Peaks corresponding to Asn, Cys, Gln, and Glu residues are missing from the spectrum because the unlabeled forms of these amino acids were used to supplement the ^{15}N-algal amino acid mixture

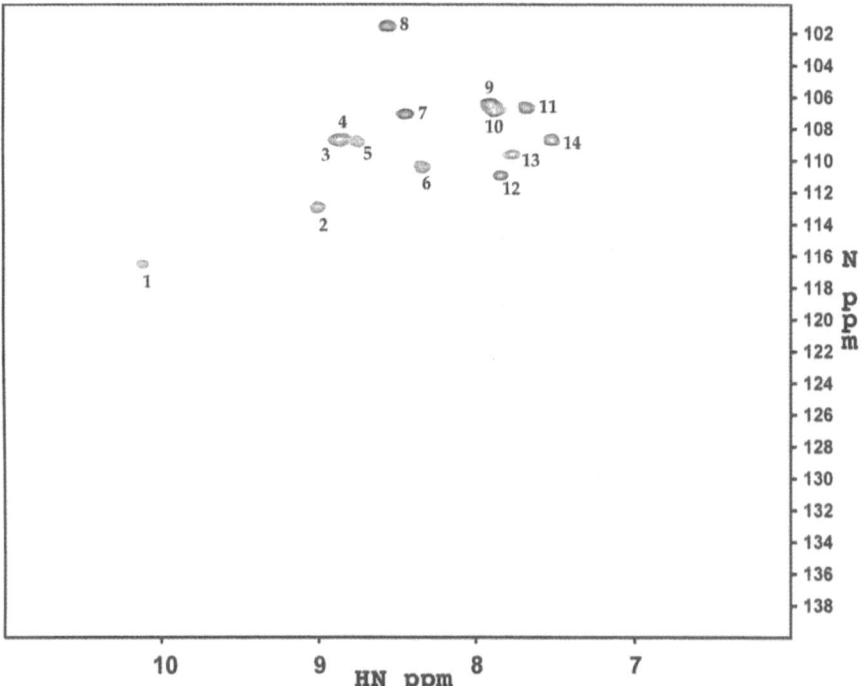

Fig. 5. ¹H-¹⁵N HSQC spectrum of PSP selectively labeled with ¹⁵N-Gly in the RTS system. The spectrum was recorded on a Bruker DRX 500 MHz spectrometer with the sample temperature set to 25°C and total recording time of 17 h. Peaks corresponding to 14 of the 15 Gly residues in PSP were observed

References

1. Wang W, Kim R, Jancarik J, Yokota H, Kim S-H (2001) Crystal structure of phosphoserine phosphatase from *Methanococcus jannaschii*, a hyperthermophile, at 1.8 Å. Structure Fold Design 9:65–72
2. Mori S, Abeygunawardana C, Johnson MO, van Zijl PC (1995) Improved sensitivity of HSQC spectra of exchanging protons at short interscan delays using a new fast HSQC (FHSQC) detection scheme that avoids water saturation. J Mag Reson B 108:94–98
3. Delaglio F, Grzesiek S, Vuister GW, Zhu G, Pfeifer J, Bax A (1995) NMRPipe: a multidimensional spectral processing system based on UNIX pipes. J Biomol NMR 6:277–93

Sequence Specific Biotinylation and Purification of Proteins Expressed in the RTS 500 System[1]

MICHAEL SCHRÄML, DOROTHEE AMBROSIUS,
JAN STRACKE, MARTIN LANZENDÖRFER[2]

Abbreviations

AviTag-PEX2	PEX2, N-terminally elongated with the AviTag
BAP	biotin-accepting peptide
BCCP	biotin carboxyl carrier protein
BirA	*E. coli* biotin holoenzyme synthetatase
BPL	Biotin Protein Ligase
MMP2	matrix associated metallo-proteinase
PEX	hemopexin-like domain of MMP2
PinPoint-PEX2	PEX2, N-terminally elongated with the PinPoint-tag
PBS	Phosphate buffered saline
RTS	Rapid Translation System
SA	streptavidin
SA-chip	streptavidin coated sensor chip
SPR	surface plasmon resonance
TIMP2	tissue inhibitor of MMP2

Introduction

The rapid development of sensitive and selective biochemical assays is a major prerequisite for streamlining the drug discovery process at pharmaceutical companies. Producing specifically labelled protein is an important parameter for achieving the desired goal. Among the established affinity systems, the streptavidin–biotin complex is one of the most frequently used principles to be applied to drug discovery processes or diagnostic assays.

Biotin combined with streptavidin is one of the strongest non-covalent interactions known in nature. This high-affinity binding pair is widely used to immobilize chemically biotinylated proteins on streptavidin-coated surfaces such as

[1] RTS is a trademark of a member of the Roche Group

[2] M. Schräml, D. Ambrosius, J. Stracke, M. Lanzendörfer (✉) (Tel.: +49–89-08856/-602843)
Roche Diagnostics GmbH, Pharmaceutical Research, Department of Biochemistry,
Nonnenwald 2, 82372 Penzberg, Germany

microtiter plates, beads, sensor chips and other matrices. The predominant areas of application are protein–protein interaction analysis (e.g. ELISA, SPR) and affinity chromatography purification. Since the K_D of streptavidin/biotin interaction is in the range of 10^{-14} M, chemically biotinylated proteins can only be eluted under denaturing conditions. This strong affinity provides an opportunity to use the interacting pair for presenting biotinylated proteins on diverse matrices for subsequent ligand fishing experiments. In this case, ligand-binding partners can be eluted under native conditions, independent of the affinity and type of interaction. Up to now, different strategies have been developed to generate biotinylated proteins [2].

In peptide synthesis, biotinylated lysins are fused to peptides in a sequence-specific manner, but the chemically synthesized peptides are limited in length and often show a lower activity than full-length proteins due to structural limitations. A very common method for post-translational chemical biotinylation of proteins in vitro is the random coupling of biotin to the ε-amino group of lysins [2]. The biotinylation rate of proteins can be adjusted by varying the ratio of biotinylating reagent to the protein of interest. However, even if proteins are being biotinylated in a 1:1 ratio, they can lose their activity due to the modification of lysins located in an active site of the protein. After immobilizing such a protein on a surface, the active site may not be accessible to interaction partners. Another reason for the loss of activity are the reaction conditions (e.g. pH value) required for a successful biotinylation, which may not be suitable for the solubility or active conformation of the target protein.

Sequence-specific biotinylation of proteins is a rare event in nature. *Escherichia coli* contains only one biotinylated protein, the biotin carboxyl carrier protein (BCCP) of acetyl-CoA carboxylase, whereas other bacteria contain one to three biotinylated proteins [3]. *Saccharomyces cerevisiae* contains four or five biotin proteins, depending on growth conditions [4], whereas mammals [5,6] and plants [7] contain four biotinylated proteins. The lysine residues of these proteins, modified at the ε-amino group, are located in amino acid sequences specifically recognized by biotin protein ligases (BPL). The best characterized BPL is BirA from *E .coli*. In the 1.3 S transcarboxylase subunit of *Propionibacterium shermanii* [8], biotin is attached by BirA to a lysine residue located 34 residues from the carboxyl terminus. Truncation of the carboxyl terminus leads to a loss of biotinylation of this protein in *E. coli* [9].The sequences surrounding the biotinylation sites of proteins from diverse biological sources are highly conserved [8, 10]. They seem to exhibit sufficient common structure to account for their specific recognition by BirA in vitro [11] and in vivo [12]. There are also shorter peptide sequences, which do not resemble the consensus sequence of these common biotinylation sites [13]. These peptide sequences are the product of in vitro selection processes out of peptide libraries [14].

We investigated the expression and biotinylation of such a commercially available 15 residue BAP (AviTag, Avidity Inc., Denver, Colo., USA), genetically fused to the amino-terminus of PEX2 (hemopexin-like domain of MMP2) [15]. The expression was performed in vitro in the RTS 500 (Roche Diagnostics GmbH, Penzberg, Germany). The biotinylation reaction was co-translationally performed in the RTS 500 by adding BirA. The biotinylated protein was chromatographically purified using a monoavidin matrix and analysed for functionality by SPR spectroscopy using a Biacore 3000 system (SPR, Biacore, Uppsala, Sweden). Here, we report the in vitro expression and biotinylation of AviTag-fused PEX2 as compared to the in vivo (*E. coli*) expression and biotinylation of another 133 residue BAP-fused PEX2 (PinPoint-tag, Promega Inc., Madison, Wis., USA).

Material and Methods

Expression of PinPoint-PEX2 In Vivo in *E. coli*

The gene, coding for PEX2 was amplified by PCR using the sense primer 5′-ATA AGA ATA AGC TTC CTG AAA TCT GCA AAC AGG ATA TCG-3′and antisense primer 5′-ATA GTT TAG CGG CCG CTT ATC AGC CTA GCC AGT CG-3′. The PCR was performed in 30 cycles with a temperature profile as follows: 1 min at 94°C, 1 min at 48°C and 1 min at 72°C. The PCR product was cloned as a *Not*I/*Hind*III fragment into the expression vector PinPoint-Xa3 (Promega). The plasmid was transformed into *E. coli* UT5600, which already contained the helper plasmid pUBS520 [1]. Cells were grown in LB-media containing 2 µM biotin, 100 µg/ml ampicillin and 50 µg/ml kanamycin. An overnight culture was used to inoculate 1 l medium of the same composition, which was incubated under vigorous shaking at 37°C. At OD_{595}=0.5, expression was induced with 1 mM IPTG for 5 h. The cells were harvested by centrifugation (2500 *g*). The cell paste was resuspended (5 ml/g cell paste) in 50 mM TRIS pH7.2, 20 mM NaCl, 5 mM $CaCl_2$, 1 mg/ml lysozyme, complete EDTA-free protease inhibitor cocktail (Roche Diagnostics GmbH, Penzberg, Germany) and subsequently incubated for 20 min at room temperature. Further cell lysis was performed by sonication on ice as recommended by the manufacturer (Branson Ultrasonic Corp., Danbury, Conn., USA) until the suspension was no longer viscous. Crude lysate was centrifuged at 10,000 *g* for 30 min at 4°C and the supernatant was filtered through a 0.22-µm Gelman filter.

Expression of AviTag-PEX2 In Vitro in the RTS 500

The PEX2 gene was genetically fused with AviTag coding DNA by add-on PCR using the primers 5′-GAAGGCATATGGGTCTGAACG-3′ (25 pmol), 5′-CTCA-GAAAATCGAATGGCACGAAGCGACCCTGAAATCTGCAAACAGG-3′ (10pmol), 5′-GCCATTCGATTTTCTGAGCTTCGAAGATGTCGTTCAGACCCATATGCC-3′ (10 pmol) and 5′-GCCGCTCGAGTCAGCAGCCTAGCCAGTCGG-3′ (25 pmol). The PCR program was performed as described above. The PCR product was digested with NdeI and XhoI, and was cloned into the expression plasmid pIVEX2.1MCS previously cut with the same restriction enzymes. The plasmid was propagated in *E. coli* DH5-α and was isolated with a commercially available kit (Qiagen, Qiafilter Plasmid Maxi Kit, Hilden, FRG). 15-µg plasmid DNA (ratio 260 nm/280 nm>1.8) and 12,500 U biotin ligase holoenzyme (EC 6.3.4.15; Avidity) were added to the reaction mixture of a commercially available RTS 500 (Roche Diagnostics). Biotin ligase activity is defined by the manufacturer: 1 U is the amount of enzyme that will biotinylate 1 pmol of peptide substrate in 30 min at 30°C using the reaction buffers provided in the manufacturers kit and peptide substrate at 38 μM. The substrate used in the enzyme assay was a 15-mer variant of sequence #85 as identified by Schatz [14]. Biotin was adjusted to 2 μM in both the reaction mixture and the substrate solution. Protein expression was performed in the RTS 500 Incubator (Roche Diagnostics) under stirring (130 rpm) for 17 h at 30°C. The RTS extract was subsequently dialysed in a 1 ml Slide-A-Lyzer dialysis chamber (MWCO 7500, PIERCE, Rockford, Ill., USA) against buffer W2 and centrifuged at 10,000 g for 30 min at 4°C.

Purification of Biotinylated Fusion Proteins

1 ml monomeric avidin sepharose resin (SoftLink, Promega) was filled in a Pharmacia HR-5 column. The column was inserted in an analytical ÄKTA Explorer HPLC system (Amersham Pharmacia Biotech, Uppsala, Sweden). Equilibration, preadsorption of nonreversible binding sites and regeneration was conducted as recommended by the manufacturer (Promega). After washing the column with 10 CV buffer W1 (50 mM TRIS pH7.2, 20 mM NaCl) and 10 CV buffer W2 (W1+5 mM $CaCl_2$), cell extract was applied with a flow rate of 0.1 ml/min. Washing with buffer W2 was done until no more protein was detectable in the flowpath of the column. To elute biotinylated protein, buffer W2+5 mM biotin was applied. The elution procedure was done as recommended by the manufacturer (Promega). The eluted protein peak was separated in 0.5-ml fractions. Fractions, containing biotinylated target protein, were pooled and free biotin was removed during ultrafiltration with the Ultrafree-0.5 Centrifugal Filter device (Millipore, USA) with buffer W2.

Detection and Quantification of the Fusion Proteins

The soluble and insoluble protein fractions were resolved by SDS-PAGE (10% BIS-TRIS SDS-polyacrylamide gel, NUPAGE, Invitrogen, Carlsbad, Calif., USA) and either stained with Coomassie brilliant blue or transferred to a PVDF-membrane by using the semi-dry Multiphor II apparatus (Pharmacia Biotech) for 70 min at 120 V and room temperature. After the transfer was completed, the membrane was blocked in PBS plus 0.2% Tween 20 (PBS-Tween) and 5% (w/v) dry milk powder with gentle agitation at 4°C. PEX2 bound to the PVDF-membrane was detected with a PEX2-specific antibody. The antibody–stock solution was 1.47 mg/ml polyclonal rabbit anti-PEX2-IgG, directed against the whole molecule. The membrane was incubated for 1 h at room temperature in PBS-Tween, 2.5% (w/v) dry milk powder, containing PEX2 antiserum (1:50.000 v/v) followed by three 10-min washes. The membrane was incubated for 1 h in PBS-Tween+2.5% (w/v) dry milk powder with 1:5000 anti-mouse/anti-rabbit-IgG-POD conjugate (Roche Diagnostics, Cat. No. 1 520 709) followed by three 10-min washes in PBS-Tween. The Western blot was developed with the BM Chemiluminescence Western Blotting Kit (Mouse/Rabbit, Roche Diagnostics, Cat. No. 1 520 709) in accordance with the manufacturer's instructions.

After the densitometric detection of PEX2 protein, the membrane was regenerated for 10 min in 0.1 M NaOH and subsequently washed 3×10 min in PBS-Tween. The membrane was blocked and washed again as described above. Biotinylated fusion protein was detected by incubating the regenerated membrane in a 1:4000 (v/v) dilution of streptavidin–POD conjugate (Roche Diagnostics GmbH, Cat. No. 1 089 153) in PBS-Tween buffer +2.5% (w/v) dry milk powder for 1 h. After washing the membrane three times for 10 min with PBS-Tween, the Western blot was developed again. Biotinylation levels of the PEX2 fusion proteins were determined by comparing the densitometric data of the two detection steps.

Densitometric quantification of the detected protein bands was performed by calibration using verified quantities of recombinant, chemically biotinylated PEX2 and the software ImageMaster 1D Prime 1D Elite (Amersham Pharmacia Biotech Europe GmbH, Freiburg, FRG).

Surface Plasmon Resonance Spectroscopy

The activity of the biotinylated PEX2 fusion proteins was measured using surface plasmon resonance spectroscopy (BIAcore 3000 technology, BIAcore AB, Uppsala, Sweden). The system was running under HBS-P-buffer. PEX2 fusion proteins were immobilized on streptavidin-coated BIAcore-SA chips in a manner recommended by the manufacturer. Various dilution steps of a 200-nM TIMP2 stock solution (0.33 mg/ml in 1×PBS-buffer) in HBS-P buffer were used to measure kinetic data in accordance to the manufacturer's instructions. TIMP2 was eluted from the chip with ImmunoPure Gentle Ag/Ab Elution buffer (PIERCE).

Results

PinPoint-PEX2

PEX2, N-terminally fused with the PinPoint-tag, was expressed and specifically biotinylated in vivo in *E. coli*. The fusion protein has a calculated molecular mass of 36 kDa. Protein identity was confirmed by N-terminal Edman degradation. Harvesting of 1 l fermentation culture resulted in 6 g of bacterial wet mass. A total yield of 0.4 mg fusion protein per gram cell paste was determined by densitometric quantification of Western blots, which was performed using PEX2-specific antibodies. Approximately 10% of the target protein was found in the soluble supernatant of the cleared cell lysate. The biotinylation yield was quantified by comparing the densitometric data as described in the Materials and Methods section. Using streptavidin–POD conjugate, no other biotinylated protein could be detected in the crude cell lysate, whereas monomeric avidin affinity chromatography enriched a second biotinylated protein in the elution fractions. Further analysis of the eluate using streptavidin-POD conjugate revealed two protein bands. The first band with an approximate mass of 40 kDa was the desired biotinylated fusion protein PinPoint-PEX2. The second protein with a size of approximately 16 kDa is either a degradation product of the target protein or, more likely, BCCP, the only biotinylated protein found in *E. coli*. Contamination with this second biotinylated protein accounted for up to 50% of the total yield. The PinPoint-PEX2 containing elution fractions were pooled. Free biotin was removed by ultrafiltration. Two samples with different degrees of purity were analysed in surface plasmon resonance spectroscopy using BIAcore technology. The activity of an immobilized ligand is indicated by the maximum analyte-binding capacity. First, the fraction of the partially purified protein concentrate was analysed. Then, 380 RU of PinPoint-PEX2 (ligand) were immobilized on a BIAcore SA-chip. Saturation of the protein ligand on the chips surface with TIMP2 (analyte) was reached at R_{max}=61 RU. Based on this data, a ligand-binding activity of 26% was calculated. In a second setting, 664 RU of biotinyl-protein were immobilized by injecting a supernatant of dialysed and cleared cell lysate in the flow-cell. Very interestingly, the streptavidin-coated biosensor captured biotinylated molecules out of the cell lysate. Saturation with the analyte TIMP2 was reached at 61 RU_{max} and the calculated ligand-binding activity was 15%. In both cases, the kinetic data of the TIMP2/PinPoint-PEX2 interaction were determined. An equilibrium constant of K_D=1.5×10^{-10} M was calculated using a numeric Langmuir simulation model of a binary complex formation.

AviTag-PEX2

PEX2, N-terminally fused with the AviTag biotin-acceptor sequence, was expressed and biotinylated in vitro in the RTS 500. Biotinylation was facilitated by adding 12,500 U of BirA-enzyme to the reaction mix. The expressed fusion protein has a molecular mass of 25 kDa and was detected by Western blotting, using the PEX2-specific antibody and streptavidin–POD conjugate. When compared to a molecular weight standard, the fusion protein shows an apparent mass of 25 kDa in a 10% Bis-Tris SDS-PAGE. Densitometric quantification showed a total yield of 72 µg AviTag-PEX2/ml of RTS 500 extract. The proportion of soluble fusion protein was 50% of the total yield. The degree of biotinylation was analysed as described in the Materials and Methods section and found to be quantitative. The biotinylated protein detected with streptavidin–POD conjugate showed no other biotinylated protein in the extract. After the affinity purification procedure using monomeric avidin, only biotinylated AviTag-PEX2 fusion protein was detected in the elution fractions. The identity of the fusion protein was confirmed by N-terminal degradation (Edman). Purified AviTag-PEX2 fusion protein as well as supernatant from the dialysed and cleared RTS 500 extract were analysed in surface plasmon resonance spectroscopy. A total of 105 RU purified AviTag-PEX2 fusion protein was attached to a BIAcore SA-chip. Saturation of the immobilized AviTag-PEX2 ligand with the analyte TIMP2 was achieved at 64 RU_{max}. Thus, an analyte-binding capacity of 70% could be detected. After injecting cleared supernatant of RTS 500 extract, 732 RU biotinylated protein were immobilized on the SA-chip surface. At R_{max}, 341 RU TIMP2 were bound, which resembles an analyte-binding capacity of 53%. Kinetics were measured showing an equilibrium constant of the TIMP2 to PEX2 interaction of $K_D=1.5\times10^{-10}$ M. The K_D was determined using the numeric model described before.

Discussion

The main purpose of this study was to produce specifically labelled protein in order to retain ligand activity after the ligands had been attached to a biosensor surface that enabled SPR interaction assays with functionally active protein. Furthermore, we tried to establish a convenient technique to provide a rapid supply of protein using the RTS 500. As a model, we chose PEX2, the hemopexin-like domain of MMP2. The interaction between immobilized PEX2 (ligand) and its natural binding partner TIMP2 (analyte) can be detected in SPR analysis, but previous studies have shown that chemically biotinylated PEX2 becomes inactivated when immobilized on a streptavidin-coated surface due to steric hindrance. Therefore, no interaction between immobilized PEX2 and the applied analyte TIMP2 could be detected, even when the PEX2 molecule is chemically monobiotinylated. Our strategy was to use specifically biotinylated fusion proteins for the attachment in a directed, highly reproducible and active orientation on a streptavidin-coated biosensor surface, whereby soluble target protein should be used in order to avoid inclusion body preparations or diverse time-consuming refolding procedures. We have fused the DNA sequences that encode the biotin attachment sites of two different commercially available BAPs (AviTag, PinPoint-tag) to the PEX2 gene in a manner designed to express two distinct N-terminally tagged PEX2 proteins. In order to compare various strategies to rapidly produce biotinylated fusion proteins, the BAP-fused PEX2 genes were expressed in different expression systems. PinPoint-PEX2 was conventionally expressed and biotinylated in E .coli, while AviTag-PEX2 was expressed and biotinylated in the cell-free RTS 500.

Expression and Purification of PinPoint-PEX2

The PEX2 gene was cloned in frame into the expression vector PinPoint-Xa3. The PinPoint-tag is derived from the well-known 1.3 S subunit of the *P. shermanii* transcarboxylase [8].

Due to the moderate expression of PinPoint-PEX2, an auxiliary co-expression of BirA was not necessary to ensure a quantitative biotinylation in *E. coli*. The affinity purification of biotinylated PinPoint-PEX2 from *E. coli* resulted in a poor enrichment of the target protein (Fig. 1), whereas a second biotinylated protein, which is likely to be a degradation product of BCCP [3, 16], was co-concentrated in the eluate. In *E. coli*, the only endogenous biotinylated protein is the BCCP subunit of acetyl-CoA carboxylase, an enzyme catalysing an essential step during fatty acid synthesis. Although BCCP could not be detected in the crude cell lysate, its affinity for the monomeric avidin matrix was kinetically favoured in

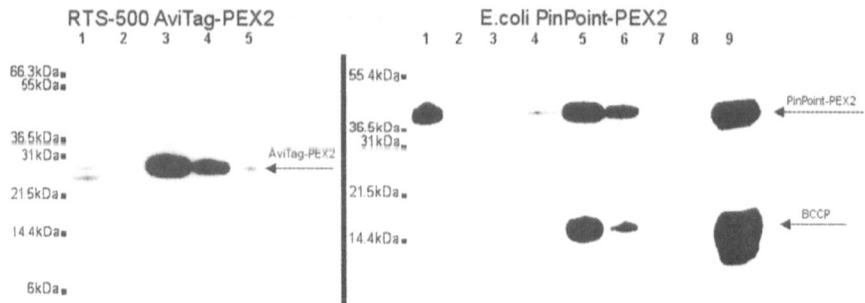

Fig. 1. Purification of the biotinyl fusion proteins AviTag-PEX2 and PinPoint-PEX2 by monomer avidin chromatography. Two Western blots are shown, where biotinylated protein was detected with Sa-POD conjugate in monomer avidin-sepharose elution fractions. *Left panel*, RTS 500 AviTag-PEX2: *Lane 1*: dialysed and centrifuged supernatant of RTS 500 extract applied to the column. A second band under the target band indicates proteolytic degradation. *Lane 2*: column wash. *Lanes 3, 4 and 5*: fractions of the 5 mM biotin elution peak. *Right panel*, E. *coli* PinPoint-PEX2: *Lane 1*: centrifuged supernatant of *E. coli* cell lysate applied to the column. *Lane 2*: column wash. *Lanes 3–8*: fractions of the elution peak containing PinPoint-PEX2, whereby the co-concentration of a proteolytic degradation product of BCCP becomes obvious in *lane 5* and *lane 6* [3, 16]. *Lane 9*: pooled elution fractions after ultrafiltration

contrast to PinPoint-PEX2 during the purification procedure. Thus, BCCP was enriched in the eluate leading to a heterogeneous 1:1 mixture of PinPoint-PEX2 and BCCP. In consideration of the low quantities of soluble protein obtained after purification, a size exclusion chromatography (SEC) to separate the target protein from the undesired biotinyl protein was omitted and the sample was directly used in SPR analysis.

Expression and Purification of AviTag-PEX2

During the expression of AviTag-PEX2 in the RTS 500, the fusion protein was co-translationally biotinylated by adding BirA to the reaction mixture, yielding 30 µg biotinylated, soluble fusion protein per millilitre of RTS 500 extract. In agreement with the biotinylation of AviTag-fused proteins in vivo in *E .coli*, which requires an addition of recombinant BirA, no biotinylation could be determined in the RTS 500 without adding this enzyme.

Quantitative biotinylation of the fusion protein was observed at a BirA concentration of 12 500 U, while lower concentrations merely led to a partial or no biotinylation of the expressed fusion protein. A Western blot analysis showed a strong enrichment of AviTag-PEX2 protein during purification and, moreover, no other biotinylated protein was detected in the eluate. Although the RTS extract descends from an *E. coli* lysate and, therefore, must contain BCCP, the protein cannot be detected by sensitive biochemical assays. This is a major advantage of the cell-free RTS 500 in comparison to *E. coli*, where BCCP is an endogenous protein and essential to cell growth. Indeed, high-level expression of biotinylated fusion proteins in *E. coli* leads to a decreased biotinylation of BCCP, which results in inhibited growth [16].

SPR Analysis of PinPoint-PEX2 and AviTag-PEX2

Two different settings were compared by analysing the analyte-binding capacities of PinPoint-PEX2 and AviTag-PEX2. Two different samples of each fusion protein – an affinity purified portion and crude material – were tested.

PinPoint-PEX2 was attached to a surface of a BIAcore SA chip from a partially affinity purified sample. The sample was contaminated in an approximately 1:1 ratio with BCCP, which makes it impossible to quantify the virtual amount of immobilized PinPoint-PEX2. Nevertheless, partially purified PinPoint-PEX2 was suitable to measure kinetics and the analyte-binding capacity was 26%. When crude material was analysed, enough PinPoint-PEX2 was captured on the biosensors surface to measure kinetics, although analyte-binding capacity deteriorated to 15% compared to the partially purified sample. In crude material, BCCP concentration was below the limit that could be detected by a Western blot. Our data do not allow any predication about the interference of BCCP nor could a putative difference between purified and crude material in ligand activity be made due to sample heterogeneity.

In contrast to PinPoint-PEX2, the analyte-binding capacities of the AviTag-PEX2 samples were significantly higher. When purified AviTag-PEX2 was attached to the biosensor's surface, 70% of the immobilized molecules captured a TIMP2 molecule, exhibiting an extraordinary analyte-binding capacity. Purified (70%, analyte-binding capacity) and crude material (53%, analyte-binding capacity) were suitable for measuring binding kinetics (Fig. 2). A purification step prior to SPR analysis was not obligatory, as the streptavidin coated biosensor served as a highly specific affinity matrix, which sufficiently captured biotinylated protein out of cell lysates. In general, purified samples showed a higher analyte-binding capacity than crude material and purifying the biotinyl-fusion proteins might be rather useful to enhance the sensitivity of an assay by upgrading the analyte-binding capacity. In our case, the ligand-binding activity of crude material was sufficient to measure kinetics with both fusion proteins.

Thus, the interaction between the immobilized biotinyl-PEX2 fusion protein and TIMP2 could be determined for the first time (Fig. 3) as the binding site of PEX2 becomes freely accessible due to the directed immobilization of the fusion protein via the biotinylated tags.

Conclusions

The goal of this study was to retain protein activity after it was immobilized on a surface. This was achieved by using specifically biotinylated fusion proteins. While it is a common procedure to enzymatically biotinylate fusion proteins tagged with biotin-accepting peptides (BAP) in vivo, it was our approach to transfer the sequence specific biotinylation to the in vitro scale using the cell-free RTS 500. This

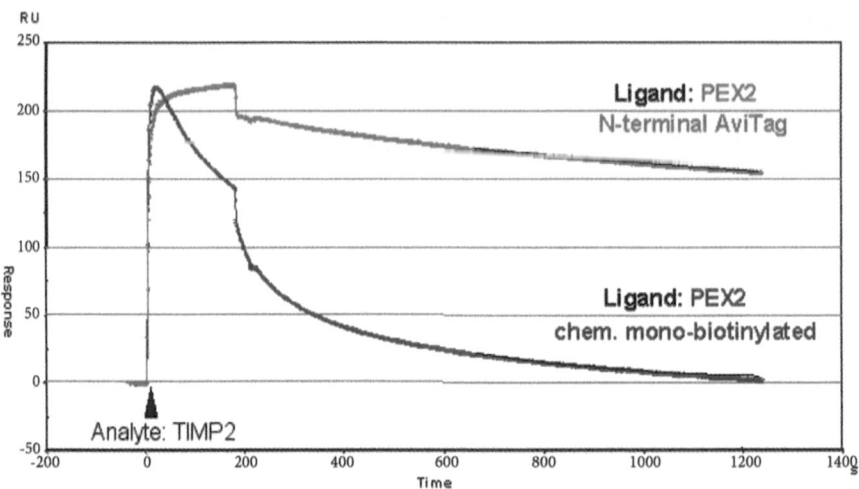

Fig. 2. Sensorgram of immobilized AviTag-PEX2 and chem. mono-biotinylated PEX2 binding TIMP2. Relative response units are defined as BIACORE response (RU).The *black arrow* indicates the injection point of 20 μl of a TIMP2 solution at 200 nM. Due to steric restrictions after immobilization, the binding site of chemically mono-biotinylated PEX2 is not accessible to the analyte TIMP2 (*blue line*). During the injection, the analyte already started to dissociate from the SA surface in a higher-order reaction, showing specific rebinding effects. Subsequently, TIMP2 dissociates ($K_D=10^{-3}s-1$) completely from the surface. In contrast, the binding site of AviTag-PEX2 becomes accessible due to the attachment of the fusion protein to the biosensor surface via the biotin-labeled AviTag. During the sample injection the surface was saturated with analyte (*red line*). TIMP2 showed a rapid binding to AviTag-PEX2 and dissociated slowly ($K_D=10^{-4}s^{-1}$) from the surface. Determination of kinetics was thus facilitated

strategy was successfully validated with PEX2, which was fused with the biotin-accepting peptide AviTag. The fusion tag proved to be superior to other BAP sequences, e.g. the 1.3 S subunit of *P. shermanii* transcarboxylase, due to its short length and decreased tendency to interfere with the protein's tertiary structure. Disadvantages of a biotinylation in vivo, such as the co-concentration of biotinylated background protein during affinity purification using monomeric avidin, are circumvented with the cell-free in vitro expression system. We demonstrated that sufficient amounts of specifically biotinylated fusion protein can be produced in vitro in the RTS 500 to supply the initial biochemical studies such as SPR spectroscopy, ELISA assays and Edman sequencing. Co-translational in vitro biotinylation has the advantage of rapidly producing specifically biotinylated protein in vitro – cell free – on a small scale and offers an alternative to commonly used protein expression strategies.

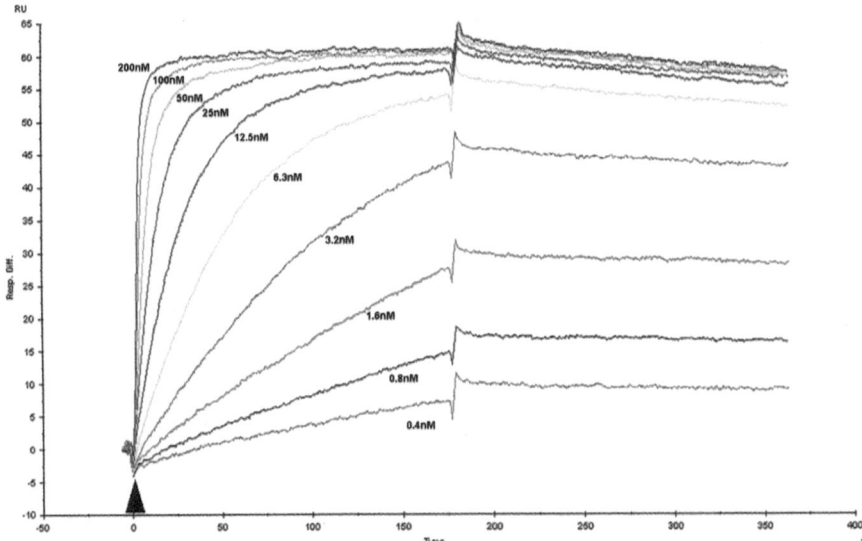

Fig. 3. Overlay plot of sensorgrams showing how TIMP2 at different concentrations interacted with immobilized AviTag-PEX2. 105 RU AviTAG-PEX2 ligand were immobilized on a BIAcore SA-chip. Response of the different units were measured against a flow cell with a blank surface. Analyte TIMP2 was injected at different concentrations in a range from 0.4 nM to 200 nM. Rmax=61 RU. A *black arrow* indicates the injection point of 60 µl analyte at a flowrate of 20 µl/min. The dissociation signal was monitored for 3 min. The regeneration procedure of the TIMP2/AviTag-PEX2 interaction was carried out by three injections of 20 µl ImmunoPure Gentle Ab/Ag Elution buffer (PIERCE) followed by a stabilisation of 2 min. Determining kinetic constants was facilitated by using a numeric Langmuir simulation model of a binary complex formation: k_a=2.5×10^6(Ms)$^{-1}$, K_D=4.7×10^{-4}s^{-1}, K_D=1.9×10^{-10} M. The Overlay plot was created with BIAevaluation Version 3.1

References

1. Brinkmann U, Mattes RE, Buckel P (1989) Gene 85:109–114
2. Bayer EA, Wilchek M (1990) Methods Enzymol184:138–160
3. Fall R. (1979) Methods Enzymol 62:390–398
4. Lim P, Rhode M, Morris CP, Occhiodoro F, Wallace JC (1987) Arch Biochem Biophys 258:219–264
5. Robinson BH, Oei J, Saunders M, Gravel R (1983) J Biol Chem 258:6660–6664
6. Chandler CS, Ballard FJ (1988) Biochem J 251:749–755
7. Nikolau BJ, Wurtele ES, Stumpf PK (1985) Anal Biochem 149:488–453
8. Samols D, Thornton CG, Murtif VL, Kumar GK, Haase FC, Wood HG. (1988) J Biol Chem 263:6461–6464
9. Murtif VL, Samols D (1987) J Biol Chem 262:11813–11816
10. Shenoy BC, Xie Y, Park VL, Kumar GK, Beegen H, Wood HG, Samols D (1992) J Biol Chem 267:18407–18412
11. McAllister HC, Coon MJ (1966) J Biol Chem 241:2855–2861
12. Cronan JE Jr (1989) Cell 58:427–429
13. Beckett D, Kovaleva E, Schatz PJ (1999) Protein Sci 8(4):921–9

14. Schatz PJ (1993) Biotechnology (NY) 11(10):1138–1143
15. Overall CM, King AE, Sam DK, Ong AD, Lau TTY, Wallon MU, DeClerck YA, Atherstone J (1999) J Biol Chem 274:4421–4429
16. Cronan JE Jr (1990) J Biol Chem 265(18):10327–10333

Synthesis of the His 6-Tagged Recombinant Protein APPC99-His and His-PS1 Using RTS 500

Miyuki Murayama*, Xiaoyan Sun, Akihiko Takashima

Introduction

Alzheimer's disease is a progressive neurodegenerative disorder associated with devastating memory loss. Major pathological features are the accumulation of Aβ called senile plaque, the aggregation of highly phosphorylated tau in neurons known as neurofibrillary tangles, and loss of neurons [1]. Genetic studies of familial Alzheimer's disease (FAD) have recently led to the cloning of three genes. The first gene identified is the amyloid β protein precursor (APP), located on chromosome 21 [2], and missense mutation in this gene appears to cause AD by generating an excess of Aβ or an extended Aβ that subsequently precipitates, leading to the formation of senile plaques. The other two are genes called presenilin 1 (PS1) on chromosome 14 and presenilin 2 (PS2) on chromosome 1 [3, 4]. Missense mutations in PS1 and PS2 can explain approximately 70% of early-onset FAD. More than 20 missense mutations of the gene in early-onset Alzheimer's disease (AD) families have been identified. These FAD-linked mutations are very likely to be involved in AD pathology [5].

Interestingly, both APP and PS1 are membrane proteins. The largest APP, comprising 770 amino acids, is a membrane protein with a single membrane spanning region. Synthesized APP transports from ER to the plasma membrane and receives proteolytic processing by α-, β-, and γ-secretase during trafficking. APP is believed to be cleaved first at the 687th or 671th amino acid by α- or β-secretase and processed into a large ectodomain form of APP, called secreted APP, and a small intracellular domain of APP, called APP C-terminal fragment; some of them contain an Aβ region. The secreted form of APP is released into the medium. APP C-100, cleaved by β–secretase, is further cleaved by γ-secretase and produces Aβ peptide.

PS1 is predicted as a membrane protein with 6–8 membrane-spanning regions. Localization of PS1 is shown in the cellular membrane (plasma, endoplasmic reticulum, and perinuclear) by immuno-electron microscopy. Using cell lines and a transgenic mouse, it has been confirmed that mutations of PS1 alter APP processing and enhance formation of amyloidgenic Aβ42. Accumulated evidence shows that PS1 is required for γ-secretase activity [6]. Furthermore, some evidence suggests that PS1 itself is a γ-secretase [7, 8]. However, there is no direct evidence that indicates that PS1 is a γ-secretase. To directly prove it, the purified PS1 and APP C-100 are required

* Miyuki Murayama, Lab for Alzheimer's Disease, Brain Science Institute,
The Institute of Physical and Chemical Research (Riken), 2-1 Hirosawa,
Wako-shi, Saitama 351-351-0198 Japan
Phone +81-48-467-9704, FAX +81-48-467-5916, e-mail: magu@brain.riken.go.jp

for detecting the influence of PS1 on γ-secretase activity. Since both proteins are membrane proteins, it was very hard to purify full-length recombinant protein from products of *Escherichia coli*. The RTS system enabled us to purify these proteins.

Materials and Methods

Materials

1. RTS500 *E. coli* Circular Template kit (Roche Molecular Biochemicals)
2. QUAGEn plasmid kit (QIAGEN)
3. Ni-NTA agarose (QIAGEN)

Construction of Plasmid

Two restriction enzyme sites, NcoI and SmaI, were added by PCR at the 5′- and 3′-terminus of the APPC99 cDNA, respectively. The PCR product was subcloned at the NcoI-SmaI site of pIVEX2.3 vector (*E. coli* Circular Template kit). This plasmid, termed APPC99/pIVE2.3, was purified by plasmid purification kit (QIAGEN). Using this plasmid, a recombinant protein APPC99-His (APPC99 tagged by His6 at its C-teminus) was produced.

The plasmid, termed PS1/pIVE2.4, was prepared in the same way as APPC99, described above.

In Vitro Synthesis of the Recombinant Protein

The recombinant protein APPC99-His or His-PS1 were synthesized as the protocol described by manufacturer. Briefly, the reaction mixture containing 25 μg of the plasmid (APPC99/pIVE2.3, PS1/pIVE2.4) was prepared in the reaction device and set the reaction device on RTS 500. The recombinant protein, APPC99-His and His-PS1, was synthesized under the following conditions:

The running conditions were: (a) stirrer speed:120 rpm, (b) temperature: 30°C, (c) running time: 20 h.

Purification of the Recombinant Protein

After centrifugation (15,000 rpm/4°C for 10 min), the supernatant was incubated with the 600 μl of the Ni-NTA Agarose (QIAGEN) overnight at 4°C. After absorption of the His-tagged recombinant protein to Ni-NTA Agarose, the Ni-NTA Agarose was washed with buffer A (20 mM Tris-HCl pH 7.5, 200 mM NaCl, 5 mM imidazole) ten times and then with buffer B (20 mM Tris-HCl pH 7.5, 20 mM imidazole). After extensive washing, APPC99-His bound to the Ni-NTA Agarose was eluted with 6 ml of the elution buffer (20 mM Tris-HCl pH 7.5, 200 mM NaCl, 200 mM imidazole). Eluates were collected at 1 ml per fraction.

Western Blotting Analysis of the Recombinant Protein

Five µl of each fraction was separated by Tris-Tricin SDS-PAGE (16.5%) and transferred to a PVDF membrane (Immobilon (MILLIPORE)). APPC99-His in each fraction was probed by anti-His rabbit antibody (SANTA CRUZ) at 1:1000 dilution and then visualized by ECL (Amersham). APPC99-His was visualized as a 14 kDa band. The highest amount of the APPC99-His synthesized was recovered in the Ni-NTA bound fraction. A peak fraction was the elute of the first fraction. A small amount of APPC99-His was observed in the flow-through fraction and the wash fraction.

Quantification Analysis

Twenty-four µl of each fraction eluted was analyzed by Tris-Tricin SDS-PAGE and Coomassie Brilliant Blue staining. Comparing with Protein Molecular Weight Standards (GIBCOBRL), the amount of purified APPC99-His was estimated densitometrically. From 1 ml of the reaction mixture, 7–8 µg of APPC99-His was purified at a concentration of 7–8 ng/µl in this experiment.

Preparation of Cell Lysates for the γ-Assay

Fibroblast cell lines derived from the presenilin-1 knockout mouse and the wild-type mouse were cultured in three different 75 cm^2 flasks filled with DMEM supplementing 10% Fetal bovine serum (Hyclone). The cells with over 90% confluency were harvested and homogenized in HEPES buffer (25 mM Hepes pH 7.2, 150 mM NaCl 2mM EDTA) containing protease inhibitors, such as pepstatin leupeptin, PH blocker, and Aprotinin. After centrifugation at 50,000 rpm for 10 min (4°C), the pellet was lysed with 1 ml of 1.0% Big-CHAPS in HEPES buffer supplementing protease inhibitors. Incubating on ice for 30 min, the lysate was centrifuged at 10,000 rpm for 5 min. The supernatant was laid on the top of the 10–40% (W/V) glycerol gradient in HEPES buffer. The lysate was fractionated by centrifugation at 24,000 rpm for 15 h (4°C).

γ-Activity

Those fractions (900 µl) were dialyzed against PBS overnight and then incubated with 100 ng of purified APPC99-His at 37°C for 2 h. To remove uncleaved APPC99-His, 30 µl of Ni-NTA Agarose was added into each sample and incubated at 4°C for 4 h after centrifugation supernatant was used for ELISA.

Aβ ELISAs

Amyloid-β proteins (Aβx-40) were quantified by using sandwich ELISAs.

Comments

Production of APPC99 Recombinant Protein

The recombinant APPC99 His was produced by the RTS system described in Materials and Methods, and the product was analyzed by Western blot using an anti-APPC antibody(APPC), which recognizes the C-terminus of APP, and an anti-His tag antibody. Western blot showed 14 kDa APPC99 and other higher molecular weight band with APPC (Fig. 1a). The His-tag antibody recognized 14 kDa doublet and 20 kDa band. (Fig. 1b) After purification using Ni-column, APPC99-His was eluted as a single band (elution fraction 2, lane 7 in Fig. 1c) detected by the anti-His tag antibody in western blot. We estimated the amount of purified APPC99 by CBB staining (Fig. 1d). We can get 7.9 μg in 900 μl elute fraction.

Production of PS1 Recombinant Protein

The recombinant PS1 was produced by the same method as previously described. After purification, PS1 was loaded in SDS/acrylamide gel and analyzed using various anti-PS1 antibodies. MKAD3.4 recognizes the amino acid residues 45–48 of human PS1. N12 recognizes the amino acid residues 1–12 of PS1, hL312 recognizes the amino acid residues 312–330 of PS1, and PS1 C directs against partial peptides of PS1 C-terminus. All antibodies recognize 50 kDa full-length PS1 protein (Fig. 2a). However, the total amount of purified PS1 was at quite a low level. To get enough of PS1, we need to improve the RTS method. In spite of small production of PS1, we could co-produce PS1 and its putative substrate APPC99 in the same reaction of the RTS system (Fig. 2b). We could find APPC99/PS1 complex but could not detect Aβ in Western analysis.

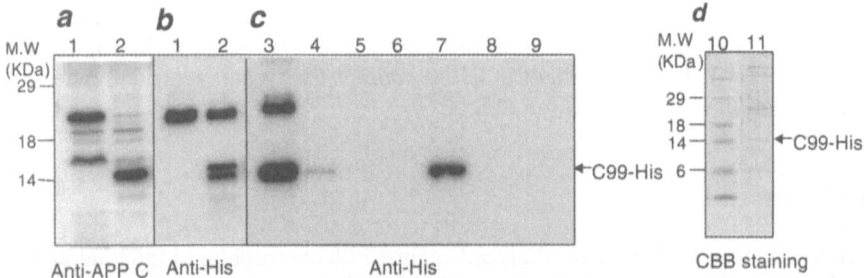

Fig. 1a–d. In vitro synthesis of the recombinant APPC99-His. **a, b** Western blot showing the synthesis of APPC99-His demonstrated by an anti APP-C antibody and anti-His antibody. pIVE2.3(1) and C99-His/pIVE2.3(2) in Western blotting. **c** Purification of C99-His. Reaction solution before purification (3), flow through (4), wash A (5), wash B (6), elution fraction no. 1 (7), elution fraction no. 2 (8), elution fraction no. 3 (9) in Western blotting. **d** Maker (10), elution fraction no. 1 (11) by CBB staining

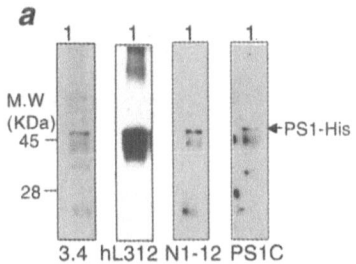

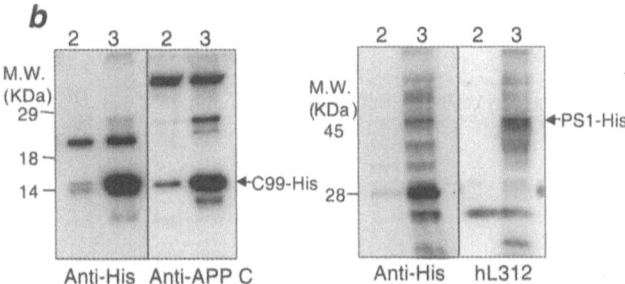

Fig 2a,b. Production of PS1-His. **a** Purification of PS1-His demonstrated by different anti-PS1 antibodies in Western blotting. PS1/pIVE2.4(1). **b** Co-production of APPC99-His and PS1-His demonstrated by anti-APP-C antibody, anti-PS1 antibody and anti-His antibody in Western blotting. C99-His (2),C99-His/Ps1-His (3)

Determination of γ-Secretase Activity Using APPC99 Recombinant Protein

Using APPC99-His, we tried to determine γ-secretase activity in mouse fibroblasts. The fibloblasts are derived from PS1 knockout and wild-type mice, which were provided by Dr. De Strooper. Cells were homogenated in buffer containing Big-CHAPS, and then protein complexes in cell homogenate were separated in ultracentrifuge by glycerol gradient method described in Materials and Methods. We determined γ-secretase activity by incubation with each aliquot of fraction and recombinant APPC99-His for 3 h at 37°C.

We could detect γ-secretase activity in cell lysate by Aβ sandwich ELISA after removing excess APPC99 by Ni-column (Fig. 3).

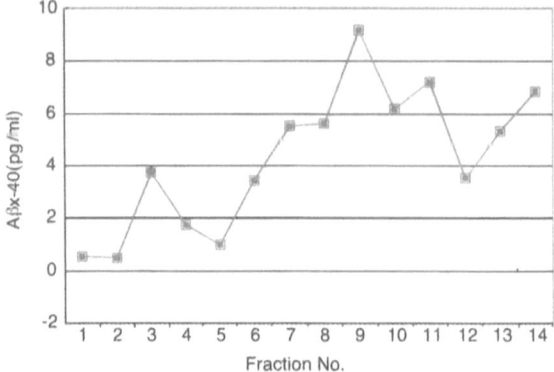

Fig. 3. The γ-secretase activity generated by co-incubation of APP99-his with PS(-/-) and PS(+/+) fibroblasts. Aβ(x-40) generation was quantified by Aβ ELISAs

References

1. Selkoe DJ (1999) Nature 399(6738 Suppl):A23–31
2. Kang J, Lemaire HG, Unterbeck A, Salbaum JM, Masters CL, Grzeschik KH, Multhaup G, Beyreuther K, Muller-Hill B (1987) Nature 325(6106):733- 6
3. Sherrington R, Rogaev EI, Liang Y, Rogaeva EA, Levesque G, Ikeda M, Chi H, Lin C, Li G, Holman K, et al. (1995) Nature 375(6534):754–60
4. Levy-Lahad E, Wijsman EM, Nemens E, Anderson L, Goddard KA, Weber JL, Bird TD, Schellenberg GD (1995) Science 269(5226): 970–3
5. Scheuner D, Eckman C, Jensen M, Song X, Citron M, Suzuki N, Bird TD, Hardy J, Hutton M, Kukull W, Larson E, Levy-Lahad E, Viitanen M, Peskind E, Poorkaj P, Schellenberg G, Tanzi R, Wasco W, Lannfelt L, Selkoe D, Younkin S (1996) Nat Med 2(8):864–70
6. Li YM, Xu M, Lai MT, Huang Q, Castro JL, DiMuzio-Mower J, Harrison T, Lellis C, Nadin A, Neduvelil JG, Register RB, Sardana MK, Shearman MS, Smith AL, Shi XP, Yin KC, Shafer JA, Gardell SJ (2000) Nature 405(6787):689–94
7. Wolfe MS, Xia W, Ostaszewski BL, Diehl TS, Kimberly WT, Selkoe DJ (1999) Nature 398 (6727):513–7
8. Herreman A, Serneels L, Annaert W, Collen D, Schoonjans L. De Strooper B (2000) Nat Cell Biol 2(7):461–2

Subject Index